Power Transducers
for Sonics and Ultrasonics

B. F. Hamonic O. B. Wilson
J.-N. Decarpigny (Eds.)

Power Transducers for Sonics and Ultrasonics

Proceedings of the International Workshop,
Held in Toulon, France, June 12 and 13, 1990

With 215 Figures

Springer-Verlag
Berlin Heidelberg New York London
Paris Tokyo Hong Kong Barcelona

Dr. Bernard F. Hamonic
Dr. Jean-Nöel Decarpigny
Acoustics Laboratory, ISEN, 41 bd. Vauban,
F-59046 Lille Cedex, France

Dr. Oscar B. Wilson
US Naval Postgraduate School, Monterey, CA 93943, USA

Program chairman
B. Tocquet, Thomson Sintra ASM, Valbonne, France

Scientific Committee
O. B. Wilson, Chairman, NPS, Monterey, CA, USA
D. Boucher, GERDSM, DCAN, Toulon, France
J.-N. Decarpigny, ISEN, Lille, France
B. F. Hamonic, ISEN, Lille, France

Organizing Committee
M. Daux, ISEN, Lille, France
B. F. Hamonic, ISEN, Lille France
J.-C. Debus, ISEN, Lille France
J.-N. Decarpigny, ISEN, Lille France

The sponsorships from:
– Direction des Recherches, Etudes et Techniques (DRET),
– Thomson Sintra ASM,
– Schlumberger Etudes et Production,
– Toulon-Var-Technologies (TVT),
– Centre National de la Recherche Scientifique (CNRS),
– Institut Supérieur d'Electronique du Nord (ISEN)
are greatly acknowledged.

ISBN-13:978-3-642-76273-4 e-ISBN-13:978-3-642-76271-0
DOI: 10.1007/978-3-642-76271-0

Softcover reprint of the hardcover 1st edition 1991

2156/3140-543210

Preface

This volume contains the Proceedings of the International Workshop on Power Transducers for Sonics and Ultrasonics, which was held in the Hotel Helios, Ile des Embiez near Toulon, France, on June 12 and 13, 1990. This Workshop was the second one in a series; the first one was held in Lille in 1987.

The main objective of this Workshop was first to expose the trends and needs in the design, analysis, construction, and testing of low-frequency, high-power transducers for sonars, for oceanographic and petroleum-finding (geophysical) applications, and for macrosonics, and second to stimulate an exchange of knowledge and experience between researchers and industrialists from various countries who are involved in this multidisciplinary field. The scientific program included 18 invited contributions distributed amongst seven sessions and a poster session, and there were 100 participants from around the world.

The main topics of the sessions were: general trends, flextensional transducers, interactions in arrays, magnetostriction transducers, other transducing mechanisms, electronics, and measurement methods.

The common theme for all sessions was summed up by three questions: What is the state of the art? What are the prospects for development? Where are we and where should we be going in the next five years? Emphasis was focused on the second and third questions.

The editors are grateful to the authors and attendees for their active participation, and they hope that the reader of these proceedings will find a lot of useful information here. They also wish to thank everyone who helped in the numerous tasks connected with this venture. Special mention has to be made of Miss M. Daux and Mrs. E. Litton (ISEN Lille) for their constant and kind help from the beginning of the project to the very end of the editing. The editors are also grateful to the publishers and their staff, and Dr. H.-U. Daniel, who exercised patience in waiting for the manuscript.

Lille, France
September 1990

B.F. Hamonic
O.B. Wilson
J.-N. Decarpigny

Contents

Introductory Lecture

B. Tocquet

Thomson Sintra Activités Sous Marines, Route des dolines,
Parc d'activité de Valbonne, B.P. 38, F-06561 Valbonne Cedex, France

1 OBJECTIVES REMINDER

The objectives are:

- to present the needs and prospects related to the study, design and reactivation of low frequency, high power sonic sources used for sonar, to ocean and oil research and to macrosonics.
- to favour exchanges between research bodies and industrialists involved in the field.

The papers presented during these two days should enable to answer, at least partially, the three following questions:

- what is the current state of the art?
- what are the development prospects?
- which improvements can be expected within the next five years?

2 TOPICS DEALT WITH

The topics to be dealt with are broken into seven sessions:

- what are the needs for low frequency sonar, oceanography, oil research and macrosonics,
- a specific solution: the flextensional transducers, the limitations involved by immersion,
- interactions between typical transducers when they are grouped to form arrays,
- the magnetostrictive materials,
- other transduction mechanisms,
- what are the needs for power electronics and systems design,
- acoustic measurement methods.

A poster session dealing with the whole field will be held on Tuesday evening.

The first workshop, organized in Lille 3 years ago, had been a real success. This promises to be as successful because:

- the topic has never been of such interest than today, in the civilian field (offshore and well oil research, ocean tomography, macrosonics) as well as in the military field (stimulation of towed or moored linear arrays, counter measurements),
- the number, the geographical origin and the quality of the participants and invited speakers witness the interest for this topic, but also the quality of the organizers,
- the rising costs of research and development (R& D) and the stability, not to say the decrease, in some budgets are elements that should incite us to collaborate tightly and to take advantage of all the possible synergies between users.

All these elements should stimulate exchanges between participants and make this workshop a real success. Now, let's hope that the weather is not going to be too fine. I think that the organizers are terribly self-confident; organizing such a 'serious' workshop in such a heavenly place, what a risk!

1. Transducer Needs for Low-Frequency Sonar

R.W. Timme, A.M. Young, and J.E. Blue

Naval Research Laboratory, Underwater Sound Reference Detachment,
P.O. Box 568337, Orlando, Fl 32856-8337, USA

The continued advances in the acoustic quieting of submarines will necessitate the use of active sonar as an adjunct to the traditional passive sonar as a means of detecting submarines. Therefore, there is a growing need for transducers that produce sound underwater at frequencies below 1000 Hz. However, reality is such that it is very difficult to design for low frequency, high power, and high efficiency and still maintain a device possessing reasonable size, weight, reliability, and cost. Different design approaches and transducer types are discussed and compared.

1.1 INTRODUCTION

The continued advances in the acoustic quieting of submarines (1.1-1.3) will necessitate the use of active sonar as an adjunct to the traditional passive sonar as a means of detecting submarines. Therefore, there is a growing need for transducers that produce sound underwater at frequencies below 1000 Hz. However, reality is such that it is very difficult to design for low frequency, high power, and high efficiency and still maintain a device possessing reasonable size, weight, reliability, and cost. Different design approaches and transducer types are discussed and compared.

1.2 FACTORS DRIVING LOW-FREQUENCY SONAR TRANSDUCER NEEDS

Given that there is a need for active acoustic projection, any competent acoustician can tell you that it generally means operating at frequencies below 1 kHz for reasons of range, resolution, and scattering. Figure 1.1 shows sound scattering intensity from a rigid cylinder at normal incidence to the side as a function of ka where k is 2π divided by the wavelength of the incident sound and a is the radius of the cylinder (1.4). Assuming a 5 m radius as somewhat typical for a submarine and seeing from Fig. 1.1 that frequencies where $ka \geq 1$ give good sound scattering, one might conclude that frequencies above 100 Hz might be of interest in submarine detection at normal incidence.

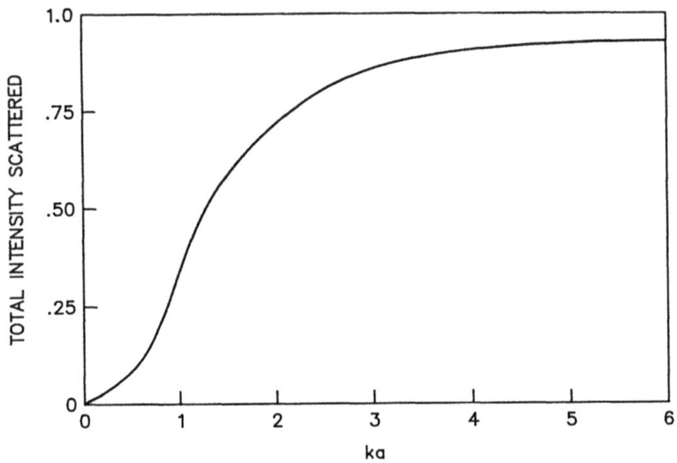

Fig. 1.1 Scattering of sound from a rigid cylinder of radius a(k = 2 π/λ)

1.3 LIMITING FACTORS ON HIGH-POWER, LOW-FREQUENCY SONAR PROJECTORS

Transducers designed to produce underwater sound at low frequencies generally have small dimensions compared to the wavelength of the sound produced (λ>>D). For illustration purposes, let us consider a pulsating sphere of radius a where λ>>a. Figure 1.2 shows the relationships that are important for low-frequency radiation.

Pressure at the surface of the sphere:	$p(a) \cong \rho c\, v_n\, (ka)^2 - i\rho c\, v_n\, ka$ v_n = velocity of surface of sphere.
Total radiation impedance:	$Z = \frac{F}{v_n} = R_r + iX_r$ $R_r \cong \rho\, c\, (4\,\pi\, a^2)\, (ka)^2, \quad X_r \cong -\omega\, (4\,\pi\, a^3).$
Source level in the water referenced to 1 meter from the sphere:	$\lvert p\rvert_{1m} = \frac{\omega\rho U}{4\pi} = \frac{\omega^2 \rho X}{4\pi}$ U = volume velocity , X = volume displacement.
Acoustic power radiated:	$P = v_n^2\, R_r = \frac{4\pi}{\rho c}\, p^2$

Fig. 1.2 Relationships for radiation from a uniformly pulsating sphere of radius a where ka << 1, ρ and c are the density and the sound velocity in water

These relationships for power and source level are plotted in Fig. 1.3. Note that at 100 Hz, a volume displacement of 1000 cm^3 (0.001 m^3) will radiate 10^4 watts of acoustic power and give a source level of 211 dB re 1 μPa at 1 m. However, as can be seen from Fig. 1.4 for λ>>a, the total input power necessary to supply the acoustic power is controlled primarily by the reactive impedance, and the resistive component of the impedance is very low. Thus, the basic problem is one of poor acoustic loading due to very long acoustic wavelengths in water at low frequencies. Simply stated, this means the transfer of mechanical power at the surface of the radiator to the water under the form of radiated acoustic power is very inefficient. Since the overall efficiency is the product of the radiation and mechanical transduction efficiencies, it will also be low even if the mechanical transduction efficiency is high.

Because high-power, low-frequency sound production requires large volume velocities, the projector must meet the conflicting requirements of counteracting large hydrostatic forces and offer a pressure release mechanism to the interior of the vibrating surface. Virtually none of the pressure-release techniques used at high frequencies are practical at low frequencies. Table 1.1 compares various pressure-release mechanisms for use in low-frequency transducers.

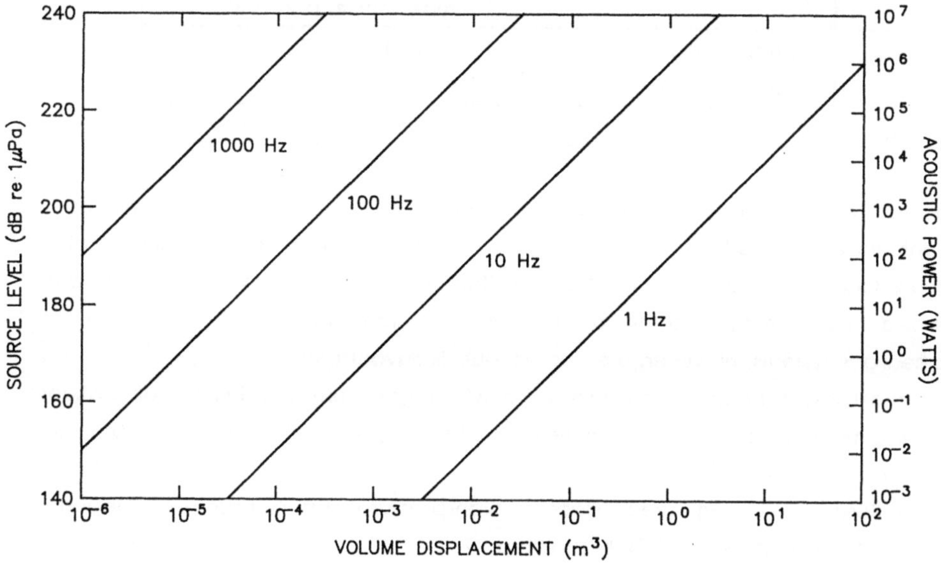

Fig. 1.3 Source level and acoustic power radiated from an harmonically pulsating sphere of radius a with ka << 1

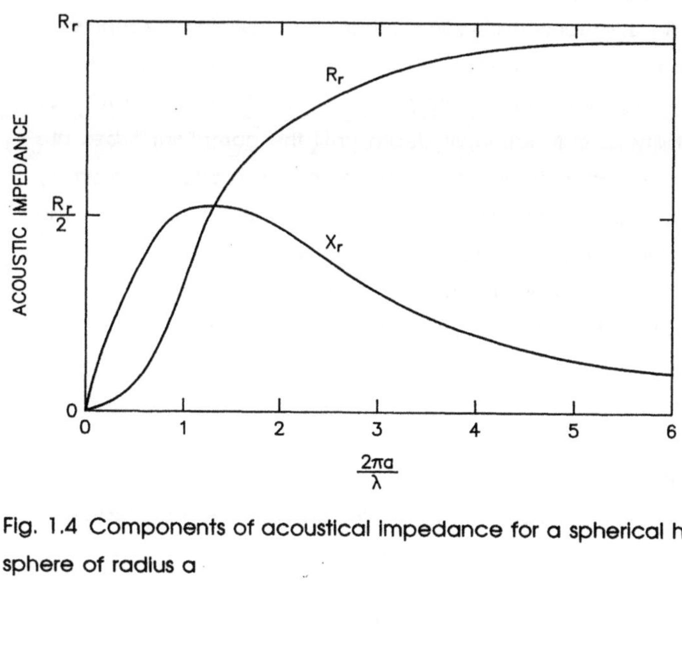

Fig. 1.4 Components of acoustical impedance for a spherical harmonic wave from a pulsating sphere of radius a

Table 1.1 Pressure release mechanisms for use in low-frequency transducers

Mechanism	Advantages	Disadvantages
Compressed Gas	Large impedance mismatch, good pressure release	Acoustic Characteristics are depth dependent, depth limited
Liquid	Independent of depth	High mechanical impedance, poor pressure
Compliant Tubes	Better pressure release than liquid alone	Depth limited
Self-supporting	Simple	Severely depth limited

The most commonly used pressure-release mechanism at low frequencies is compressed gas. If the interior of a transducer is filled with gas at the same pressure as the surrounding water, the transducer is obviously balanced against the forces due to hydrostatic pressure and the large impedance mismatch provides an excellent pressure release.

Compressed gas systems, however, are not without disadvantages:

- since most low-frequency transducers have a large internal volume, applications requiring many depth changes may require a large high-pressure gas storage volume, and,
- for applications at very great depths, high-pressure gas systems can become complicated and pose a reliability problem.

Transducers can be made essentially independent of operating depth by filling the enclosed volume with liquid. To do so, however, provides essentially no pressure release. For a given frequency and acoustic output, liquid-filled transducers will be larger, heavier, and require larger driving forces than will transducers using some other compensation mechanism.

As a compromise usable to moderate depths, sealed, air-filled, oval metal tubes can be inserted into the liquid-filled cavity to increase its compliance. This technique provides decreasing pressure-release capability as a function of depth until the compliant tubes are collapsed by the hydrostatic pressure. Some transducer designs can be made to be simply self-supporting by filling the internal cavity with air at some predetermined pressure. The primary disadvantage, of course, is the severely limited depth capability.

To summarize (see Table 1.2) from the transducer designer's point of view, cost, size, weight, and level of difficulty in design of low-frequency sources increases with:

- decreasing frequency,
- increasing sound pressure level,
- increasing bandwidth,
- increasing operational depth.

Most design techniques used at higher frequencies do not scale well to low-frequency applications and each new requirement is essentially a new design problem.

Table 1.2 Characteristics of low-frequency C-W acoustic sources

Characteristic	Attributable To:
Large size, low overall efficiency	Low radiation efficiency
Heavy, poor reliability	Large forces required
Expensive	Poor acoustic loading requires large amount of reactive power

1.4 TRANSDUCER TYPES

Before discussing any of the specific transducer types, it should be noted that there is no general way to quantitatively compare different transduction mechanisms. The ratio of output power to total weight is frequently used, but it neglects the effects of bandwidth, transduction efficiency, and reliability.

Since all of the available low-frequency transducers cannot be treated here, only those most likely to be encountered will be addressed. The most common transduction mechanism in use is, of course, piezoelectric ceramics. Design techniques used at higher frequencies, such as the longitudinal resonator, are not feasible at low frequencies because of the size that would be required to generate the required volume velocities. Low-frequency ceramic designs attempt to take advantage of vibrational modes not normally used at higher frequencies.

1.4.1 Ceramic Flexural Disc

The trilaminar configuration of the ceramic flexural disc transducer shown in Fig. 1.5 lends itself reasonably well to the high-power, low-frequency application. In the trilaminar configuration, an inactive disc (normally steel or aluminum) is laminated between two ceramic disc composites; when the two ceramic discs are driven in opposition, a flexing motion is produced in the trilaminar structure. To keep the size of the ceramic within reasonable limits, the ceramic discs may be assembled in a mosaic instead of one piece.

In a common configuration, two trilaminar structures are mounted back to back with a spacer ring forming the housing between them which is the compliant annular cavity. The volume between the discs may either be gas filled or oil filled with compliant tubes inserted to provide the necessary pressure release. The enclosed volume can simply be oil filled at the expense of reduced acoustic output.

The ceramic flexural disc can offer a good power-to-weight ratio for some applications over a bandwidth of 1 to 1.5 octaves. Its primary disadvantage is the sensitivity of the resonance frequency to the impedance of the internal cavity.

The maximum input power is limited by the electrical field and maximum stress that the ceramic can withstand.

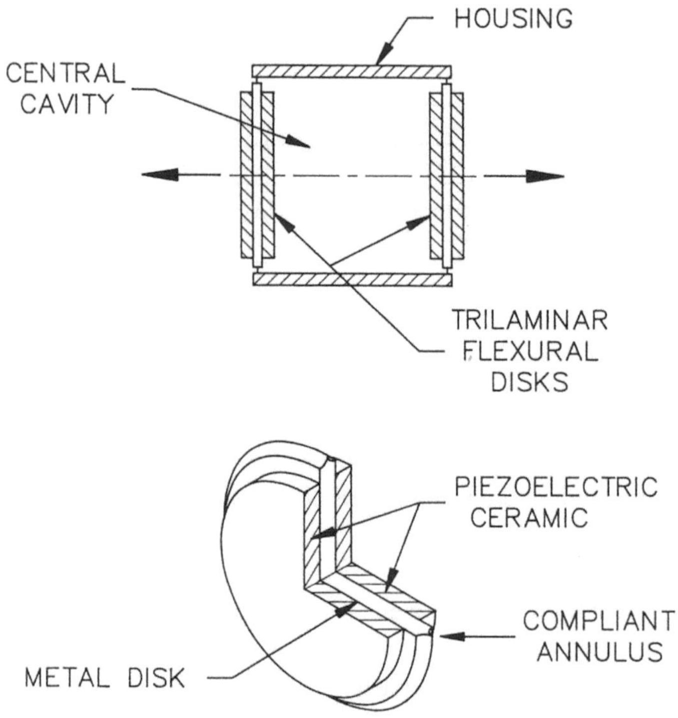

Fig. 1.5 Flexural disk transducer

1.4.2 Flextensional Transducers

In its common form, the flextensional transducer shown in Fig. 1.6 consists of an elliptically shaped housing, or shell, with a longitudinally vibrating ceramic stack mounted along its major axis. Unlike the flexural disc transducer, the housing (not the ceramic) forms the radiating surface. The ceramic stack is compressively prestressed by the shell to assure that it does not go into tension and fracture at high drive levels.

A single large shell may be used or several small shells may be stacked together in a line configuration. In either case, the open ends of the shell are sealed and the resulting internal volume may be either gas filled or oil filled and compliant tubes inserted. For relatively shallow depths, the transducer can be made self-supporting by filling the cavity with air at atmospheric pressure. To do so, however, means that the prestress on the ceramic and, therefore, the safe driving voltage decreases as a function of depth.

The flextensional transducer does offer a good power-to-weight ratio, but they are resonant devices and have a Q higher than most non-ceramic designs. It can also be highly efficient. Its primary disadvantage compared to other low-frequency ceramic transducers is difficult design, particularly for low-resonance frequencies.

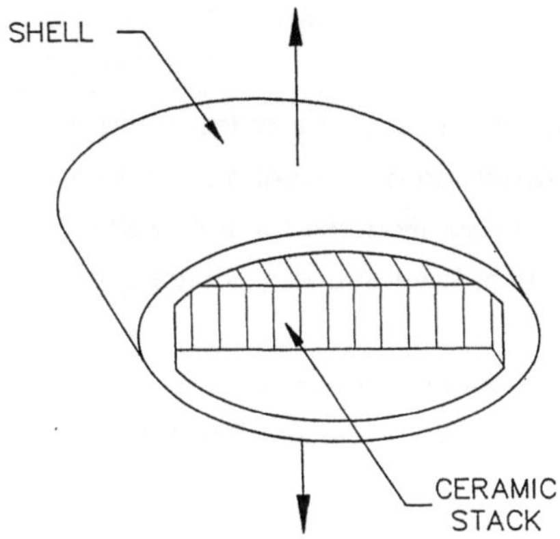

Fig. 1.6 Flextensional transducer

1.4.3 Ceramic Bender Bar Transducers

The ceramic bender bar transducer shown in Fig. 1.7 typically consists of multiple 'bars' arranged in a 'barrel stave' configuration around a cylindrical housing. Each bar consists of two segmented stacks of ceramic and is 'hinged' at each end. When the stacks are driven in opposition, a bending motion is produced in the bars.

The barrel stave configuration of the transducer results in a central cavity which is normally oil filled to compensate for hydrostatic pressure. Compliant tubes are inserted into the cavity to increase its compliance and to provide the necessary pressure release mechanism for radiation from the inner surfaces of the bars. Some very low frequency designs do, however, use compressed air as the pressure-release mechanism.

The transducer is capable, however, of producing moderately high output power levels over a frequency range of an octave or so and at depths to several hundred meters; it does have the advantage of proven reliability. However, its design is such that it uses a very large amount of ceramic, thus it is heavy and expensive.

As in the case of the flexural disc, the input power is limited by the electric field and maximum stress that the ceramic can withstand.

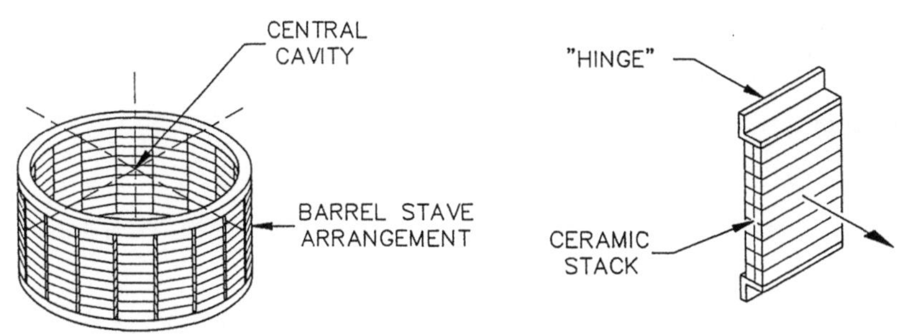

Fig. 1.7 Bender bar transducer

1.4.4 Moving-Coil Transducer

The electrodynamic, or moving coil, transducer shown in Fig. 1.8 is one of the oldest designs still in use and derives its driving force from the interaction between an ac current moving in a conductor and a large magnetic field. In the most common configuration, the force is used to drive a rigid piston radiator. When applied to the requirements of low-frequency sources, the moving coil offers some distinct advantages:

- it can be, and usually is, designed to have a very low resonance frequency,
- being a typically large compliance system, it can accommodate large linear displacements,
- wide operating bandwidths are relatively easily achieved.

It does, of course, also have several disadvantages:

- the moving coil is typically an inefficient transduction mechanism,
- it is a relatively small force device,
- compressed gas is used as a pressure release mechanism.

Being a large compliance system with a gas pressure-release mechanism, low-frequency, moving-coil transducers typically exhibit large changes in performance as a function of depth.

As a relatively low-force device, the moving-coil transducer is capable of producing moderate output power levels when used as a single surface radiator. For higher acoustic output requirements, it is normally used in arrays.

The limitation upon the maximum input electrical power is determined by how well the heat generated in coil can be dissipated.

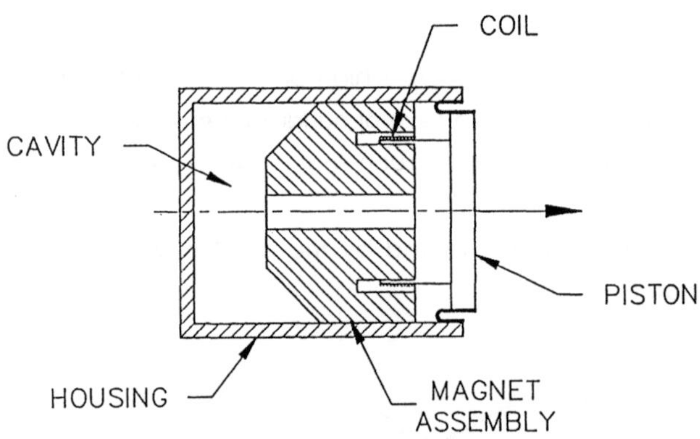

Fig. 1.8 Moving coil transducer

1.4.5 Hydraulically Actuated Transducers

In a common form, the hydraulically actuated transducer as shown in Fig. 1.9 consists of two opposing flexural discs driven by a central hydraulic amplifier. A low-level electrical signal with the desired waveform is used to control the hydraulic amplifier while the hydraulic power is supplied by an electrically driven pump. The hydraulic system can essentially be housed within the transducer module, eliminating the need for handling high-pressure hydraulic lines from the surface. The transducer is normally designed to be self-supporting to depths of several hundred meters and may be gas compensated to go deeper.

Hydraulic transduction seems ideally suited for low-frequency broad bandwidth applications because of the ability to produce very large mechanical forces from a relatively small package and yet allow for the required large linear displacements. A hydraulically actuated source typically produces moderately high acoustic output levels over bandwidths of two octaves or more.

The power-to-weight ratio is comparable to or slightly less than that for some of the ceramic sources.

The primary disadvantage of hydraulic transduction is reliability; it is a relatively complex system and, as such, must be maintained to a greater extent than other transduction mechanisms.

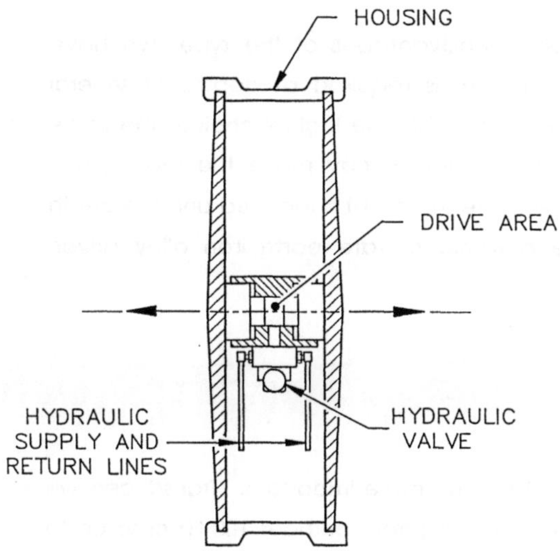

Fig. 1.9 Hydraulically actuated transducer

1.4.6 Other Sources

There are source types other than C-W, and the restrictions on the use of impulse type sources (sparkers, air guns, "water hammer" devices, etc.) are primarily set by the requirements of the experiment. If the repeatable complex waveform from a C-W source is not essential, the use of impulse sources should be considered since they are generally easier to design and operate. There is no general basis for comparing impulse sources with other types.

In the discussion above on flexural discs, flextensionals, and bender bar transducers, no mention was made of utilizing rare earth-iron alloys as the transduction mechanism in place of ceramic. These alloys can, of course, be used, and the potentially greater transduction energy density will provide certain advantages. However, the overall efficiency of the transducer will not be appreciably changed because of the dominance of the radiation efficiency which results from the fact that ka is small.

There are other transduction mechanisms, such as the tow-powered source and the thermoacoustic source, which have not been discussed here because they are still very much in the development phase. They are radically different from the conventional methods and will be described later in these proceedings.

1.4.7 Summary

Table 1.3 compares some of the advantages and disadvantages of the types we have discussed. If, for example, broadband acoustic power is required at depths of several hundred meters, the hydraulically actuated source is probably the logical choice; the same bandwidth and acoustic requirements at a shallower depth may make the moving-coil transducer the most attractive; and yet, if the bandwidth requirement is reduced and the depth requirement retained or increased, one of the ceramic or rare earth iron alloy driven transducers could become the best choice.

1.5 CONCLUSION

We are entering an era in which active sonar may become more important. Transducers will be needed which operate over a frequency range extending from perhaps 100 Hz or so up to about 1000 Hz, although it is unlikely to expect one transducer to operate over this entire band. The required source levels will likely be high - of the order of at least 200 dB re 1 μPa at 1 m. Emphasis will also be on efficiency and reliability, and cost will always be a factor. Knowledgeable users will not insist upon small size because of the greater efficiency concern. It is also likely that for the sonar application, these transducers will be operated in close-packed arrays for the gain and directivity.

In conclusion, there is no general answer to the question of "which type of transduction mechanism is best suited for low-frequency applications?" The user and the transducer designer must decide which transducer type best suits the requirements of a particular experiment. The transducer designer cannot make the decision alone since the user defines the restrictions as well as the requirements.

The idea of designing single sources to cover a wide range of applications may not be as attractive as one might initially think. The transducer designer has no choice but to design for the worst case requirement, which usually means the maximum required volume velocity. If the required bandwidth is appreciable, the result will probably be a transducer which is grossly overdesigned for the higher frequency end of the band. One last thought concerns array interactions; if the projector is to be used in an array configuration, then consideration had best be given to interactions between a projector and its neighbors. Its performance will very likely be different than free-field.

Table 1.3 A comparison of the advantages and disadvantages of the several types of low-frequency sources

Type	Advantages	Disadvantages
Ceramic Flexural Disc	Relatively simple device, good power to weight ratio	Resonance frequency is sensitive to impedance of internal cavity
Flextensional	Good power to weight ratio	Difficult design
Bender Bar	Reliability	Power to weight ratio
Electrodynamic	Large linear displacements Low resonance frequency Wide bandwidth Low harmonic distortion	Typically inefficient Small force device Compressed gas compensation
Hydraulically Actuated	Wide bandwidth high force device	Complicated design Maintenance schedule required Typically inefficient

REFERENCES

1.1 T. Sedgwick, ed., "Soviet submariners close gap on U.S. rivals," Navy News and Undersea Technology, **7**, 1-3 (12 Feb 1990).

1.2 T. Sedgwick, ed., "New soviet anechoic tiles use layers for sound control," Navy News and Undersea Technology, **7**, 1-3 (5 Mar 1990).

1.3 T. Sedgwick, ed., "Sub threat from Soviets and Third World increasing," Navy News and Undersea Technology, **7**, 8 (5 Mar 1990).

1.4 P.M. Morse, and K.U. Ingard, *Theoretical Acoustics*, (McGraw-Hill, New York, 1968).

2. Transducer Needs for Oceanography

R. Person

IFREMER, 29280 Plouzane, France

Acoustical devices are commonly used in oceanography to investigate the oceans. Transducer requirements are overviewed for various applications such as acoustical tomography, subsurface drifting, floats tracking, remote sensing networks for seismic studies, multibeams echo-sounder, side scan sonar and fish detection. New developments are considered to improve these different instruments.

2.1 INTRODUCTION

The whole acoustical frequency spectrum is used by oceanographers to investigate the oceans. Biologists, physicists, geologists, geophysicists, etc, resort to various acoustical equipments. We shall essentially describe the frequency range below 200 kHz. Higher frequencies fit with very particular applications such as high resolution sonars or ultrasonic current meters; transducer requirements are then similar to those used in non-destructive control or medical imaging. We shall distinguish two ranges:

- the frequency range 80-5000 Hz that we shall call "low frequencies", excluding lower frequencies only used by geophysicists,
- the frequency range 5-200 kHz that we shall call "medium frequencies". They are at present the most widely used in oceanography.

2.2 THE LOW FREQUENCY DOMAIN

Up to a recent time, these frequencies were rarely used in oceanography. With few exceptions, only sub-bottom profiling systems operated in this frequency range. 3.5 kHz was the most often encountered frequency. It seems, today, that they are expected to play a more important role in the future.

2.2.1 Physical Oceanography

It had been thought for a long time that the general oceanic circulation could be represented mainly by currents at the surface, deep water being at rest. Recent research works have proved that, on the contrary, subsurface circulation exists on large spatial and temporal scales as well as mesoscale. The quantities of energy involved in these fluctuations are ten times bigger than those in the mean circulation. So, they take an essential place in the dynamics of oceans and numerous, often international, research programs are led to improve our understanding of these phenomena.

Classical measurement systems are not suitable for monitoring subsurface circulation patterns:

- hydrological stations give detailed vertical profiles but, because of temporal spacing, are not able to infer an instantaneous image of a studied area,
- instruments fixed on mooring lines give an excellent temporal resolution but the required spatial coverage needs too many of them;
- satellites can only have access to surface phenomena.

Two complementary techniques have been proposed to monitor oceanic variability on the mesoscale (spatial order 100 km, temporal order 100 days) and on the large scale (spatial order 1000 km, temporal order 1 year):

- ocean acoustic tomography of which concept was proposed by W.K. Munk and C. Wunsch (2.1),
- tracking of subsurface Lagrangian drifters (2.2).

2.2.2 Ocean Acoustic Tomography

Ocean acoustic tomography is a method of observing the interior of an ocean volume by transmitting acoustic energy through it from its edges without having to probe within it. Appropriate mathematical techniques are used to transform the travel time variation data, measured in the different vertical planes defined by the array of sources and receivers, into a three dimensional image of the sound speed field. Acoustic tomography is a synoptic tool and yields a direct measure of sound speed and currents over long distances. In some circumstances other fields (temperature, density, vorticity) can be inferred.

Gradients in the speed of sound in water cause refraction and acoustic rays are bent. Between two points, A and B, it is generally possible to identify many multipaths. Let T(+) be the travel time along a path from A to B and T(-) the travel time in the opposite direction. The speed of sound $C(x,z)$ can be written as the addition of a mean value $C^{\circ}(z)$ and a perturbation $\delta C(x,z)$. Let T° be the travel time without currents and perturbations. The expressions:

$$\delta T = (T(+) + T(-))/2 - T^{\circ}$$

$$\alpha T = (T(+) - T(-))/2$$

which can be expressed as a function of $C(z)$, $\delta C(x,z)$ and the currents, are the basis of the inverse problem. For a 300 km path, with 0.2 m/s as mean current and 6 m/s as sound speed anomalies, δT and αT are respectively of the order of 5 ms and 60 ms.

For IFREMER tomography programs, an instrument called ERATO (Emetteur Récepteur Acoustique pour la Tomographie Océanique) has been developed jointly by IFREMER, WHOI (Woods Hole Oceanographic Institution) and WRC (Webb Research Corporation). ERATO is a self contained apparatus moored during many months. The depth of sources and receivers are selected between 500 and 2000 meters to achieve good vertical sampling of the water column by providing a variety of resolved multipaths. Spatial scales are of the order of 100 to 400 km.

Arrival times are measured with high accuracy. The overall time keeping accuracy of the system must be about 1 ms for the duration of the deployment so that ocean-induced travel time variations can be distinguished from instrument clock fluctuations and drifts. Excursions of the moorings in response to internal tides and currents alter the distances between sources and receivers. A 150 m excursion is not uncommon and results in a travel time change of 100 ms. So, each ERATO includes an acoustic pulse position navigation system for measuring mooring displacement. Three acoustic transponders are anchored on the bottom about the mooring base. Before each transmission time, the navigation module simultaneously interrogates the three transponders and stores the position. Pressure and temperature are also recorded.

As acoustic sources, ERATO employs 2 stagger-tuned resonant tubes, approximately $1/4^{th}$ of the acoustic wavelength in water, driven at one end by a flat piezoelectric element and open at the other. They are fabricated from 12" diameter, 1/4" wall aluminium tube. The overall bandwidth is about 100 Hz around 400 Hz. The nominal source level is 180 dB (re 1μPa at 1m). A pseudo random PSK sequence is transmitted.

As an acoustic receiver, ERATO is equiped with a vertical 2x4 element hydrophone array (Benthos, type AQ-1). The sensitivity is -85 dB (re 1μPa/1V). Only one beamforming is accomplished: beamwidth is ±17° and directivity index is 5 dB.

Other experiments were conducted at other frequencies (224 Hz, 800 Hz, etc.) The choice of the operating frequency results of a compromise between range and time measurement precision given by the bandwidth.

Test long range ocean acoustic tomography experiments were conducted some years ago. Explosives were used as sound sources in the South Indian Ocean, receivers being deployed on the East Coast of the United States. In this kind of experiment, very low frequency sound sources powered by electrical cable from shore may be used in the future.

New developments may improve different aspects of instruments:

- increasing source efficiency: for the same deployment, it would be possible to reduce the size of the batteries, and thus to have a less bulky instrument;
- increasing source bandwidth,
- increasing angular resolution with multi-beamforming. Multipath identification would be easier.

2.2.3 Subsurface Lagrangian Drifters

In this method, neutrally buoyant floats are used. Float weight is adjusted to neutral buoyancy corresponding to a certain density surface (isobaric or isopycnal operation). The compressibility of the float is less than seawater. The control of variations of compressibility as a function of depth and temperature gives isobaric or isopycnal characteristics.

SOFAR (Sound Fixing And Ranging) floats have been regularly used since the MODE experiment in the spring 1973. The floats are tracked acoustically by means of the deep sound or SOFAR channel. They emit acoustic signals on a regular schedule. These signals are received at autonomous listening stations. Then, the float position can be determined as a function of time.

But SOFAR floats are expensive, large and heavy. Because of their size and weight, they require specialized launching equipment. Moreover, only a limited number of sources can operate simultaneously in a given area.

The RAFOS floats (SOFAR spelled backwards) was proposed by T. Rossby et al five years ago (2.3). RAFOS and SOFAR floats serve identical purposes. But, in the RAFOS concept, the floats listen to moored sound sources. Arrival times of signals from sources, temperature and pressure are stored. At the end of the mission, the float activates a release circuit which drops a ballast weight and returns to the surface. Then an ARGOS transmitter is turned on and stored data are transmitted. Typically, at least three days are required to transfer the complete data set. An unlimited number of floats listening to the same sound sources can be deployed. The RAFOS float consumes less power and, fitted out with hydrophones only, is smaller and cheaper than the SOFAR float.

Pop-up floats are also used. At regular intervals they return to the surface, transfer stored data via the ARGOS system and then submerge and drift again.

According to the type of studies, various frequencies are used: 1560 Hz, 800 Hz, 260 Hz, 80 Hz etc. As tomography sources, SOFAR sources employ a resonant tube closed at one end where it is excited at its first natural frequency by a disc of piezoelectric elements. Standard SOFAR sources built by WRC operate at 260 Hz. They transmit an 80 sec CW pulse centered near 260 Hz during which time the frequency increases linearly from 259.376 Hz to 260.898 Hz. Typically, sources transmit every 8 hours with a 20 minutes offset for each source. They provide an acoustic source level of roughly 175 dB (re 1μPa at 1m). The ambient noise conditions are a strong function of weather and shipping. Usable signals are commonly obtained to at least a 1200 km range. Sources are buoyed up from bottom at the depth of the sound channel axis (500 - 2000 m).

A similar tuned projector for 80 Hz would be prohibitively long (6 m) and three times as wide. A modified Helmholtz resonator solves the problem through the use of twin flexural disks. The dimensions of the projector do not exceed the typical 260 Hz float. The twin disks are driven 180° out of phase with each other to set the system free from wavelength-tuning considerations. The loss of efficiency of the system is compensated by the use of the two disks. The peak sound source level is again roughly 175 dB (re 1μPa at 1m).

Improvements are to be made to increase acoustic source level and efficiency. It would also be interesting to reduce the transducer size.

2.2.4 Long-Range Data Transmissions

At the present time long range acoustic data transmissions are not commonly used. But a variety of human activities within the ocean addresseses the need for wireless communications. Typical examples are oceanography and environmental studies or underwater industrial installation monitoring. Many projects using long range data link have been proposed. The bit rates would be between 1 and 100 bits/s at a range between 2 km and 50 km, depending on environmental conditions. In all the cases, the acoustic data link is required to avoid maintaining a vulnerable surface buoy.

We describe as an example, the ACHAIOS (Acoustical CHAnnel for Information On submarine Seismicity) project submitted for EEC funding by FORTH (FOundation for Research and Technology Hellas-Heraklion). In order to increase the capability for seismic risk prediction in Southern Europe, an extension of existing land networks to the sea is proposed. An ocean bottom seismometer is moored in deep water (4000 m). Collected data are transmitted by an acoustic data link (26 kHz) to a subsurface float buoyed up from the bottom at the depth of the channel axis. The acoustic link is implemented to avoid the noise at the sensors due to the cable connecting the seismometer and the float. Data are processed in the float and then relayed by a low frequency data link to a listening station at shore. Range is roughly 50 km. A 2 kHz omnidirectional beam pattern transducer, roughly 400 Hz bandwidth, would be used. A sound source level of about 180 dB (re 1μPa at 1m) would be required. Maximum efficiency will be wanted to reduce battery power.

2.2.5 Fishing Industry

Suggested uses of low frequency sources in the fishing industry include long range fish finding sonars operating at about 3 kHz. This frequency may give good acoustic returns and overcome target aspect problems associated with current equipment. Detection range would be 5 to 10 km.

At lower frequencies, it has been suggested that sensitivity of some species of fish to these frequencies may make low frequency transducers useful as acoustic fence for offshore fish farms or as deterrent to mammal predators. But the efficiency of such devices must be demonstrated.

Non linear acoustics could, in some cases, solve the problem of directivity control. A highly directional beam can easily be obtained. But, low efficiency of such a technique limits its use. The ATLAS Parasound is an echo-sounder system for geological and hydrographic research. In sub-bottom profiling mode, two primary frequencies, very close together (18 kHz and 20.5 to 23.5 kHz), are transmitted simultaneously and generate a low secondary frequency of 2.5 to 5.5 kHz. Nominal acoustic sound level is comprised between 203 and 210 dB (re 1μPa at 1m).

The transducer array consists of four replaceable modules. Its size is 1.20 x 0.95 m and its weight is 160 kg.

2.3 THE MEDIUM FREQUENCY DOMAIN

This frequency range is the most widely used in oceanography. Bathymetry is determined with echo-sounders. Echo-ranging systems are used for detection, location and, to some extent, classification of fish shoals. Passive listening and communication devices as well as acoustic markers and positioning systems have many applications. A large variety of specialized acoustic transducers, produced by many manufacturers is available. Recent developments show some trends, sometimes conflicting:

- transducer arrays are more and more sophisticated to satisfy both the angular resolution of the principal beam and the requirements for the relative level of secondary lobes. Multibeam devices are often required,
- universal tendencies in scientific instruments is towards wideband operation. Beamwidth must be controlled over the whole bandwidth,
- new echo-sounders can operate simultaneously as many as three different transducers. Multifrequency transducers are needed for these systems,
- low cost and efficient transducers have appeared, based on the use of piezoelectric composite materials. The two main advantages of these materials, in comparison with dense ceramics, are their high electromechanical coupling coefficient and the ability to match their specific acoustic impedance with sea water.

The first trend can be illustrated by the ATLAS HYDROSWEEP mapping echo-sounder used for large area hydrographic surveying in shallow and deep sea zones. A special feature of this system is the large coverage of 2x45°. A swath width equal to twice the depth of the water is covered on the sea bed. 59 beams are preformed by means of two transducer arrays, each consisting of three replaceable modules. The arrays are mounted in a "T" configuration in the bottom of the ship. The operating frequency is 15.5 ± 0.6 kHz; acoustic source level is 220 dB (re 1μPa at 1m) for sectorial direction transmission (deep water > 1000 m) and 237 dB for omnidirectional transmission. The beamwidth per preformed beam is 2.2° with 1.5° separation. The input sensitivity is 45 dB/μPa with reception directivity higher 27 dB. The weight of each transducer array is 393 kg.

Another example is provided by the advanced acoustic imaging towed system SAR developed jointly for IFREMER by Thomson-Sintra ASM and ECA. This equipment combines simultaneously or separately a side scan sonar, a bottom penetrator echo-sounder and various underwater sensors. All data are transmitted to surface through a single coaxial cable. The operating depth is 6000 m and height is adjustable over the sea bottom. The main acoustic characteristics are summed up in Table 2.1:

Table 2.1 Advanced acoustic imaging towed system

side scan sonar	
range	2 x 750 meters
vertical beam	80°
horizontal beam	0.5°
range resolution	30 cm
transducer size	914 x 63 x 43 mm
transducer weight	6 kg (in air)
sub-bottom profiling:	
operating frequency	3.5 kHz
transducer size	ø 600 x 350 mm
transducer weight	30 kg (in air) 20 kg (in water)

The main problem was encountered with the sub-bottom profiler. Modifications in the beam pattern with depth were recorded, due to variations with pressure of the acoustic properties of the materials used to baffle the piezoelectric elements.

High capacity, long range acoustic transmission systems require high performance transducers. The TIVA system, developed by ORCA Instrumentation, allows 9600 bit/s data rate transmissions up to 6000 m depth with a good error rate. The system operates at 25 kHz with a 5 kHz bandwidth, 6000 m operating depth transducer. A low cost transducer would widen its domain of application.

Another example of a sophisticated transducer is given by the transducers proposed for the new SIMRAD EK 500 scientific echo-sounder. Three different transducers can be operated simultaneously, single beam or split beam. It is expected that the use of multiple frequencies in acoustic surveys of fisheries will aid the identification of different kinds of scatterers by studying systematically differences with frequency of the echo integral, target strength distribution, echo trace, acoustic density in shoals, etc.

To aid the comparison process, it could be convenient for each transducer to sample identical volumes. To accomplish this, K.G. Foote (2.4) proposes to arrange transducer elements in concentric patterns. The highest frequency beam would be formed by a central circular pattern of elements, the two other beams with two rings surrounding the central one. The outer diameter of these rings would be in rough proportion to the respective wavelength in water. For example, 38, 120 and 200 kHz could be chosen. Such a complicated transducer is not commercially available, but the standard 38 kHz splitbeam transducer (SIMRAD ES38B) is already not simple. Its basic configuration is 10x10 identical square elements of side length 30 mm and center-to-center distance along rows and columns of 32 mm. The nominal acoustic level source is 232 dB (re 1μPa at 1m) and the beamwidth 8°.

2.4 CONCLUSION

Many kinds of transducers are used to satisfy the needs of oceanographical activities. As more and more precise measurements are required, more and more sophisticated transducers have to be developed. Improvements are possible, but the high cost of complex transducers limits their number. One of the specific needs for oceanographic transducers is a large operating depth. It is probably in this area of deep operating low frequency transducers, that maximum effort will be made in the next several years.

REFERENCES

2.1 W. Munk, and C. Wunsch, "Ocean tomography: a scheme for large scale monitoring," Deep-sea Res., **26A**, 123-161 (1978).

2.2 C. Swallow, "A neutral-buoyancy float for measuring deep currents," Deep-sea Res., **3**, 74-81 (1955).

2.3 T. Rossby, D. Dorson, and J. Fontaine, "The RAFOS system," J. of Atmospheric and Oceanic technology. **3**, 672-679 (1986).

2.4 K.G. Foote, "Wideband beam pattern," ICES C.M. 1989/B:2 (1989).

3. Transducer Needs for Petroleum Acoustics

B. Froelich

Etudes et Production Schlumberger, 26 rue de la Cavée, 92140 Clamart, France

A variety of acoustic techniques are used for geophysical measurements in oil wells. The frequency used ranges from 10 Hz for the lower end of the seismic range to 1 MHz in the ultrasonic range. Except for the very low frequencies, the sources are located in the well and, as such, are subjected to a harsh environment. The available room and electric power are limited, the pressure and temperature are high (up to 1000 bars and 175° C), and the mud filling the borehole can have unfavorable acoustic properties. Moreover, the time allowed for a measurement is restricted for economical reasons.

Low frequency (below 1 kHz) acoustic wave propagation between two wells can be used to image the rock properties through tomography. To overcome the attenuation encountered over distances of several hundreds meters, a high power, efficient source is needed. Bandwidth is required to get resolution. In the medium frequency band (1 kHz to 50 kHz), propagation along the borehole over distances of about 1 m is used to measure compressional and shear speed of the formation, and wideband monopole and dipole transducers must be used. Ultrasonic applications (100 kHz to 1 MHz) include nondestructive testing of the steel casings or high resolution imaging of the formation, and one of the main challenges is to overcome the mud attenuation.

3.1 INTRODUCTION

The process of detecting and producing oil in subterranean rock formations requires a detailed physical investigation of these formations from the borehole which has been drilled through the earth. This is the objective of the 'wireline' industry, and it is achieved by means of sondes carrying down various sensors and their associated electronics (Fig. 3.1). These sondes are connected to the surface through an electrical cable which provides the power to operate the sensors, and carries uphole the acquired data, where they are displayed as 'logs'. These data must answer a variety of questions: in the open hole, what is the lithology of the rock matrix (sandstone, carbonate or clay for example), what is its porosity, and what is the fluid filling this porosity (water, oil or gas). In a producing well, the hole is generally cased with a steel pipe cemented in place, and the problems can be the evaluation of the integrity of this cement, or the measurement of the fluid flow. To face this variety of problems, a wide spectrum of physical principles are called upon. The main sensors today in use fall into the following three categories: electromagnetism, nuclear physics or acoustics. The field of acoustics ranges from seismic to ultrasonic frequencies. However, conventional seismic, where at least the source is located at surface, will not be considered in this review of borehole transducers Some of the techniques presented are mature and the measurements are routine, while others are experimental or in early development stages. They all share a common and difficult environment which dictates the technical solutions.

Fig. 3.1 Schematic of the borehole environment

3.2 BOREHOLE ENVIRONMENT

The borehole environment puts severe restrictions on acoustic transducers. The first one is the size constraint, which will be deeply felt when designing low frequency sources. The usual borehole diameter limits the sonde diameter to be less than 100 mm. In cased holes, it is often desirable to restrict the diameter to less than 43 mm.

Because acoustic transducers require a direct mechanical contact with the borehole fluid, they cannot be shielded from the hydrostatic pressure. The fluid column must hold the formation fluids in place. This is achieved by using muds loaded with heavy mineral particles, so that its density can be twice the water density. An oil well reaching depths of several kilometers, downhole devices must be able to withstand more than 1000 bars. At the same time, the temperature increases with depth, and the usual specification is 175° C, although temperatures higher than 200° C can occasionally be encountered.

The cable linking the downhole and surface equipments has a limited cross section and a length of several kilometers. This limits to less than 1 kW the total electrical power available to the sonde. This power not only drives the sources, but also performs other functions, such as data acquisition and transmission to the surface. Moreover, the small size of the sonde makes downhole energy storage rather unattractive. These energy considerations severely penalize inefficient sources.

The data acquisition concerns depth intervals of several hundreds to several thousands of meters and must be done in the minimum amount of time for economical reasons. Thus, the preferred mode of operation is the continuous logging mode, where the sonde is moved continuously, with a typical speed of 1000 m per hour. However, some techniques may require stationary measurements, which increase their cost.

3.3 SEISMIC IMAGING

Surface seismic waves have been used for decades to image subsurface reflectors. The seismic waves launched and received at surface have to travel twice the distance to the target, and cross twice the attenuative weathered zone, located just below surface. As a consequence, only low frequencies can be used (10-100 Hz), and the resolution is in the range 30 to 300 m. This kind of resolution is inadequate when small features need to be seen within a reservoir. Such features which can be critical for the oil company, are faults, continuity of a layer or the position of a steam front. This situation recently prompted the idea of borehole to borehole seismic (Fig. 3.2): both source and receivers are located downhole in two different wells, taking advantage of the multiplicity of wells drilled to drain a known reservoir. A typical well-to-well distance is several hundred meters, an order of magnitude less than the surface to reservoir distance. The transducers are located in the immediate vicinity of the target zone and the waves do not have to cross the weathered zone. Higher frequencies can be used, with the benefit of higher resolution. A resolution of roughly 3 m can be expected at 1 kHz.

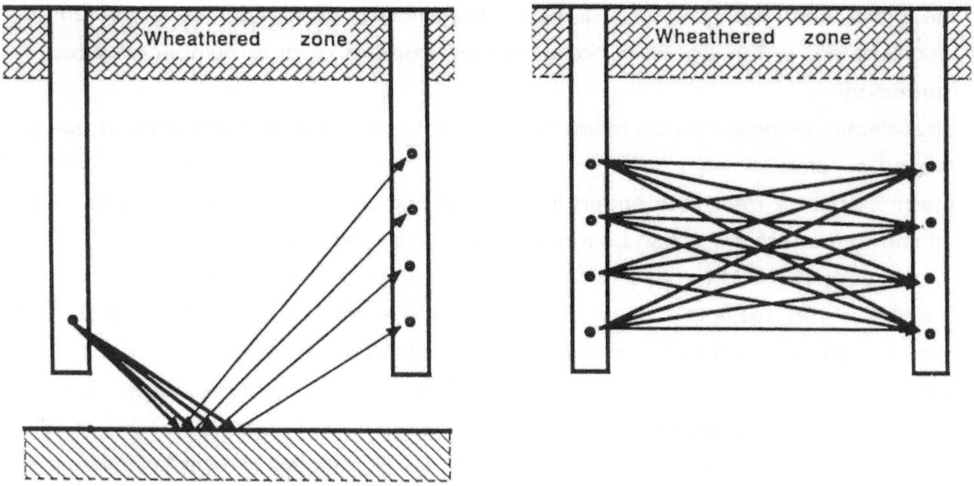

Fig. 3.2 Crosswell seismic source and receivers configuration, showing either reflected path or transmitted path

The imaging techniques include the classical reflection technique used in surface seismic techniques, if one can separate the direct and the reflected arrivals. This solution is the only one when the target reflector is located below the bottom of the wells, but leads to signals at least one order of magnitude weaker than the direct arrival. On the contrary, tomography makes use of these direct arrivals. With travel time tomography, the P wave times of arrival at a multiplicity of source and receiver locations are inverted to reconstruct an image of the formation velocity between the two boreholes (3.1,3.2).

Receivers are either simple hydrophones suspended in the fluid, or three axis geophones or accelerometers clamped to the borehole wall. This last solution allows the recovery of the complete wave field, at the cost of a more complex mechanical design. For the source, a variety of physical principles have been used experimentally. This includes explosives, air guns (3.3), electric arc discharges (3.4), hydraulic or pneumatic devices (3.5), and piezoelectric devices (3.6). Depending on the configuration, these sources, which are generally small compared to the wavelength, can behave either as monopoles or as dipoles. Monopoles involve a volume change, while dipoles represent horizontal or vertical forces embedded in the formation. Among these sources, the piezoelectric monopole has a number of interesting features:

- it is suited to the generation of pulses in the kHz range, with a low enough intensity not to damage the well,
- the frequency content is easily controlled by the shape of the electrical excitation. By using frequency modulated sweeps, it is possible to increase the energy level of one shot, while keeping the peak energy low,

- the signal from such a source is highly repeatable, and thus may allow further improvement in the signal to noise ratio by stacking a large number of shots, if necessary,
- piezoelectric devices can be made very efficient and this is an advantage in power limited systems,
- the monopole is freely suspended in the fluid and does not need a complex wall clamping mechanism. It can be moved quickly from one position to the other.

However, a number of problems need to be addressed by the designer:

- the size is very restricted. The diameter of the sonde is much smaller than the wavelength, and so are at least two dimensions of the source,
- the fluid filled borehole is a waveguide, and the low frequency mode, called the tube wave, carries away most of the energy radiated by a monopole (3.7). Only a very small fraction of the total energy (0.1 to 1 % at 1 kHz) is radiated into the formation, and the lower the frequency, the worse this energy partitioning,
- another consequence of this guided wave problem is that the radiation impedance as seen by the source is predominantly real and does not tend toward zero with frequency. This situation is very different from the free field situation, where the low frequency impedance is mostly imaginary and decreases with frequency,
- because of the high pressure encountered, acoustic transducers need to be pressure compensated with fluids. As a consequence, a volume change source has to work against this fluid stiffness, and its performance is lowered,
- at low frequency, piezoelectric devices are displacement limited, and the pressure response has a 12 dB per octave slope which cuts down the lower part of the spectrum,
- to separate the possible different structures, a short pulse signature and a wide bandwidth is required. Resonances within the frequency band of interest would obscure the interpretation of the traces.

Imaging the structure between two wells can be extended to imaging the structure around a single well (3.8). In this case, both the source and the array of receivers are located on the same sonde. This geometry (Fig. 3.3) is very similar to the marine seismic geometry, where a boat supports the source and tows a long array of hydrophones. The sonde is moved continuously along the borehole and the source is fired every 15 cm, which is also the inter-receiver spacing. These receivers are omnidirectional hydrophones, and the reconstructed image has no azimuthal information. The optimum frequency is a few kHz, and the depth of investigation reaches tens of meters, depending on the rock attenuation and the impedance contrast. The receivers record reflections from interfaces which are not perpendicular to the borehole, and the imaging process relies on back-projection of the data, taking advantage of the large number of source and receiver positions to increase the signal to noise ratio. Experimental applications of this technique are targeted towards imaging dipping beds, or horizontal beds from deviated wells, nearby salt domes, or monitoring the reservoir cap position when drilling a horizontal well in such a reservoir.

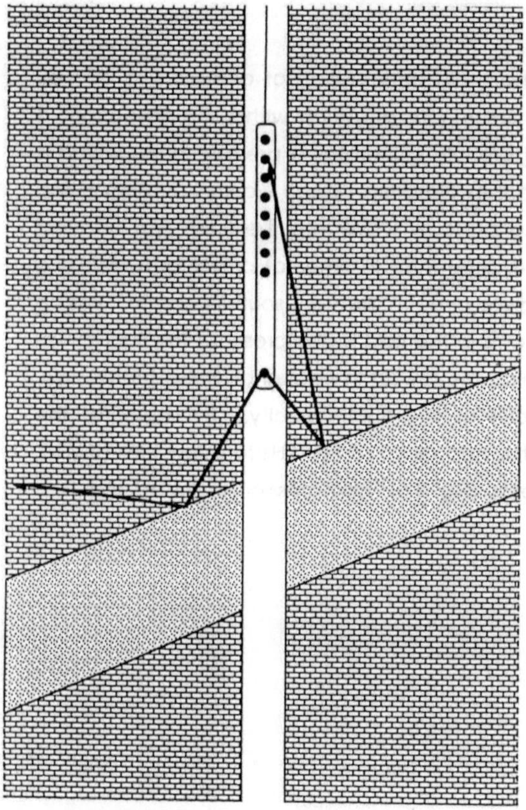

Fig. 3.3 Configuration used to image the vicinity of a borehole with source and receivers located in the same borehole

The source requirements in this case are quite different from the crosswell imaging. Here, the predominant noise at the receiver level is the direct tube wave arrival, propagating through the fluid as a guided mode. Increasing the source power does not change the signal to noise ratio, and the power constraint is less severe. In fact, regular sonic transducers, as described in the next paragraph, can be used. Elimination of tube wave is achieved through velocity filtering: the tube wave has a low velocity, close to the fluid velocity, while reflected events have high apparent velocity along the array.

3.4 SONIC LOGGING

Measurement of the rock velocities has been the main contribution of acoustics to oil well logging for decades. The principle (Fig. 3.4) is based on refraction within the formation: a monopole source located at the center of the borehole generates a pressure pulse which is converted into compressional and shear waves in the formation. These waves travelling along the solid/fluid interface, continuously reradiate a pressure wave into the fluid. An array of centralized hydrophones detects this signal. The difference in times of arrival between two receivers is independent of the fluid path and gives directly the rock sound velocity. Beside the refracted arrival, a direct arrival propagating only in the fluid is also present. This fluid arrival is in fact a combination of guided modes, of which the lowest in frequency is the tube wave. Although dispersive, their group velocity is close to the fluid velocity. By selecting the transmitter to receiver distance (about 1 m) and the frequency (a few kHz to a few tens of kHz), it is possible to separate in time the three events (Fig. 3.5): compressional, shear and fluid arrival.

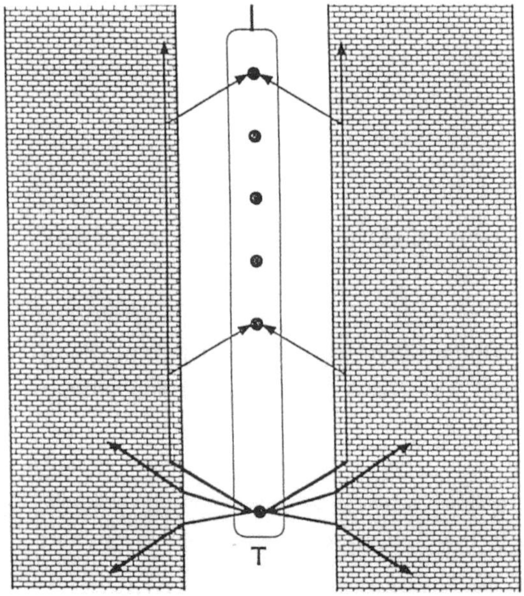

Fig. 3.4 Schematic of the ray paths from transmitter to receivers with a sonic sonde used to measure rock velocities

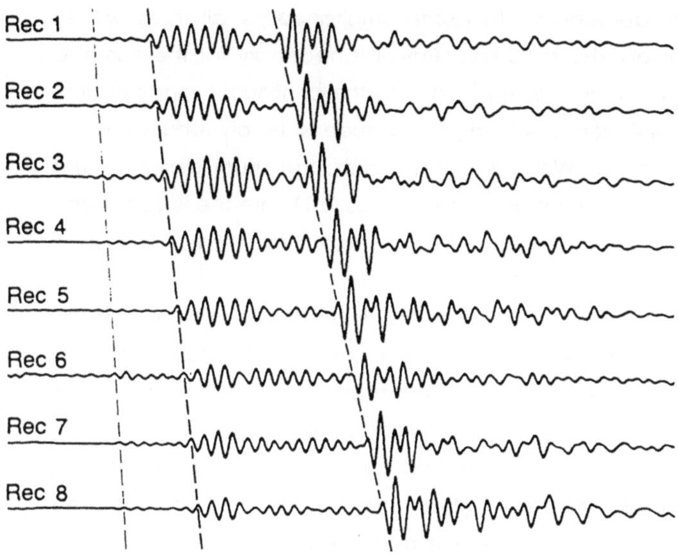

Fig. 3.5 Waveforms received along the array of receivers with a sonic sonde

The determination of compressional velocity has a number of applications (3.9). The first one is to correlate time and depth on surface seismic traces. Seismic record signal versus time, and to get the depth of a given reflector, the transit time between surface and this reflector is required. It is obtained by integrating the local compressional slowness (inverse of velocity) along a well. Laboratory experiments also show that the compressional velocity V_p is a function of rock porosity for a given mineralogy. Under certain conditions, useful estimates of these critical data can be obtained from the sonic velocity. V_p is also very sensitive to small amounts of gas in the pores, and can be used as a gas zone indicator. Adding the shear velocity measurement provides a way to estimate the mechanical properties of the rock, such as bulk and shear modulus. These properties are helpful in the prediction of fracture initiation or potential sanding problems. Fractures are unwanted during drilling, because it can cause caving of the borehole wall, but can be useful during production to drain a low permeability reservoir. Sanding may occur during production in poorly cemented sandstones and quickly corrodes well equipments. Finally, the tube wave can give an estimation of the rock permeability, one of the most important and difficult parameters to measure downhole. The low frequency tube wave is basically a pressure pulse. If the formation is permeable, fluid will be forced into the pores of the formation. The wave is simultaneously attenuated and dispersed (3.10), and although these effects are small, the measurement of either attenuation or dispersion allows an estimation of the order of magnitude of the permeability.

The main transmitter requirements are bandwidth and power. Bandwidth is necessary to separate in time the various events. Moreover, the different waves have preferential frequency domains of excitation. These differences are due to the cylindrical geometry and to the fact that the borehole diameter is roughly comparable to the wavelength. Maximum compressional and shear wave excitation occurs at respectively 10 kHz and 7 kHz. Tube wave excitation increases when frequency decreases. To excite simultaneously all three waves requires a bandwidth ranging from a few kHz to 20 kHz. Power is necessary for the signal to overcome the noise level. Noise at the receiver is mainly due to the continuous sonde motion (flow noise, shocks against the wall), while refracted arrival amplitude is sharply reduced when the fluid/solid impedance contrast is high. With these constraints, the usual design of the transmitter is a simple free-flooded ring. The three main resonances (3.11) are the length, hoop and fluid cavity resonances. With length and diameter of the order of 5 to 10 cm, it is possible to cover the above bandwidth.

When the formation shear wave velocity is smaller than the fluid velocity (slow formation), the simple picture of Fig. 3.5 is altered, since no shear wave can be refracted along the borehole. In this case, the shear arrival disappears from the traces, and the monopole sonde does not provide an answer. Unfortunately, this situation of unconsolidated rocks is the situation where shear velocity and mechanical properties are the most valuable. To solve this problem, it has been proposed to use horizontal dipole (or higher order) transmitters and receivers (Fig. 3.6). The dipole source centered in the borehole acts as a point force which shakes the formation and radiates shear waves in the vertical direction. This radiation is picked up by an array of dipole receivers. If the frequency is low enough, the measured velocity is the shear velocity. However, beyond a few kHz, a guided mode (borehole flexural mode) is excited, and its velocity is lower than the true shear velocity (3.12). Measurements are thus restricted to the low kHz frequency band. Taking advantage of the shear wave polarization, it is possible to use two sets of transducers oriented at 90 degrees and detect formation anisotropy in the horizontal direction. Such anisotropy can be related to vertical fractures oriented in a preferred direction, which are important in reservoir drainage.

The first requirement for the source/receiver combination is to be free of the monopole component. Imperfect dipoles generate compressional and tube wave arrivals, which complicate the detection of the shear event. Then, the source must be able to overcome the noise level. This noise is again due to sonde motion and is difficult to avoid. Finally, a bandwidth ranging from 1 to 5 kHz is necessary to efficiently generate shear waves in various formations.

Various transduction principles have been used for a dipole source: the electrodynamic transducer (3.13) combines several attractive features, such as the absence of monopole component and a flat frequency response. It is very similar to audio loudspeakers (Fig. 3.7.a), with a moving coil excited by a current pulse in a static magnetic field. This coil drives a pair of plates which exerts a horizontal force on the fluid. This force is transmitted to the formation and generates shear waves propagating in the vertical direction. Even more simple is the piezoelectric bender (3.14), where two ceramic plates are bonded together with opposite polarity and deflect laterally when a voltage is applied across them (Fig. 3.7.b). These transduction mechanisms are reciprocal and can also be used for the receivers. Another design takes advantage of the small size of hydrophones. A horizontal pair of monopole hydrophones is matched in sensitivity and electrically connected in opposition. Because the wavelength is much larger than their separation distance, the pair acts as a horizontal dipole.

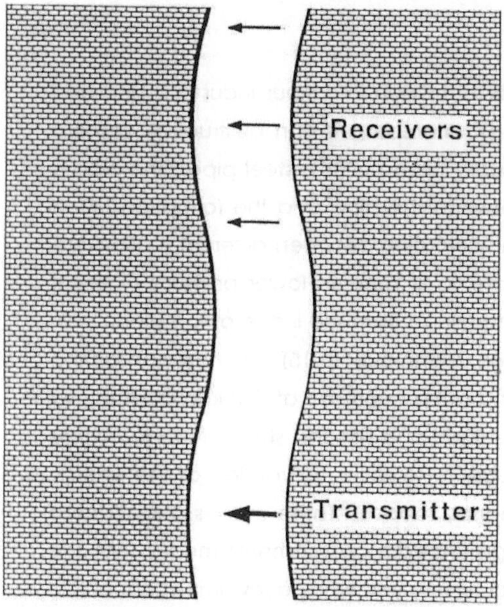

Fig. 3.6 Dipole transmitter and receivers configuration used to measure shear velocities in rock

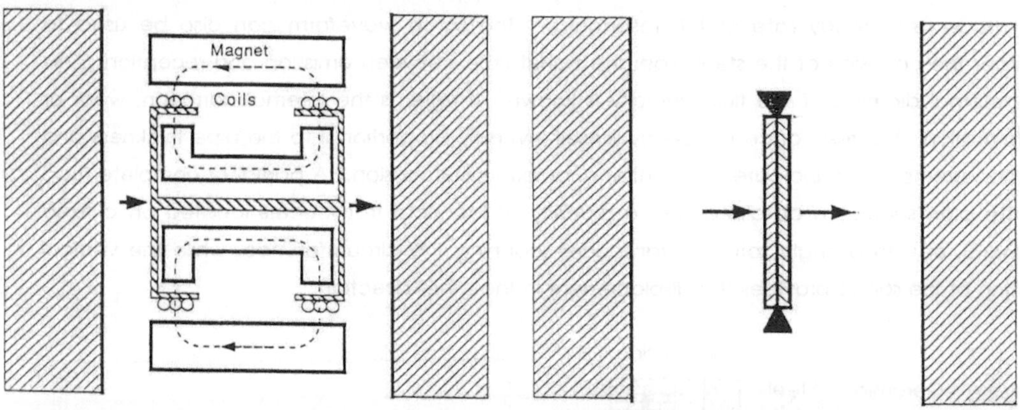

Fig. 3.7 Examples of dipole transducers showing electrodynamic and piezoelectric bender transmitters

3.5 ULTRASONIC LOGGING

Ultrasonic waves are widely used as a non destructive technique in various industries, and also found some applications in oil wells. One of the most interesting is non destructive cement evaluation. After a borehole has been drilled, it is usually cased with a steel pipe. This pipe is held in place by pumping cement in the annulus between the pipe and the formation. One critical function of the cement is to prevent fluid communication between different rock layers. For example, valuable hydrocarbons must not migrate to a zone of lower pressure where it would be lost. Then, the cement integrity must be determined from the inside of the pipe.

This goal is achieved with an ultrasonic reflection technique (3.15). A non contacting ultrasonic transducer (Fig. 3.8.a) launches a broadband pulse centered at 500 kHz, and normal to the pipe surface. Such a pulse propagates in the fluid, and hits the steel interface, where most of the energy is reflected back to the transducer. However, a fraction of the energy enters the steel layer and excites the thickness resonance modes. These modes are affected by the outside medium acoustic impedance. A high impedance (cement) means that the resonance is quickly damped by transmission to the outside layer, while a low impedance fluid means that the resonance decay is slow, because of the high acoustic impedance contrast between steel and fluid (Fig. 3.8.b). The outside medium impedance is deduced from the amplitude and decay rate of this resonance. The same waveform can also be used to monitor the corrosion of the steel pipe: the transit time between emission and reception gives the internal diameter if the fluid velocity is known. It reflects the internal corrosion, wear or deformation. The resonance frequency is itself inversely proportional to the pipe thickness and its measurement can indicate either internal or external corrosion. A partial or complete map of the pipe surface is obtained with a plurality of identical transducers located on a fixed mandrel, or with a single rotating transducer scanning the circumference, while the vertical motion of the sonde provides the displacement in the other direction.

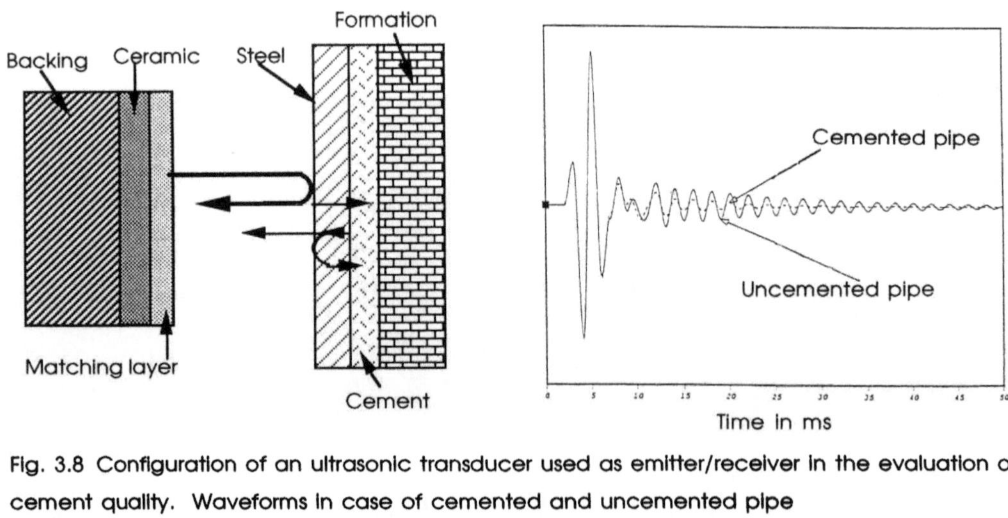

Fig. 3.8 Configuration of an ultrasonic transducer used as emitter/receiver in the evaluation of cement quality. Waveforms in case of cemented and uncemented pipe

A similar, although less sophisticated technique can be used in an open hole, in which case a single interface is present between fluid and rock. The echoes transit times and amplitudes are measured and displayed as images. The first gives the borehole shape, while the second is sensitive to both surface geometrical defects and acoustic impedance. These images are helpful for example in fracture detection, which might be an important component of the reservoir porosity or permeability. When intersecting the borehole, these fractures create small rock breakouts which are easily detected as ultrasonic amplitude reduction, even if the fracture width is less than a millimeter and much smaller than the beam width.

The transducer requirements are similar for both cased hole and open hole applications. The frequency of operation is controlled by the necessity to excite the pipe resonance, to have a sufficient lateral resolution and to travel through about 10 cm of mud without excessive attenuation. This attenuation is a severe problem since it can reach several decibels per cm at 500 kHz . Such an attenuation is due to the nature of the mud which is usually a suspension of solid barite particles in a viscous fluid. Barite content controls the density and the pressure at bottom, while the viscosity stabilizes the suspension and helps carry to the surface the rock cuttings created by the drilling process. The practical frequency range is limited to below 1 MHz. A broad bandwidth transducer is also required to separate the main reflected pulse from the pipe resonance in cement evaluation. Moreover, pipe thickness and the resonance frequency may vary in a wide range.

The active piezoelectric material of the transducer is preferentially lead metaniobate ceramic. Its acoustic impedance is lower than PZT and makes broadband operation easier to achieve. To increase bandwidth and efficiency, one or two matching layers may be inserted between the ceramic and the fluid medium. The critical component of the transducer is the backing material, which must combine high impedance and high attenuation and retain its properties over a broad range of temperature and pressure. Acoustic impedance should in principle match the ceramic impedance to get the highest bandwidth. Energy emitted in the backing can be reflected back towards the ceramic and is the main source of noise. To lower this noise level, and because the available room is limited, a highly absorbing material is needed.

3.6 CONCLUSION

In the past ten years, the use of acoustic waves in oil wells has been extended far beyond the traditional audio frequency range by new applications. In the low frequency domain (1 kHz), the gap between seismic and acoustic is aimed at mapping structures around one or several wells. This objective and the unusual environment are prompting the development of innovative source designs. The use of the high frequency range (500 kHz) is mostly non destructive testing of pipes and cement annulus. It requires high temperature, high pressure and low internal noise transducers. Even the routine measurement of rock wave velocities has been renewed by dipole transmitters and receivers which allow the generation and detection of shear waves in any kind of formation.

REFERENCES

3.1 N.D. Bregman, R.C. Bailey, and C.H. Chapman, 'Crosshole Seismic Tomography,' Geophysics, **54**(2), .200-215 (1989).

3.2 N.D. Bregman, P.A. Hurley, and G.F. West, 'Seismic Tomography at a Fire-Flood Site,' Geophysics, **54**(9), 1082-1090 (1989).

3.3 M.W. Lee, A.H. Balch, and K.R. Parrot, 'Radiation from a Downhole Air Gun Source,' Geophysics, **49**(1),.27-36 (1984).

3.4 T.E. Owen, W.T. Balogh, and W.R. Peters, 'An Arc Discharge Pulse Source for Borehole Seismic Applications,' Presented at the 58th Annual International Meeting of the Society of Exploration Geophysicists, Anaheim, California, (1988).

3.5 H.C. Hardee, G.J. Elbring, and B.N. Paulsson, 'Downhole Seismic Source,' Geophysics, **52**(6), .729-739 (1987).

3.6 W.T. Balogh, T.E. Owen, and J.M. Harris, 'A New Piezoelectric Transducer for Hole-to-Hole Seismic Applications,' Presented at the 58th Annual International Meeting of the Society of Exploration Geophysicists, Anaheim, California, (1988).

3.7 M.W. Lee, and A.H. Balch, 'Theoretical Seismic Wave Radiation from a Fluid-filled Borehole,' Geophysics, **47**(9),.1308-1314 (1982).

3.8 B.E. Hornby, 'Imaging of Near-Borehole Structure with the Array Sonic Tool,' Presented at the 58th Annual International Meeting of the Society of Exploration Geophysicists, Anaheim, California, (1988).

3.9 D. Ellis, *Well Logging for Earth Scientists*, (Elsevier, New-York, 1987).

3.10 J.E. White, *Underground Sound, Application of Seismic Waves*, (Elsevier, New-York, 1983).

3.11 P.H. Rogers, 'Mathematical Model for a Free-Flooded Piezoelectric Cylinder Transducer,' J. Acoust. Soc. Am., **80**(1),13-18 (1986).

3.12 A.L. Kurkjian, 'Theoretical Far-field Radiation from a Low-frequency Horizontal Acoustic Point Force in a Vertical Borehole,' Geophysics, **51**(4), 930-939 (1986).

3.13 C. Kitsunezaki, 'A New Method for Shear Wave Logging,' Geophysics, **45**(10), 1489-1506 (1980).

3.14 J. Zemanek, F.A. Angona, D.M. Williams, and R.L. Caldwell, 'Continuous Acoustic Shear Wave Logging,' Trans. SPWLA 25th Annual Logging Symposium, (1984).

3.15 B. Froelich, D. Pittman, and B. Seeman, 'Cement Evaluation Tool. A New Approach to Cement Evaluation,' presented at the 56th Annual Fall Technical Conference and Exhibition of the Society of Petroleum Engineers, San Antonio, Texas, (1981).

4. Transducer Needs For Macrosonics

J.A. Gallego-Juarez

UEI Ultrasonidos, Instituto de Acústica del CSIC,
Serrano 144, E-28006-Madrid, Spain

Macrosonics is the term used for the application of high intensity acoustic energy for producing some kind of permanent effect in a treated medium. Most practical work in macrosonics is done at ultrasonic frequencies, generally within the range of 20 to 100 kHz, using acoustic intensities from about 0.1 W/cm^2 to several kW/cm^2. The main effects produced in macrosonic applications are heat, cavitation, agitation, streaming, interface instabilities and friction, diffusion and mechanical rupture. Transducers for macrosonics are devices of various types with working frequencies within the range of 10 to 100 kHz, power capabilities from hundreds of watts to several kilowatts and large vibration amplitudes. At present, most macrosonic transducers are of the piezoelectric type and have structures based on the classical sandwich arrangement. The main problems for their successful application to industrial processes are related to obtaining uniform amplitude distribution and improving the design and the calculation methods. This paper presents a review of basic piezoelectric transducers for macrosonics and discusses the specific needs for different applications.

4.1 INTRODUCTION

The applications of acoustic waves can be classified into two main groups, which respectively deal with the use of low and high intensities. The term macrosonics covers the application of high-intensity acoustic energy for producing permanent effects in the treated medium. Most practical works in macrosonics are done at ultrasonic frequencies, generally within a range of 20 to 100 kHz, using acoustic intensities from about 0.1 W/cm^2 to several kW/cm^2. Using ultrasonic frequencies permits very high accelerations to be reached easily with small displacements and moderate particle velocities.

The effects produced by high-intensity sonic or ultrasonic waves are mainly due to finite-amplitude pressure variations and radiation pressure. These basic phenomena lead to a series of effects, the most important of which are: heat, cavitation, agitation, acoustic streaming, interface instabilities and friction, diffusion and mechanical rupture. These effects are used to enhance a wide range of processes, such as machining, welding, metal forming and powder densification in solids; cleaning, emulsification, atomization of liquids, acceleration of chemical reactions, degassing, defoaming, drying, aerosol agglomeration, etc., in fluids.

A certain number of macrosonic prócesses have already been introduced in industry, but many of them still remain at a laboratory stage, and have not been commercially developed. Transducer and power generation technology problems prevent these processes from progressing in this direction.

At present, most high power sonic or ultrasonic transducers are based on the classical sandwich transducer. The main points to be considered for a successful application are the increase in power and efficiency, the improvement of design and calculation, the enlargement of the working area and the amplification of the range of useful frequencies. This paper presents a review of basic transducers for macrosonics, and discusses specific needs for different applications.

4.2 TRANSDUCERS FOR MACROSONICS

In the last ten years, most macrosonic transducers have been of the piezoelectric type. Therefore, the present description will focus on this kind of transducer. Nevertheless, it is interesting to point out recent developments of new and promising magnetostrictive materials (rare earth compounds), which show a great potential for high-power transducers (4.1).

Transducers for macrosonics are narrow-band transducers with a resonant working frequency within the range of 10 to 100 kHz, power capabilities of a hundred watts to several kilowatts and large vibration amplitudes.

4.2.1 Transducer Materials

Transducers are typically composite devices in which the core is a piezoelectric (or magnetostrictive) element, which changes dimensions in response to an electric (or magnetic) field. In modern transducers, the piezoelectric materials generally used are piezoelectric ceramics. It can be shown that piezoceramics present the highest electromechanical conversion and efficiency, and have, in general terms, the most favorable properties for macrosonic transducers (4.2). At present, the most suitable and popular ceramic materials are lead zirconate titanates (PZT), various commercial versions of which are manufactured in different countries. The leading position of the PZT ceramics is due to their strong piezoelectric effect and high Curie point, as well as to the wide range of properties which they offer by changing slightly their composition. Lead titanate and lead zirconate form a solid solution with an approximately equal molar concentrations of titanium and zirconium. The advantage of such compositions is that the mixtures richer in zirconium are rhombohedral, while those richer in titanium have tetragonal crystal symmetry. Dielectric constant, compliance and electromechanical coupling reach their maximum value at the phase boundary. Properties may be changed by replacing small fractions of the constituents with other elements such as calcium, strontium, barium, niobium, etc. Calcium, strontium or barium may be substituted for a fraction of the lead, and tin may be used instead of zirconium, resulting in a lowered Curie point and an increased permittivity (piezoceramic PZT-4). When the substituted atom has higher or lower valence, more profound changes occur. This is the case of the ceramic PZT-8, in which the electron-donor-doped lead titanate zirconates with, for instance, Fe^{3+} replacing $(Ti, Zr)^{4+}$, are characterized by decreased piezoelectric coupling, permittivity, compliance, mechanical and dielectric losses (4.3). The PZT-4 and PZT-8 compositions are particularly adapted for high-power transducers. They are both on the tetragonal side of the phase boundary, and the latter has remarkably low dielectric and elastic losses at high drive levels (Fig. 4.1).

The effect of internal losses on transducer performance can be severe with high-Q transducers and/or when operated at high amplitudes. Mechanical losses are more important with high-Q and dielectric losses with low-Q. For example, suppose that the power P per unit volume delivered to the mechanical and acoustical elements is set at 25 W/cm^3 at 10 kHz, and consider two different Q values: Q = 3 and Q = 30. Table 4.1 shows the values of dissipated power for several piezoceramic transducers (4.3). The advantages of PZT-8 material are clear.

Table 4.1 Dissipated power in W/cm^3 for piezoceramic transducers with Q = 3 and Q = 30 and P = 25 W/cm^3 at 10 kHz

Material	Electrical dissipated power (W/cm^3)		Mechanical dissipated power (W/cm^3)	
	Q = 3	Q = 30	Q = 3	Q = 30
PZT-8	0.10	0.01	0.09	3.0
PZT-4	0.20	0.01	0.23	9.9
PZT-5	1.35	----	3.75	---

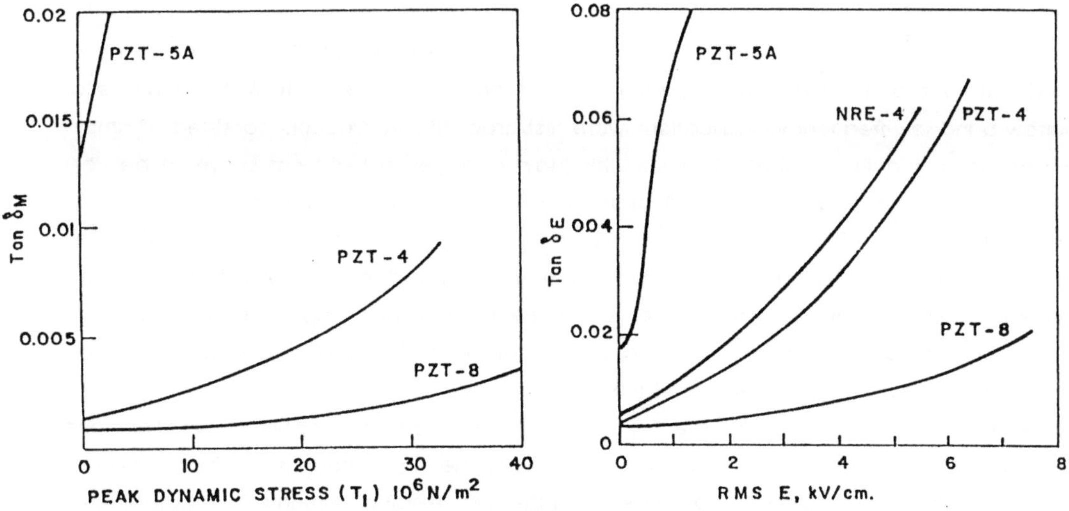

Fig. 4.1 Dielectric and mechanical losses for several piezoelectric ceramics

The overheating due to internal mechanical losses in the piezoelectric elements can produce, if it is excessive, a decrease in their electromechanical properties or even a depolarization.

As already mentioned, transducers are composite devices composed of active piezoelectric materials and other passive materials. In macrosonic transducers, the passive materials are generally metals. The metal choice affects the transducer power capacity and efficiency. Titanium alloys, specially Ti-6Al-4Va, are the best for high-power. Other useful metals are duralumin, aluminium bronze, naval brass and tool steel. Table 4.2 shows some properties of materials used in macrosonic transducers.

Table 4.2 Properties of materials used in macrosonic transducers

Material	Young's modulus (10^{11}N/m^2)	Density (10^3 kg/m^3)	Velocity of sound (m/s)	Fatigue strength (10^7 N/m^2)	Maximum strain (10^{-3})
Titanium alloy (Ti-6Al-4Va)	1.06	4.42	4900	72	6.80
Duralumin	0.74	2.80	5130	19	2.57
Aluminium bronze	1.43	8.50	4070	37	2.59
Naval brass	0.89	8.45	3240	15	1.69
Tool steel	2.20	7.80	5200	55	2.52

4.2.2 The Sandwich Transducer

The most characteristic piezoelectric transducer for high-power applications is the well-known sandwich transducer, which is reminiscent of the Langevin transducer (4.4). When ceramics for narrow-band low-frequency applications were first used, the transducers consisted of simple piezoceramic blocks or tubes. However, this plain arrangement did not prove to be very useful, especially for high-power applications, due to the low tensile strength of the ceramic and to the physical dimensions of the single-piece needed for such a low frequency. Because of these difficulties, Langevin's design was re-investigated and adapted to the new circumstances. The sandwich transducer is a half-wave resonant length-expander structure, which, in its simple version, consists of a disc, or of paired discs, of piezoelectric ceramic sandwiched between two identical metal blocks (Fig. 4.2).

When used in pairs, the piezoelectric ceramics are polarized in opposite directions and separated by an electrode (Cu-Be alloy) connected to the high-voltage lead. The electrode is therefore located at a node. Coupling between the piezoelectric elements at the metal end-sections and increase of the tensile strength are achieved by mechanically prestressing the assembly in the axial direction by means of a bolt.

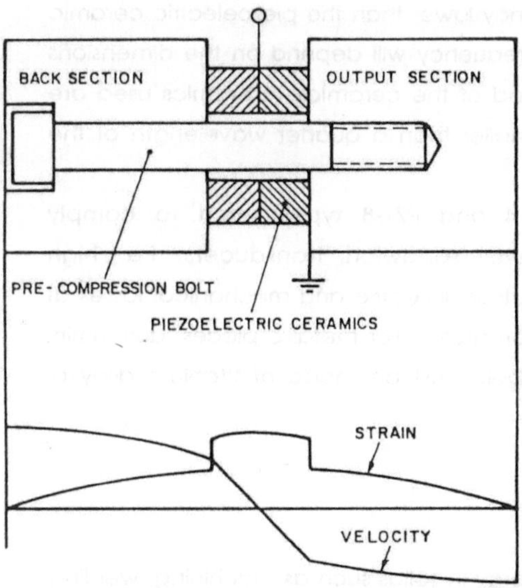

Fig. 4.2 Basic structure of prestressed sandwich transducer

In other designs, the central bolt is replaced by several peripheral bolts, eliminating the need for drilling the ceramics and the metal pieces (4.5). A typical compressive prestress may be in the order of 300 kg/cm^2. Evidently, the materials used for these bolts must have a high fatigue strength and low mechanical losses.

In practice, the sandwich construction does not generally follow the symmetrical structure just described. In most applications the output and back sections are made of metals of different density in order to increase the vibration amplitude on the radiating face and to improve matching to the load. Other cases require the node, which is the supporting point, to be in one of the metallic sections. The result in all these cases is an asymmetric sandwich structure as schematically shown in Fig. 4.3 (4.6).

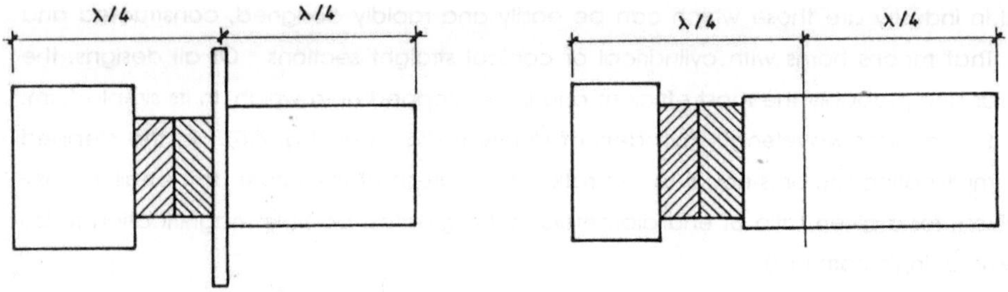

Fig. 4.3 Asymmetric sandwich transducers

The sandwich transducer will resonate at a frequency lower than the piezoelectric ceramic alone, when unloaded. The value of the resonant frequency will depend on the dimensions and acoustic characteristics of the metal sections and of the ceramics. Ceramics used are generally circular plates or rings, with a diameter smaller than a quarter wavelength of the extensional mode to avoid lateral coupling.

As mentioned above, piezoceramics of PZT-4 and PZT-8 types seem to comply adequately with the requirements of high-power sandwich transducers, i.e. high electromechanical coupling factor, high Curie point, low dielectric and mechanical losses at high drive and stable properties over time and temperature. For metallic pieces, duralumin, stainless steel or titanium are used. The prestress bolt must be made of titanium alloy or special steels.

4.2.3 Transducers for Applications in Solids

High-intensity applications of sonic and ultrasonic energy in solids such as machining, welding, metal forming, etc. are based on mechanical effects as a result of particle motion. In these processing applications, the sandwich transducer is also used, but it also includes a metallic transmission line of special shape, which produces a displacement amplification at the working end. These transmission lines are formed by half wavelength resonant elements, called mechanical amplifiers or horns, which are generally stepped, conical, exponential or catenoidal (Fig. 4.4). The horn must be designed to resonate at the same frequency as the transducer which is to drive it. A schematic drawing of a typical processing transducer with one exponential horn is shown in Fig. 4.5. A successful application of these transducers depends on obtaining the correct amplitude of movement at the horn tip. Amplitude depends on shape and dimensions. It can be difficult to design a horn having the correct shape and required gain. In such cases more than one horn section is used in order to obtain some amplification stages (Fig. 4.6). The intermediate horns, which are inserted between the transducer and the work horn are sometimes called booster horns. In general practice, the horns used in industry are those which can be easily and rapidly designed, constructed and modified. That means horns with cylindrical or conical straight sections. Of all designs, the most popular and probably the most efficient one is the stepped horn which, in its simple form, consists of two quarter wavelength cylinders of different diameter (Fig. 4.5). In the stepped horn, the amplification factor is equal to the ratio of the areas of the larger and smaller cross-section, which, for a given ratio of end diameters, is the greatest possible magnification to be obtained with a single horn (4.7).

In the horns described in Fig. 4.4, the cross-sectional dimensions were assumed to be smaller than one-quarter or one-third of the wavelength, in order to obtain a pure extensional mode of vibration. Nevertheless, most applications require wide horns or horns of large cross-sections. In these cases, one of the main problems is to obtain a uniform amplitude distribution on the radiating face avoiding radial or lateral vibration modes. For this purpose, large horns are constructed with slots running parallel to the direction of longitudinal motion (Fig. 4.7). This way, the large horn can be considered as an array of narrow horns in which the apparent cross-sectional dimension of each element is less than a quarter of a wavelength. The slots are also useful for heat dissipation. Large horns may be of different geometrical shapes and the procedure for designing must take into account that the bridging section joining the elements is an additional mass which presents a mechanical impedance to the longitudinal element of the horn (4.8). Recently some methods for modal vibration control of large horns have been proposed (4.9).

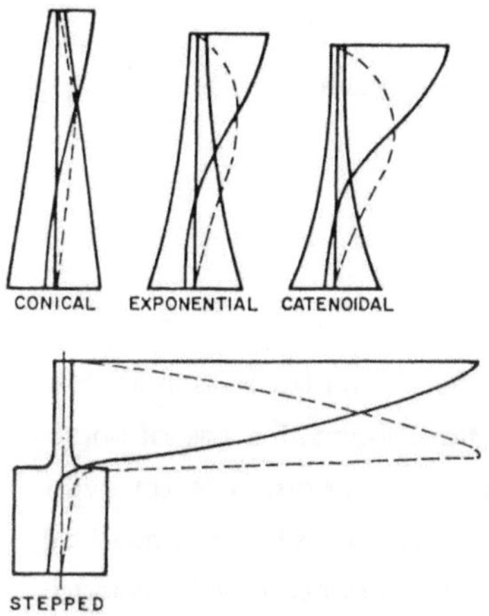

Fig. 4.4 Basic horn shapes and amplification effect

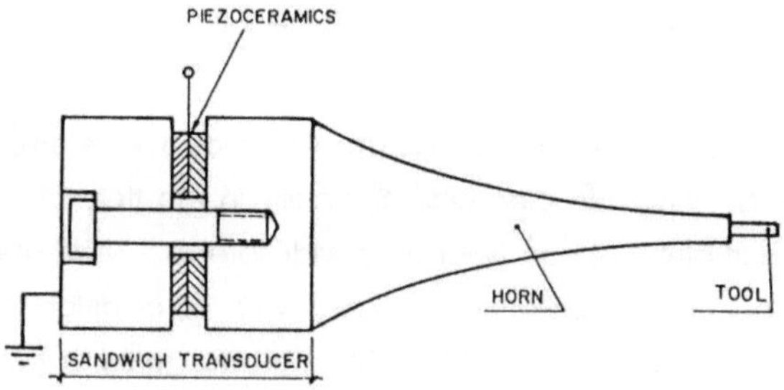

Fig. 4.5 Schematic structure of a typical ultrasonic processing transducer

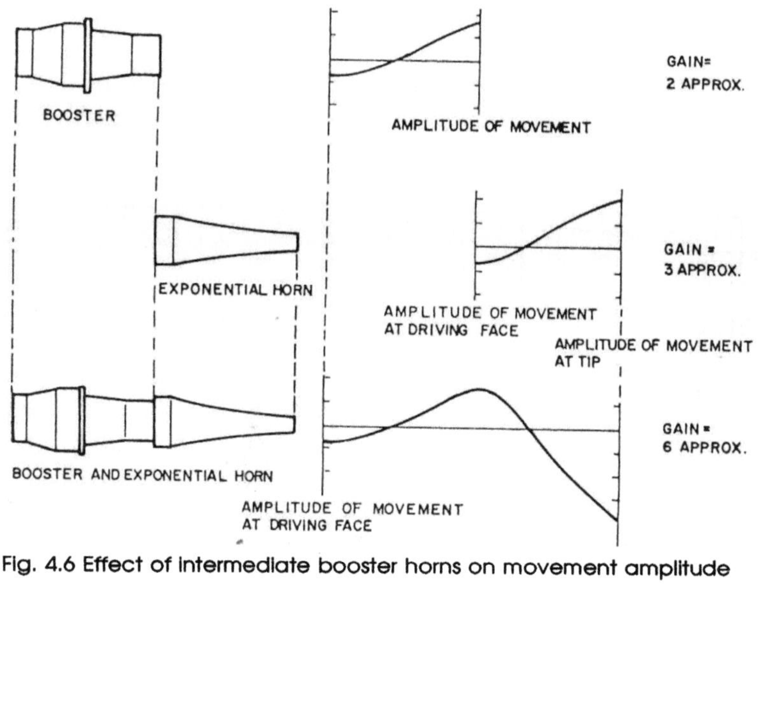

Fig. 4.6 Effect of intermediate booster horns on movement amplitude

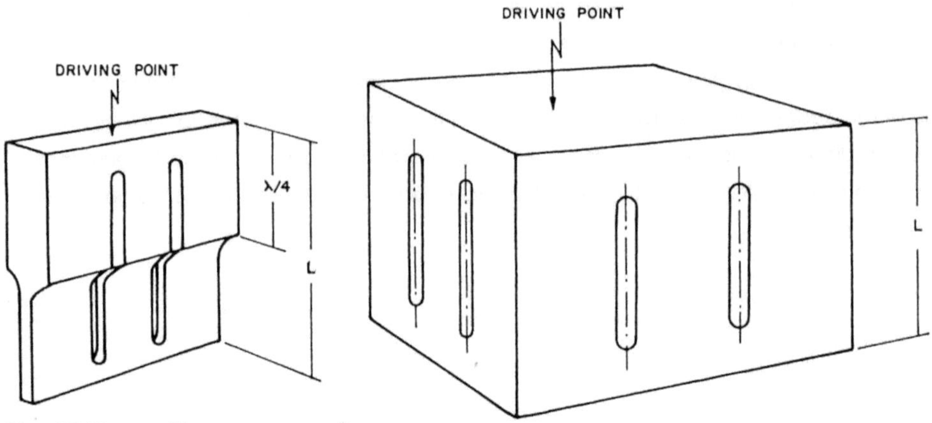

Fig. 4.7 Horns of large cross-section

4.2.4 Transducers for Applications in Fluids

The sandwich transducer, as described above, is employed directly for applications in liquids, such as ultrasonic cleaners. Specially in the field of underwater transducers, the use of single or multi-element sandwiches is widespread. Multi-element sandwiches incorporate an even number of piezoceramic discs (Fig. 4.8). In addition, the work-face area can be increased by tapering the output section. This can be useful in the matching of the transducer to its load. Most sonars are designed on the basis of different kinds of multi-element sandwich transducers and they are constructed very often in arrays of a number of elements in order to obtain the desired directivity and total power.

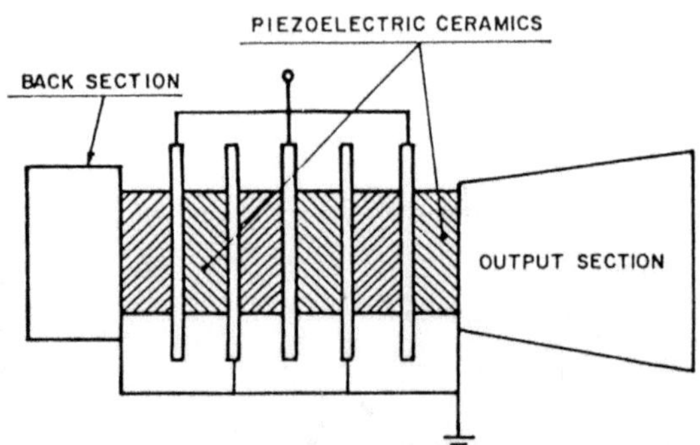

Fig. 4.8 Multi element sandwich transducer

Generation of ultrasonic energy in gases presents problems related to the low specific acoustic impedance and high absorption of the medium. Therefore, in order to obtain an efficient ultrasonic transmission and to produce high pressure levels, it is necessary to achieve good impedance matching between the transducer and the gas, large amplitudes of vibration and highly directional or focused beams. There are few airborne piezoelectric transducers and almost none of them seems to cover all the above-mentioned requirements. In recent years, a new type of stepped-plate transducer, in which these prerequisites have been attained (4.10), has been developed. It consists essentially of an extensive circular flexural vibrating plate of stepped shape driven at its center by a piezoelectrically activated vibrator (Fig. 4.9). The vibrator, formed by a sandwich transducer and a solid horn, is similar to the one used for applications in solids. The extensive surface of the plate increases the radiation resistance and offers the vibrating system good impedance matching with the medium. The special shape of the plate permits, despite its flexural vibration, a piston-like radiation to be obtained. As is well known, a flat-plate radiator vibrating in its flexural modes presents a poor directivity, due to phase cancellation. The procedure to design these radiators consists in alternately shifting along the acoustic axis the surface elements vibrating in counterphase on both sides of the nodal circles, half a wavelength of the radiated sound. The radiation produced will be in phase across the whole beam. Following this procedure, it is possible, with adequate modifications of the plate surface, to obtain any acoustic field configuration. Focused radiators have also been constructed (4.11). Stepped-plates vibrating up to seven nodal circles have been designed and constructed. In the latest version of this transducer, efficiencies of around 80%, beam width (at 3 dB) of 1.5 degrees and power capacities of 400 watts have been attained with a resonant frequency of around 20 kHz.

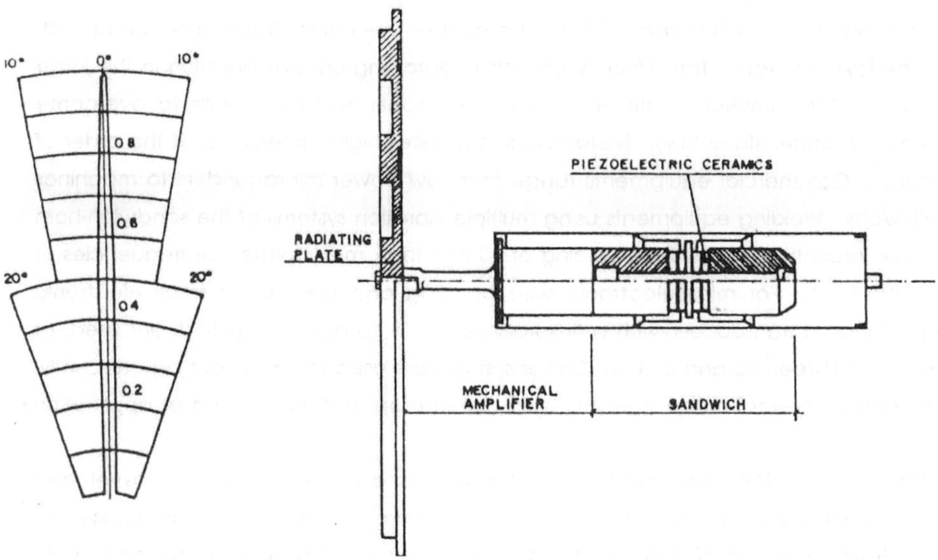

Fig. 4.9 Stepped-plate airborne transducer with a directional radiator and the corresponding directivity pattern.

Following the same principle used to modify the emission pattern in air of a flexural vibrating plate, but using a special delaying liquid or helical waveguides, a new type of directional underwater projector has also been designed and constructed (4.12).

4.3 TRANSDUCER NEEDS FOR APPLICATIONS

As mentioned above, many difficulties in making macrosonic applications to industry are related to transducer technology and coupling of the acoustic energy from the source to the process. The main general problems with transducers concern power limitations, efficiency, working area and designing methods. Concerning the latter point, much important progress has been made during the past few years, in particular with the application of the finite element method (4.13). Power limitations and efficiency generally depend on the materials and structure of the transducer: contributions can be expected from new titanium alloys with special thermal treatments, better qualities of piezoceramic materials and manufacture, and new designs. Reaching a uniform distribution of vibration amplitude on an extensive working area is a general aim linked to transducer structure and vibration modes (4.9).

4.3.1 Needs for Applications in Solids

The most popular and industrially established application concerns probably solids, plastic and metal welding. Macrosonic welding is a fast and clean process characterized by low heat. Frequencies used for plastic welding range from 10 to 40 kHz and the intensities needed from about 200 to 1000 W/cm^2. Commercial plastic welders operate predominantly at about 20 kHz and power outputs between 100 and 3000 W. Transducer thermal limitations are not critical, because of the low duty rate. The difficulty consists in obtaining uniform vibration in the large horns used to provide, simultaneously, several welding spots; some methods to overcome these problems are under study (4.9). Metal welding requires higher intensities, of the order of several kW/cm^2. Commercial equipments range from low power microbonders to machines of several kilowatts. Welding equipments using multiple vibration systems of the sandwich-horn type have been recently proposed for welding of 10 mm thick metal plates, at frequencies of 19 and 27 kHz (4.14). For microelectronic wire bonding and welding of small electronic components, standard transducers, with a flexurally resonant tip as bonding tool, are used, at frequencies lying between 40 and 60 kHz. Similar systems are used for metal cutting. Recently, a new bolt clamped flexural mode transducer has been proposed for this kind of application (4.15).

Metal forming, specially wire and tube drawing, is also an industrially developed application. The ultrasonic vibration is supplied in the same direction and/or transversely to the drawing. Transducers are of standard structure, but they must be adapted to the specific forming operation, e.g., for wire drawing, the transducer and the horn are perforated lengthwise. Frequencies of 20-30 kHz and powers from 0.4 to 14 kW are needed.

Machining of hard and brittle materials is a process which has been used for many years. At present, ultrasonic vibration is particularly applied to assist turning and drilling processes, increasing the cutting rates and extending the tool life. The frequency used is generally 20 kHz and vibration amplitudes between 10 and 40 microns are needed.

The ultrasonic vibration powder press is now used more frequently to obtain denser and more uniform compacting of metal or ceramic powders. Standard sandwich-horn transducers are used with different type vibration dies and punches. Longitudinal, radial and coupled modes are employed. The frequencies are within the range of 20 to 30 kHz and powers between 1 and 3 kW are used (4.14).

4.3.2 Needs for Applications in Fluids

Ultrasonic cleaning is the oldest and most widespread macrosonic industrial application. The main phenomenon responsible for cleaning is cavitation. The powers needed are of about 5 to 25 W/liter. At present, the most used operating frequencies are 25, 40 and 50 kHz. At 25 kHz, a more efficient cleaning is attained, because the shock intensity from the cavity collapse is higher, but a frequency of 40 or 50 kHz is better for the cleaning of delicate parts as in the electronic industry. Recently, a new technology, called microacoustic, has been developed for the cleaning of surface mount printed circuits working at 400 kHz. This new method does not rely on high transient forces produced by cavitation, but on the rapidly changing pressure gradient produced by an acoustic array, which generates a narrowly focused field. In this technology, the peak pressure is less than 50 psi and the displacement amplitude is one-tenth of the amplitude associated with cavitation (4.16). Transducers used for cleaning are generally sandwich transducers. Ultrasonic cleaning works well on sound-reflecting materials such as metals, glasses and plastics and is less efficient on sound-absorbing materials, such as rubbers or cloths.

Atomization of liquids is one of the few macrosonic applications which has been developed in a very wide range of frequencies: from 20 kHz for metal spraying to 3 MHz for medical nebulizers. Therefore, the transducers used belong to different types depending on the application.

Sonochemistry is a fast developing field for macrosonics. Frequencies from 20 kHz to several hundred kHz are used and ultrasonic intensities of 2-10 W/cm^2 are needed.

Gas removal from high-temperature metal and glass melts by using high-power ultrasonics is a process which has been demonstrated, but which is still under study. Degassing of carbonated drinks, photographic solutions and other liquids has been more successful in industry. Ultrasonic cleaning tanks working within a range of 20 to 40 kHz are normally used.

Macrosonics in gases is the least exploited field of application. This was due to the lack of efficient gasborne transducers. The new stepped-plate transducer previously described represents a major step in the industrial development of applications such as aerosol agglomeration and precipitation (4.17), defoaming (4.18), drying, separation of gaseous mixtures, pest control, sterilization, etc. For these applications, transducers within the range of

10 to 40 kHz, with power capacities from one hundred watts to kilowatts, are needed. Different versions of the stepped-plate transducer for frequencies between 20 and 40 kHz and power capacities of up to about 400 W have been constructed and tested. At present, a new model for 10 kHz with a radiator of 670 mm in diameter and an expected power capacity of about 1 kW is under development. This transducer will be useful for treating gas volumes of several cubic meters within a range of 10 to 20 m. This will probably lead to new indoor and open air applications.

4.4 CONCLUSION

As a conclusion, it is to be noted that the word macrosonics refers to a very wide range of processes, the majority of which are still far from being successfully applied at the industrial level. To increase applications of these processes, an essential point is the development of the right type of transducer to couple and distribute the acoustic energy in the medium to be treated.

REFERENCES

4.1 A.E. Clark, "Highly magnetostrictive rare earth compounds for high power acoustic projectors," in *Power Sonic and Ultrasonic Transducers Design*, Edited by B. Hamonic and J.N. Decarpigny, Springer-Verlag Berlin Heidelberg, 41-99 (1988).

4.2 J.A. Gallego-Juárez, "Piezoelectric ceramics and ultrasonic transducers," J. Phys. E. Sci. Instrum. **22**, 804-816 (1989).

4.3 D. Berlincourt, "Piezoelectric crystals and ceramics," Ultrasonic Transducer Materials, (Edited by O.E. Mattiat), Plenum Press New York, 63-124 (1971).

4.4 P. Langevin, French patents Nr. 502913 (29.5.1920), Nr. 505703 (5.8.1920), Nr. 575435 (30.7.1924).

4.5 S. Van Randeraat, and R.E. Setterington, *Piezoelectric ceramics*, 2nd edition, (Mullard Ltd, London, 1974).

4.6 E.A. Neppiras, "The prestressed piezoelectric sandwich transducer," Ultrasonics International Conference Proc., 295-302 (1973).

4.7 E.A. Neppiras, "Very high energy ultrasonics," British Journal of Applied Physics, **11**, 143-150 (1960).

4.8 D. Ensminger, *Ultrasonics*, 2nd edition, (Marcel Dekker Inc. New York 1988).

4.9 K. Adachi and S. Ueha, "Modal vibration control of large ultrasonic tools with the use of wave-trapped horns," J. Acoust. Soc. Am. **87**, 208-214 (1990).

4.10 J.A. Gallego-Juárez, "High power ultrasonic transducers for use in gases and interphases," in *Power Sonic and Ultrasonic Transducers Design*, Ed. B. Hamonic and J.N.Decarpigny, Springer-Verlag Berlin, 175-184 (1988).

4.11 G. Rodriguez-Corral, J.L. San Emeterio, and J.A. Gallego-Juárez, "Focused high-power ultrasonic transducer with stepped-plate radiator for industrial application in gases," Ultrasonics International Conf. Proc. 794-799 (1987).

4.12 R. Montero de Espinosa, and J.A. Gallego-Juárez, "A directional single-element underwater acoustic projector," Ultrasonics **24**, 100-104 (1986).

4.13 P. Tierce, and J.N. Decarpigny, "Design of high power transducers for use in macrosonics," in *Power Sonic and Ultrasonic Transducers Design*, Ed. B. Hamonic and J.N.Decarpigny, Springer-Verlag Berlin, 185-207 (1988).

4.14 J. Tsujino et al., "Some recent applications of high-power ultrasonic using multiple transducers and vibration system," IEEE Ultrasonics Symp. Proc., 593-598 (1986).

4.15 E. Mori et al., "New bolt clamped flexural mode ultrasonic high-power transducer with one dimensional construction," Ultrasonics International Conf. Proc., 257-261 (1989).

4.16 C. Schmidt, "New technology and trends in the industrial application of ultrasonic cleaning," Ultrasonics International Conf. Proc., 334-339 (1989).

4.17 E. Riera, F. Sarabia, and J.A. Gallego-Juárez, "Ultrasonic agglomeration of micron aerosols under standing wave conditions," J. Sound Vib., **110**, 413-427 (1986).

4.18 G. Rodriguez, J.A. Gallego-Juárez, A. Ramos, J.L. SanEmeterio, and F. Montoya, "High-power ultrasonic equipment for industrial defoaming," Ultrasonics International Conf. Proc., 506-511 (1985).

5. Class IV Flextensional Transducers

G. Bromfield

Admiralty Research Establishment, Portland, Dorset, DT2 2JS United Kingdom

The class IV flextensional transducer has attracted considerable interest over the past two decades and the emerging technology has been broad based and highly innovative. Progress has been made in solving the fundamental design limitations and comprehensive well proven analytical and finite element models have been developed and published. The evolving practical technology has often been cloaked in commercial confidentiality but Defence Technology Enterprises (DTE) decided to market ARE flextensional technology to industry three years ago. Companies have also invested in R&D and some are now publishing the often unique attributes of their designs as well as acoustic performance information. It is intended to use this information as the basis of a review of current state of the art technology. Initially design limitations are identified, followed by a more detailed discussion on the choice of shell materials and their effects on performance and cost. An overview of novel depth compensation methods, and the design of flextensional transducers for low cost expendable applications are also included.

5.1. INTRODUCTION

The class IV flextensional transducer has attracted considerable interest over the past two decades and the emerging technology has been broad based and highly innovative. Progress has been made in solving the fundamental design limitations and comprehensive well proven analytical and finite element models have been developed and published. The evolving practical technology has often been cloaked in commercial confidentiality but Defence Technology Enterprises (DTE) decided to market ARE flextensional technology to industry three years ago. Companies have also invested in R&D and some are now publishing the often unique attributes of their designs as well as acoustic performance information. This paper is divided into five parts:

- Simple theory - use ARE 350 design to illustrate design limitations - stack stress changes with depth, shell stress, and maximum electric field,
- Shell materials - Glass Composites, Carbon Graphite Composites, Aluminium, Steel, MMC. Practical construction methods, advantages and disadvantages, cost,
- The published performance of different shell type flextensionals and some comparison with FLEXT model,

- Low cost flextensionals for expendable applications,
- Depth compensation techniques.

5.2 DESIGN LIMITATIONS

The ARE 350 Hz flextensional (5.1) will be used to illustrate a typical set of practical design limitations for a class IV construction. A mechanical amplification of the small but high force displacement of the major axis is achieved by exciting a flexural resonance in the shell. The magnitude of these displacements for the ARE 350 Hz flextensional at full power together with the major axis dimension are shown in Fig. 5.1.

During operation the shell is normally subjected to relatively slow changes in hydrostatic pressure which magnify the tensile force applied to the drive stack. Both piezoelectric and magnetostrictive rare earth drive stacks are weak in tension and have to be biased with a compressive load capable of withstanding the dynamic stress excursion of the stack. For the flextensional, additional compressive bias is required to balance the tensile force that will be exerted at maximum operating depth. If a ceramic driver is used, the maximum allowable stress excursion is in the region of 80 MPa and this translates into a corresponding depth limitation of 150 m for the 350 Hz flextensional. The application of the bias stress can impose a further limitation as it can involve loading the shell across the minor axis and inserting a stack between the extended major axis (Fig. 5.2).

Under this severe loading condition, composite shell flextensionals are particularly prone to interlaminar shear failure and any fracture of a metal shell could be dangerous. Note that a compressive bias stress of 25 MPa is required for the dynamic stress excursion of the ceramic stack and this requirement progressively reduces output power as the depth is increased from 150 m to 220 m when the stack and shell become detached.

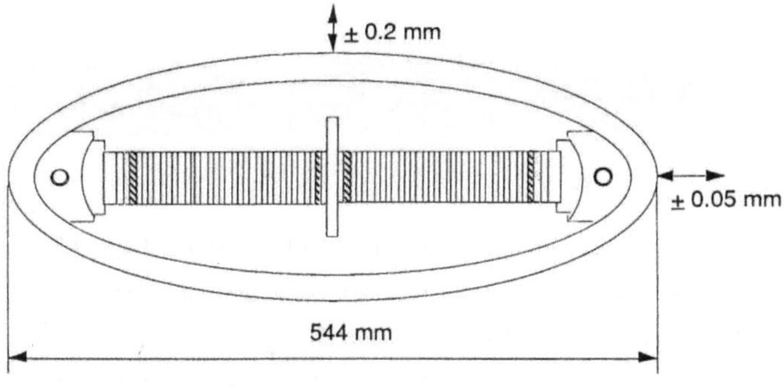

Fig. 5.1 ARE class IV flextensional transducer

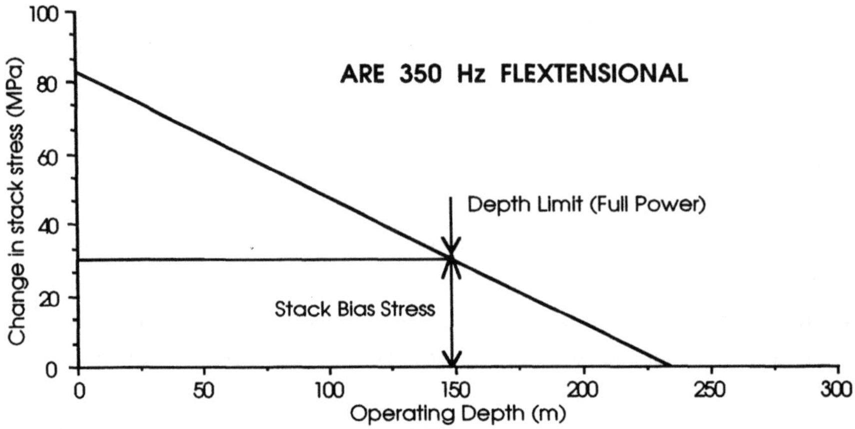

Fig. 5.2 Depth limit of ARE class IV flextensional

For a typical design the shell stress at the maximum operating depth and full power are initially estimated using the FLEXT5 computer program (5.2) as part of an iterative procedure. The final candidate designs are then analyzed using the finite element method to more accurately evaluate the principal stresses in the shell and stack. The FLEXT5 estimate for peak maximum stress (static plus dynamic) is 450 MPa. The PAFEC finite element analysis of the stack insertion stress given a value of 300 MPa and these values have to be conservatively compared with the maximum value specified for uni-directional 'E' glass composite of 760 MPa to decide if the design is viable. The application of high level electric field is the principal design limitation associated with a ceramic drive stack although heat dissipation can also be a limiting factor in some designs. Using a combination of careful electrode design, stack coatings and a SF_6 (sulphur hexafluoride) or Freon fill gas, fields of 6 KV rms/cm have been applied to experimental ARE and British Aerospace designs. However, a maximum field of 4 KV/cm is recommended for long term reliability but this could be increased for expendable transducers with a short life span.

5.3 SHELL MATERIALS

The choice of shell material depends on the performance requirements and cost of the flextensional. Metal matrix composite materials have been included because of their long term potential although current production costs are prohibitively expensive.

Table 5.1 can be used as a quick guide but there are a number of other factors that should be carefully considered before selecting a shell material. For example, the composites are susceptible to high levels of interlaminar shear during the stack insertion procedure and to both static and dynamic creep thereafter. Metal shells are generally more expensive and have potentially dangerous failure modes due to the energy stored in the pre-compressed stack. Composite shells are normally filament wound on a removable mandrel resulting in a high initial tooling cost but potentially low unit cost production. 'E' glass composite is very inexpensive but carbon graphite is considered relatively expensive at about 15 times the material cost. However, unless MMC is used the base material cost of the shell does not significantly effect the unit cost of the flextensional. Aluminium shells are machined from solid using a conventional milling or spark erosion technique while steel shells can be machined from solid or alternatively fabricated from rolled sheet with electron beam welded end pieces. As a general rule, it is more economical to construct the larger low frequency metal shell flextensional from a number of narrow shell segments. Each segment usually has one stack inserted across its major axis and typically three of these units are cascaded together to form one flextensional. In the FLEXT program, shell materials are characterized by their modulus and density and the effect of this change to the existing ARE 350 Hz graphically demonstrates the choice of shell in Fig. 5.3.

Table 5.1 Shell materials and their effects

MATERIAL	SOURCE LEVEL	BANDWIDTH	COST	SIZE/WEIGHT
'E'GLASS	LOW	MEDIUM	LOW	SMALL
'S'GLASS	MEDIUM	MEDIUM	LOW	MEDIUM
GRAPHITE	MEDIUM	WIDE	MEDIUM	LARGE
ALUMINIUM	MEDIUM	MEDIUM	MEDIUM	MEDIUM
STEEL	HIGH	LOW	MEDIUM	LARGE
MMC	VARIABLE	VARIABLE	HIGH	VARIABLE

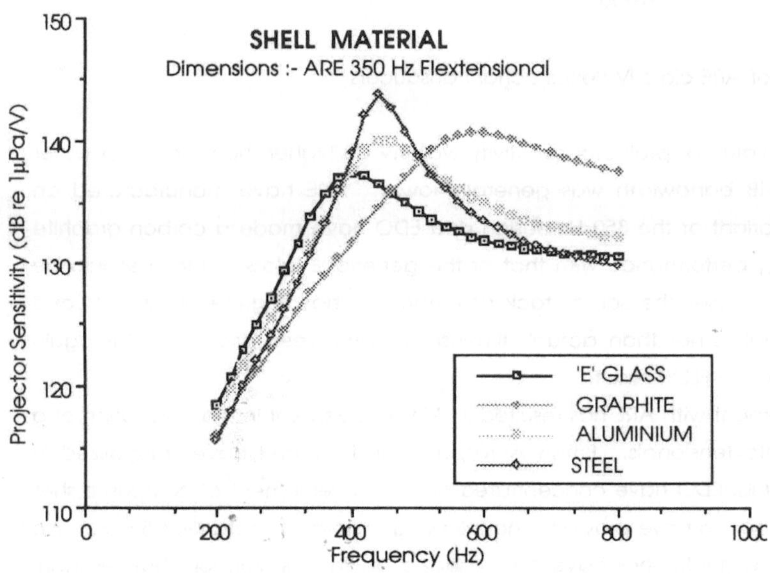

Fig. 5.3 Importance of shell material

5.4 PUBLISHED PERFORMANCE DATA

Published calibrations of low frequency transducers have to be carefully interpreted before realistic comparisons can be made. The most common ambiguity arises from a difficulty in achieving ideal freefield conditions, similar test depths, water temperature, test distance and measurement techniques. The ARE 350 flextensional has been calibrated in deep freefield conditions and the results (Fig. 5.4) compared with the shallow (18 m) Crystal facility located in Portland Harbor.

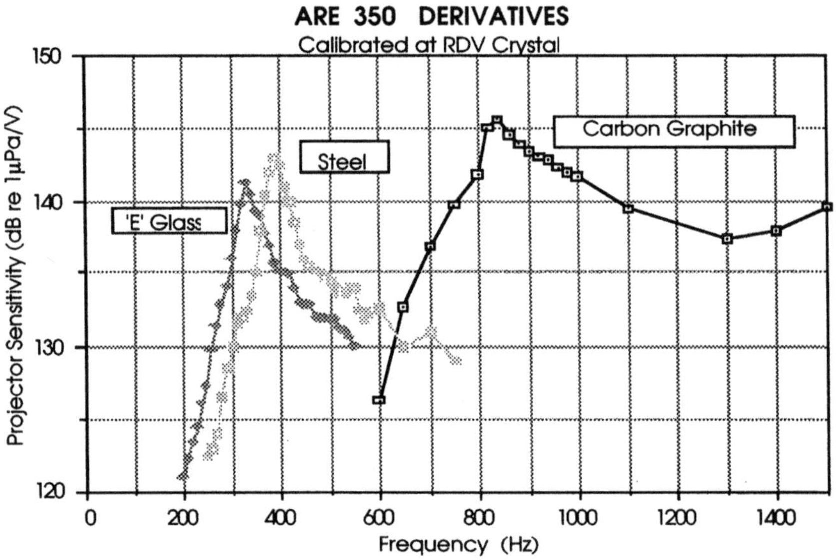

Fig. 5.4 Calibration results of ARE class IV flextensional transducers

The Crystal measured value of projector sensitivity was 1.9 dB higher than the deep water calibration whilst the 3 dB bandwidth was generally lower. ARE have manufactured an experimental steel shell variant of the 350 Hz design and EDO have made a carbon graphite version. When comparing performance with that of the generic 'E' glass model it should be noted that the EDO design uses the same stack assembly but has a generally smaller and thicker shell. Also nominal rather than actual dimensions have been used and this again reflects in the accuracy of the FLEXT result.

The DTE licence agreement with ARE has resulted in ARE involvement in the calibration of a number of commercial flextensionals. British Aerospace and Ferranti have specialized in aluminium shell designs whilst EDO have concentrated on the development of composite shell flextensionals. These companies have provided the following calibration data (Figs 5.5, 5.6 and 5.7) to which performance guide lines have been added on the assumption that resonant frequency can be scaled by transducer size.

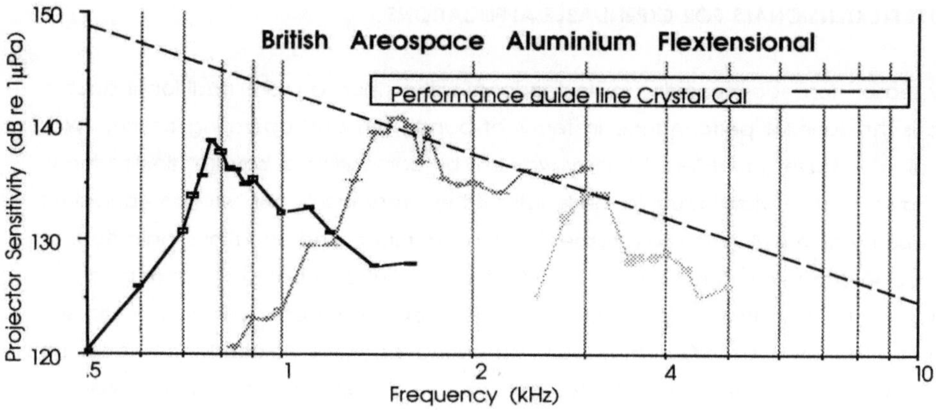

Fig. 5.5 Calibration data and performance guide lines (BAe)

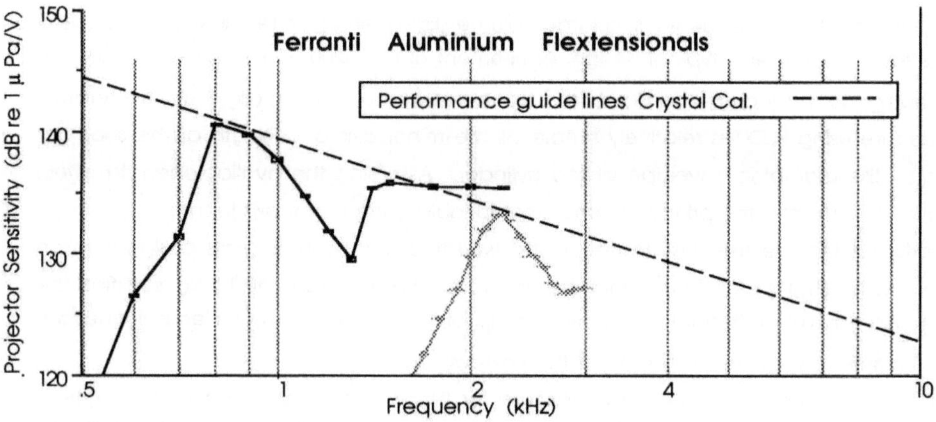

Fig. 5.6 Calibration data and performance guide lines (FERRANTI)

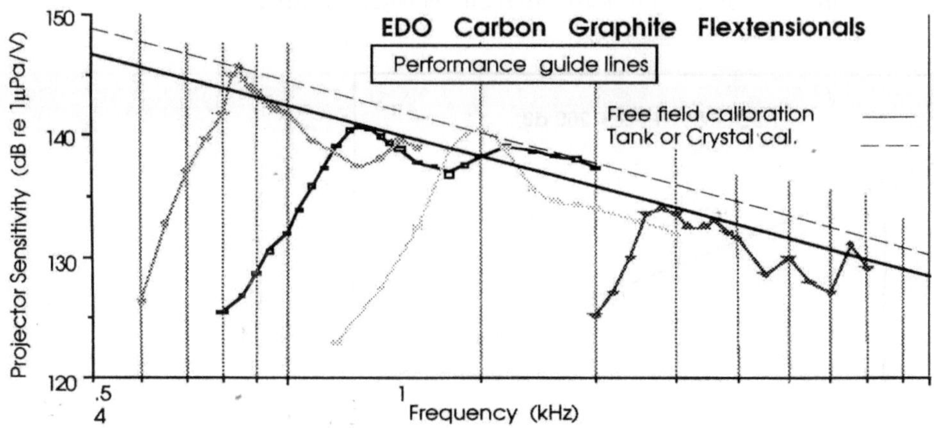

Fig. 5.7 Calibration data and performance guide lines (EDO)

5.5 LOW COST FLEXTENSIONALS FOR EXPENDABLE APPLICATIONS

The primary reason for recommending a flextensional rather than a more traditional double flexural disc is the superior performance in terms of bandwidth and operating depth. W.J. Marshall et al. (5.3) have quantified the improvement by comparing a 'square' flextensional, that is, one having a shell width equal to the length of the ellipse major axis, with an equivalent size flexural disc by normalizing size with respect to wavenumber. The equations indicate that the mechanical Q of a flextensional is lower by a factor of about 2 for small units and up to 4 or more for large units. The derived relationship between depth capability versus acoustic size also indicates that for a specified maximum operating depth a flextensional would be approximately half the size of a flexural disc. The size weight and cost of a transducer are particularly important in the design of expendable devices. A comparison with a tonpilz longitudinal transducer gives an increasing advantage to the flextensional at frequencies below about 4 kHz. At low frequencies, a large volume displacement is required to produce a given source level. Consider a typical design requirement of F_r = 1200 Hz, source level = 200 dB, operating depth = 300 m and sized to fit within a 150 mm diameter housing (Fig. 5.8). An iterative design procedure using FLEXT is relatively simple as the minor axis and length of the shell are constrained by the diametric envelope of the cylinder. Assuming the available length within the housing is not restricted the principal remaining variable is the major axis length.

Other variables include the shell material and wall thickness. An 'E' glass design with an overall major axis length of 0.33 m and an estimated total weight of 11 kg satisfies the requirement. An equivalent 'tonpilz' design with a Q_m of 4, would be in the order of 4 times the weight and 1.5 times the cylindrical length of the flextensional.

A potentially low cost method of manufacture for flextensional transducers was originally conceived at ARE and further developed by EDO under their licence agreement. The EDO trade name of FLEXWRAP relates to the assembly procedure which involves wrapping the shell around the stack rather than inserting a stack into a pre-formed shell (Fig. 5.9).

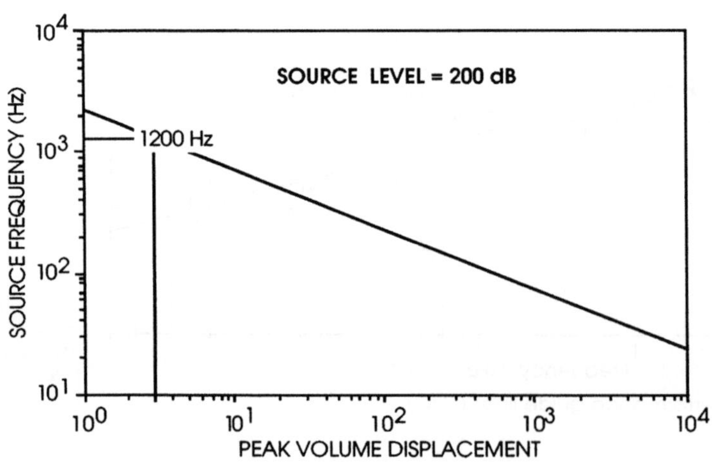

Fig. 5.8 Source frequency vs. peak volume displacement expressed in cm^3

Fig. 5.9 Expendable flextensional

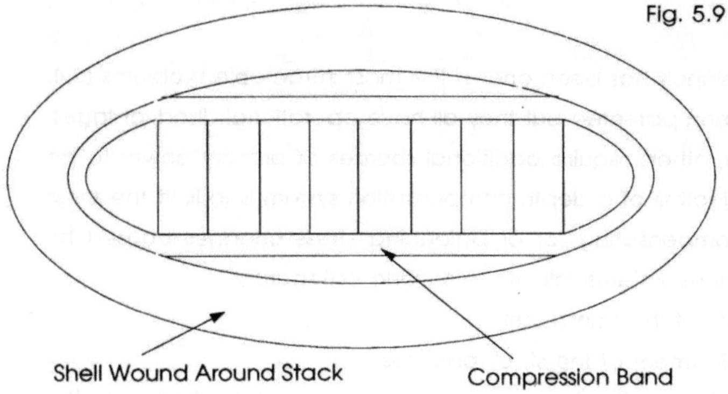

The cost savings are achieved by building a stack or stacks that do not require post assembly machining, the use of a universal rather than customized shell winding mandrel, and no additional ceramic prestressing procedure. The compression band is wound at high back tension and bonded to both the ceramic and to the shell. Two removable semi-elliptical formers are attached to the stress band before winding the outer shell. This type of construction is only suited to higher frequency flextensionals, above about 3 kHz as the thicker compression bands required for lower frequency designs would have a significant compressive modulus and thus reduce coupling to the shell. The change of stack stress with hydrostatic pressure has been measured for an EDO 3500 Hz design and shows (Fig. 5.10) that this transducer could operate at a depth of 300 m with a dynamic stress excursion of at least 8 MPa.

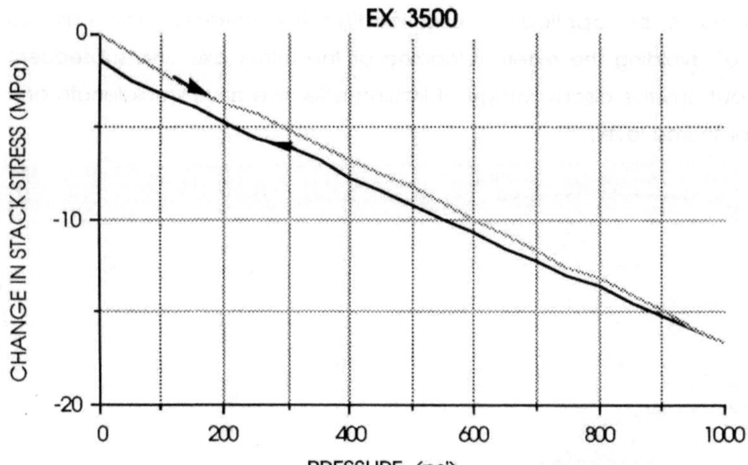

Fig. 5.10 Change of stack stress with hydrostatic pressure

5.6 DEPTH COMPENSATION

The depth compensation of flextensionals has been one of the most intractable problems (5.4, 5.5). Several novel methods have been patented but they all have operational disadvantages. Some require complex engineering, others require additional sources of primary power to be supplied to the transducer. The objective of a depth compensation system is to limit the stress excursion in the drive stack by compensating for or balancing stress changes caused by changes in hydrostatic pressure. These systems fall into two basic categories:

- pressurizing the internal cavity of the transducer,
- mechanical control and adjustment of the stack prestress.

The primary disadvantage of pressurizing the internal cavity is the resulting change in the flexural resonant frequency due to the decrease in the compliance of the fill gas as the pressure is increased. Practical examples of this method of compensation include Scuba divers valve pressure balanced system using compressed air, pumped gas to and from external pressure vessel developed by ARE and the Hymatic Engineering Co, dual air/water bladder developed by DREA Canada (5.6).

The mechanical control of the stack prestress has to be capable of maintaining good dynamic coupling between shell and drive stack and yet de-couple or reduce the steady state bias stress induced by changes in hydrostatic pressure. Ultimately, the design of non pressure balanced models will be limited by the maximum value of induced tensile shell stress, but interlaminar shear could be the limiting factor for composite shells.

British Aerospace (5.7) have patented a sliding wedge (Fig. 5.11) that is incorporated within the nodal plate and can be adjusted during operation by a closed loop external power source and depth sensor. The design of the wedge is substantial enough to allow the full value of compressive stack prestress to be applied by end loading the wedge. This has the considerable advantage of avoiding the external loading of the minor axis and subsequent over-stressing of the shell but has the disadvantage of limiting effective drive stack length and thus reducing the potential source level.

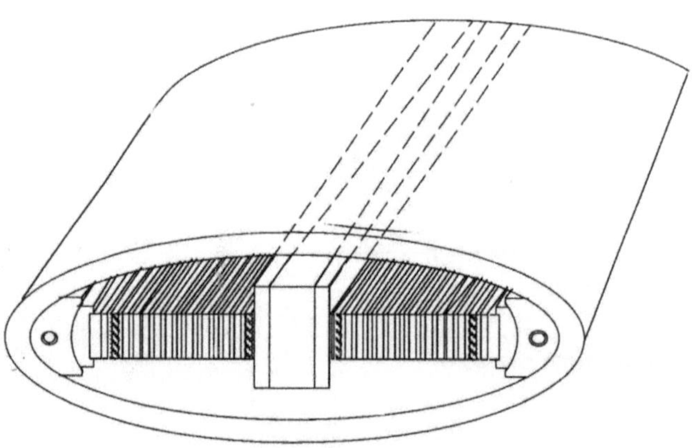

Fig. 5.11 Wedge for stack prestress adjustment

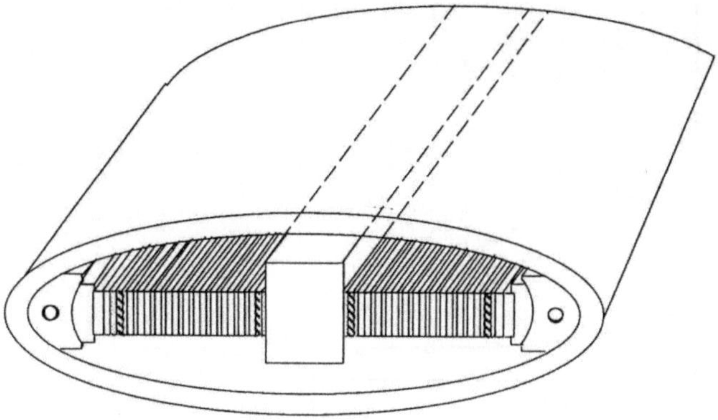

Fig. 5.12 Dynamic filter incorporated nodal plate

Thomson-Sintra ASM (5.8) have developed a dynamic filter (Fig. 5.12) that is also incorporated into the middle of the drive stack. The detailed performance information has not been published but a diagram shows that considerable space is required and this again reduces the effective length of drive stack. The compliance of the filter has to be low compared with the stack over the operating frequency band in order to reduce the degradation in the effective coupling coefficient. Conversely it has to be high at the very low frequencies associated with changes in operating depth.

EDO have patented a self locking wedge band (5.9) that is attached to the apices of the elliptical shell and locates the stack by gripping across the 'D' shaped couplers (Fig. 5.13). The stack prestress is applied by either a center bolt or a composite band wound at high back tension. The value of the compressive bias stress is calculated from the dynamic peak to peak stress at full power plus an extra amount to allow for the stretching friction between shell stack assembly.

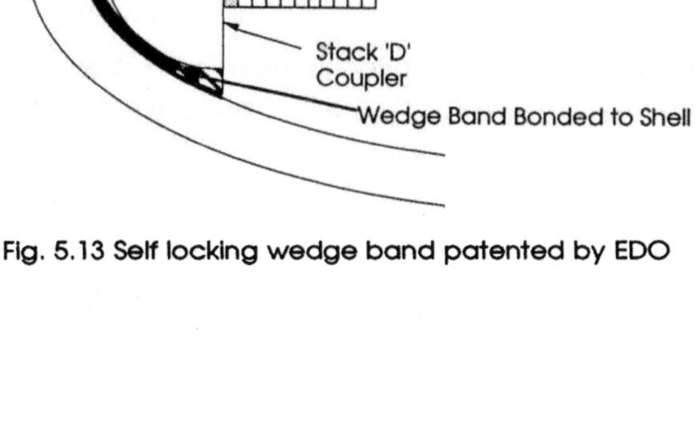

Fig. 5.13 Self locking wedge band patented by EDO

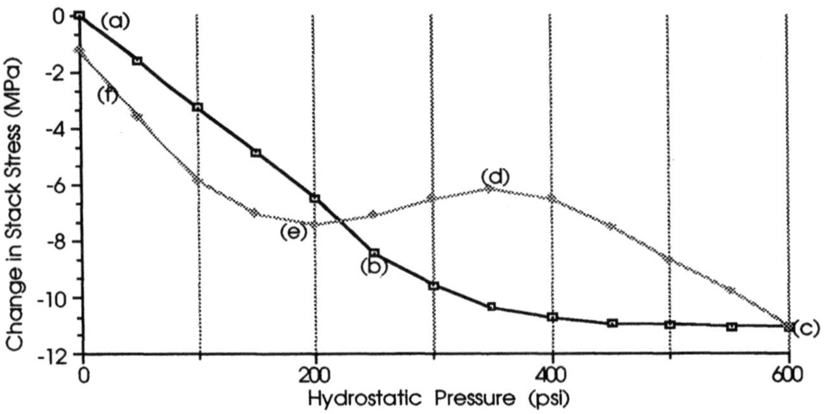

Fig. 5.14 Change in stack stress

A graph of change in stack stress (Fig. 5.14) for an experimental flextensional demonstrates the mode of operation over a 600 psi depth cycle. Initially, a linear change in stress with increasing depth is associated with the wedge locked in position ((a) to (b) on the graph). The stack is stretched until the wedge starts to slip and the sliding friction is progressively reduced as maximum depth is approached ((b) to (c)). A reduction in pressure locks the wedge and a linear change occurs until the wedge starts to slip at point(d).

The wedge continues to slip until sufficient friction and change of wedge angle again locks it at point (e). The wedge then remains locked as the pressure is returned to ambient at point (f). By careful design of wedge angle and 'D' coupler geometry the change in stress associated with the decreasing pressure between (c) and (d) could be adjusted to handle the peak dynamic stress excursion of the stack. For example, the experimental flextensional illustrated could be operated at a pressure of 500 psi by initially applying a pressure of 600 psi, and reducing the pressure by 100 psi to lock the wedge. This is clearly an operational disadvantage but transducers of this type are capable of operating at depths limited only by shell stress or wall thickness, have relatively small changes in resonant frequency and source level over the depth cycle and do not require a secondary power source.

5.7 CONCLUSION

The selection of a shell material for a flextensional transducer is dependent upon a number of factors including acoustic performance, method of manufacture and cost. Theoretical and practical performance data are presented for different shell materials.

The published performance of flextensionals has been collated for three manufacturers and gives useful practical performance guide lines for aluminium and carbon graphite shells. The design of a low cost flextensional for expendable applications is presented. A variety of depth compensation methods are reviewed.

REFERENCES

5.1 G. Bromfield, and G. Ash, 'The ARE 350 Hz Flextensional Transducer,' UDT Conference Proceedings (1988).

5.2 J.L. Butler, and T.J. Pierce, 'A Desktop Computer Program For a Flextensional Transducer,' Proc. Inst. Acoust., **9**, part 2 (1987)

5.3 W.J. Marshall, J.A. Pagliarini, and R P White, 'Advances In Flextensional Transducer Design,' Trans. IEEE Oceans 124-129 (1979).

5.4 G. Brigham, and B. Glass, 'Present Status in Flextensional Technology,' J Acoust Soc Am **68**, 1046-1052 (1980).

5.5 J.R. Oswin, and A. Turner, 'Design Limitations of Aluminium Shell Class IV Flextensional Transducers,' Proc Inst Acoust. **6**, part 3, 94-100 (1984).

5.6 B.A. Armstrong, and G.W. McMahon, 'Discussion of the Finite Element Model Performance of Ring-Shell Projectors,' IEE Proceedings, **131**, part F, No 3. (1984).

5.7 K.J. Ponchard, BAe Patent for wedged nodal plate. GB 8522652 dated 12/9/85

5.8 Y. Lagier, Th. Mazoyer, D. Guyomar, Ph. Dufourcq, and P. Ventura, 'Flextensional Transducer Modelling,' UDT Conference Proceedings (1990).

5.9 G. `Bromfield EDO Patent Application filed 5 may 1988 for self locking wedge. USA 4,845,687 (Class 367/158)

6. The Ring-Shell Flextensional Transducer (Class V)

G.W. McMahon

Defence Research Establishment Atlantic, P.O. Box 1012, Dartmouth,
Nova Scotia, B2Y 3Z7, Canada

The ring-shell flextensional transducer comprises a piezoelectric ceramic ring sandwiched between two spherical shell segments that are driven in flexure. Research on the ring-shell projector (RSP) at DREA began in the mid 70's with a concave shell configuration for relatively shallow operation, and has progressed to the present high-power design that uses convex shells and an internal water bladder for depth compensation. The DREA finite-element program MAVART has been a valuable tool in RSP research. New designs include a versatile unit that features removable, interchangeable shells, and a robust, shock-hardened design. Ring-shell transducers are marketed by Sparton of Canada Limited, who have been actively involved in their development for several years.

6.1 INTRODUCTION

The ring-shell transducer is a versatile flextensional design for generation of low frequency, high power underwater sound. It can include an internal, passive pressure compensation system that allows greater operating depths at lower frequencies than uncompensated flextensionals of similar size and weight. Ring-shell transducers have been under investigation at Defence Research Establishment Atlantic since the mid-70's and are now manufactured and marketed by Sparton of Canada Limited. This paper will present an historical review of the ring-shell projector (RSP) development, briefly describe finite-element modeling of RSP's, discuss some array interaction effects, and introduce two new RSP design configurations.

6.2 HISTORICAL REVIEW

The original idea for the ring-shell transducer was conceived by Frank Abbott in 1955 (6.1) at the US Naval Electronics Laboratory (now Naval Ocean Systems Center) in San Diego, and a patent was issued in 1959 (6.2). It was called the Compliant Membrane transducer and consists of a radially-poled barium titanate ring sandwiched between two concave copper shells (or membranes). It was re-invented at least twice since then. The patented Multidyne™ hydrophone (6.3), used by Seismic Engineering of Texas in their towed arrays, employs the same principle. Then at DREA, being at first unaware of this earlier work, we began our studies in 1972. Apparently, no follow-on work had been conducted at NEL, so we continued our research.

Our first low frequency ring-shell projector was the 605 Hz unit, shown in Figs. 6.1 and 6.2, which used a tangentially-poled, lead zirconate-titanate ring, driving concave aluminum shells. Electrical connections to the ceramic staves were buried in a fiberglass/epoxy wrapping, which was applied under tension to provide a compressive bias to the ring. The ends of the electrodes on the ceramic staves were insulated from the metal shells by plates of alumina. Its measured source level was 201.5 dB re 1μPa at 1m at the 605 Hz resonance and its -3dB bandwidth was 107 Hz. This transducer had a diameter of 36 cm and a weight of 19 kg. An attractive feature of this design is the favorable stress imparted to the ceramic ring by hydrostatic pressure due to submersion.

Transducers of a similar design, but larger and resonant at 205 Hz, were built in 1978-79, and were used in a two-element vertical array in propagation and reverberation experiments in the North Atlantic during 1980. A source level of 206 dB was achieved from a single transducer at a depth of 30 m, with an electro-acoustic efficiency of 90% at the 205 Hz resonance.

The 205 Hz RSP's were the first transducers that were designed with the aid of our finite-element program MAVART, which was delivered to us in 1977. The shells were manufactured on the basis of MAVART predictions for a resonance frequency of 200 Hz. Since then, MAVART has proven to be a powerful tool for transducer design and is discussed in more detail in a later section.

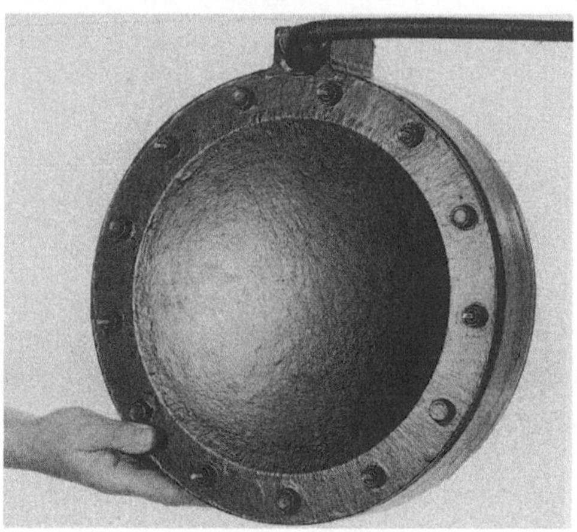

Fig. 6.1 DREA's first high power ring-shell projector. It delivered 1200 W of acoustic energy into the water at its 605 Hz resonance, with an electro-acoustic efficiency of 83%

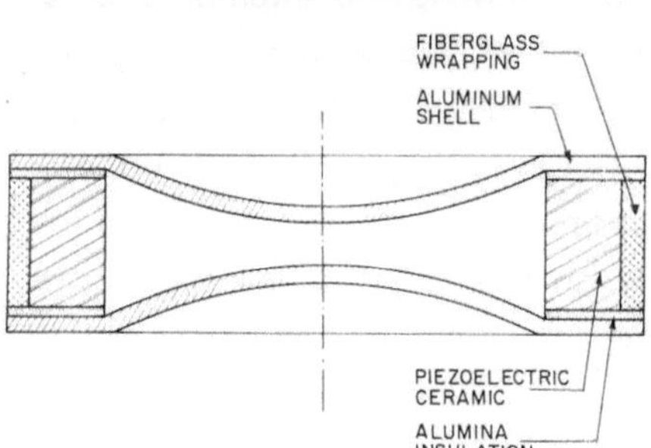

Fig. 6.2 Cross-sectional diagram of the RSP shown in Fig. 6.1

Further research on RSP's during the 1980's was aimed at improving their reliability, increasing their output power, and extending their depth capability. The first depth-compensated design, shown in Figs. 6.3 and 6.4, appeared in 1980 (6.4). It features an internal water bladder for passive depth compensation and uses a convex configuration for the aluminum shells. The shells are bolted and bonded to the ring, the bolts passing through several aluminum staves in the ring. A valve, not shown in the diagram, allows dry compressed gas to be introduced into the transducer to extend the depth range. Its resonance frequency is 220 Hz and bandwidth 14 Hz, measured at a depth of 20 m. For a low frequency, thin-shelled unit such as this, there is some increase of resonance frequency with increasing depth, due to the decreasing compliance of the internal gas.

In 1981, a technology transfer contract was negotiated with Sparton of Canada to begin manufacture of ring-shell transducers by a Canadian company. Two units, resonant at 340 Hz, were delivered to DREA in 1982. They were used in sea trials during 1983 in a vertical line array, which delivered a source level of 205 dB. Since that initial contract, a number of design improvements have been introduced and patents obtained (6.5,6.6). Sparton has continued to be heavily involved in the development, manufacturing, and marketing of ring-shell transducers. Their RSP's have so far been made in two standard ring sizes, (**A**) 46 cm and (**B**) 88 cm in diameter. The fundamental resonance frequencies have ranged from 340 to 1000 Hz, and source levels up to 207 dB from the smaller units and 214 dB from the larger units have been attained. Design details will be presented later.

Fig. 6.3 The first depth-compensated ring-shell projector

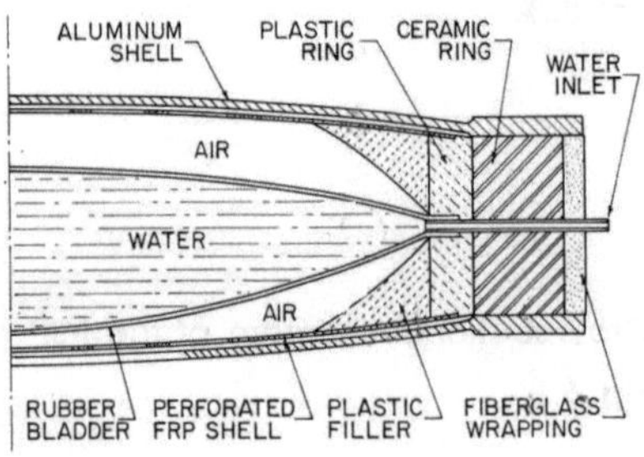

Fig. 6.4 Cross-sectional view of the projector in Fig. 6.3, showing the internal bladder partially filled with water, as it might appear near the middle of its operating depth range. Not shown are vents at the edge of the bladder support that allow the air pressure on both sides of the bladder to equalize

6.3 FINITE-ELEMENT MODELING

The name MAVART is an acronym for 'Mathematical Model for Analysis of the Vibration and Acoustic Radiation of Transducers'. MAVART can model the steady-state response of an axially-symmetric, piezoelectrically-driven structure immersed in an acoustic fluid of infinite extent. The solid structure is enclosed in a sphere of the fluid, and both are discretized into finite elements. Figure 6.5 shows a typical finite-element grid for a ring-shell transducer. Only one-quarter of the cross-section is required in the model because of the symmetry. The internal depth compensation system is normally not modeled, as it has only a small effect on the dynamic performance.

The pressures on the outer boundary of the sphere are matched to spherical harmonic solutions of the Helmholz integral equation in the infinite fluid. The solution returned by MAVART consists of complex values of the degrees-of-freedom (DOF), which are: displacements at solid nodes, pressures at fluid nodes, and voltages at nodes that have an electrical DOF.

The complex solution is postprocessed to yield farfield transmitting and receiving responses, directivity patterns, electrical admittance or impedance, stresses and strains in the solids, and nearfield pressures in the fluid. The structure need not be driven symmetrically, in which case the solution in the external fluid involves spherical harmonics of higher order than zero in the azimuthal coordinate.

MAVART assumes that the structure has infinite-fold axial symmetry, but it can accommodate moderate deviations from this symmetry, such as in the RSP, if the appropriate material properties are used. For example, a ring comprising alternating wedges of metal and plates of ceramic is modeled as a composite material having homogeneous properties.

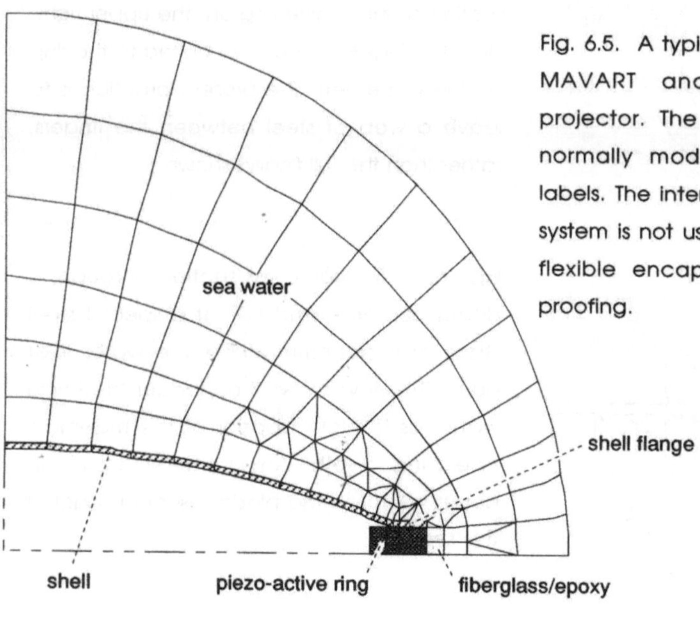

Fig. 6.5. A typical finite-element grid for MAVART analysis of a ring-shell projector. The five materials that are normally modeled are indicated by labels. The internal depth compensation system is not usually modeled, nor is the flexible encapsulant used for water-proofing.

6.4. DESIGN DESCRIPTION

The ring-shell transducer that has evolved includes several design improvements over the earlier transducer shown in Fig. 6.3. A typical RSP is illustrated in Figs. 6.6 and 6.7, which show a plan view and a radial cross-section, respectively, of an A-size, 350 Hz unit. Some of the most important new features incorporated into this design are as follows:

6.4.1 The Depth Compensation System

The bladder support structure is decoupled from the ring to reduce damping and improve efficiency, which is now over 90% at resonance. The bladder is made in a toroidal shape to allow free dynamic flow of the internal gas.

6.4.2 The Composite Ceramic/Metal Driving Ring

The active ring is made with one metal wedge for every pair of ceramic plates, and the wedge material is steel. The wedges are slightly longer axially than the ceramic plates to remove the point of high dynamic stress at the shell edge from the ceramic. A major advantage of the ring-shell over other flextensional designs is the excellent thermal coupling of the driver to the water. Dielectrically generated heat in the ceramic is quickly dissipated, and the full-power duty cycle can be 100%.

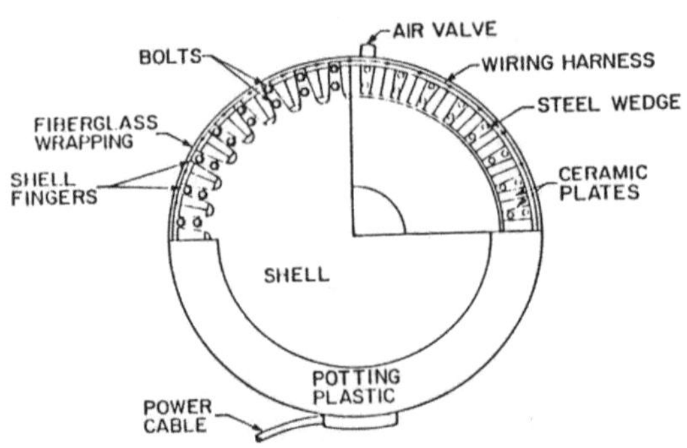

Fig. 6.6 A partially cut-away plan view of a 46 cm diameter RSP, showing a cross-section of the active ring on the upper right, and the fingered steel shell bolted to the ring on the upper left. The present practice is to leave a web of steel between the fingers, rather than the full fingers shown

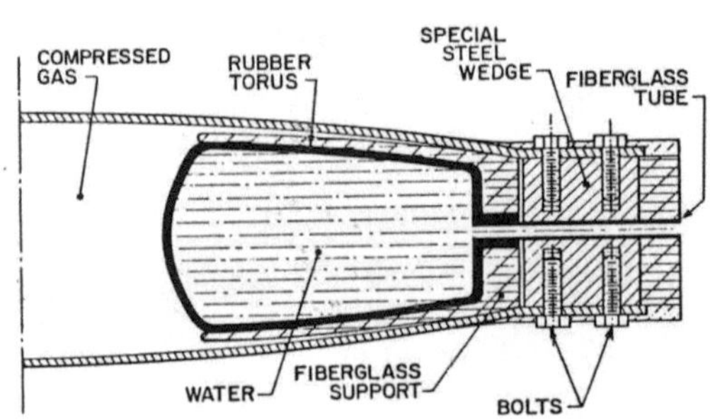

Fig. 6.7 Radial cross-section through a depth compensated RSP at a special steel stave that accommodates the water inlet port. This view shows the bladder filled with water, as it might appear at the maximum operating depth. Any further increase in depth will force the bladder support against the shells

6.4.3 Fingered Steel Shells

The formerly solid shell flange has been largely cut away between each bolt, leaving only a thin web of metal (webbed fingers). This reduces the hoop stiffness of the flange and hence, reduces its clamping effect on the ring driver. The resulting increased coupling of strain energy from the driving ring to the shells increases the source level. Also, the cut-away increases the electrical break-down path from the shell flange to the high voltage electrode of the ceramic. Steel shells provide greater strength and fatigue resistance than aluminum shells, and MAVART modeling indicates that only a small sacrifice in source level and bandwidth will result. Some designs use only one bolt per wedge instead of the two shown in Figs. 6.6 and 6.7.

6.5. PERFORMANCE

Ring-shell transducers have been manufactured for DREA by Sparton of Canada in both standard sizes. The essential characteristics of these transducers are as follows: (**A**) 350 Hz resonance, 35 Hz bandwidth, 204 dB source level, and 30 kg weight in air (6.7); (**B**) 610 Hz resonance, 160 Hz bandwidth, 212 dB source level, and 220 kg weight (6.8). The **A**-size projector has been described above as typical of the RSP design. Figure 6.8 is the transmitting voltage response for a 350 Hz unit, showing the measured radial response compared to the MAVART predictions.

An eight-element vertical line array of the **A**-size projectors, spaced one half-wavelength apart, has been assembled for DREA underwater acoustics research and is shown suspended from a crane in Fig. 6.9. It is normally housed in the canister seen at the top and deployed from the canister after launching from the research ship. The array is then drawn back into the canister before recovery.

It is about 20 m long when deployed and has been named the "Jellyfish" array. Eight driving transformers are housed in the upper part of the canister.

Each element can be driven separately, allowing the directivity pattern to be steered vertically. Computed directivity patterns are shown in Fig. 6.10 for two steering angles. The maximum source level of the array with broadside steering is 220 dB at 355 Hz.

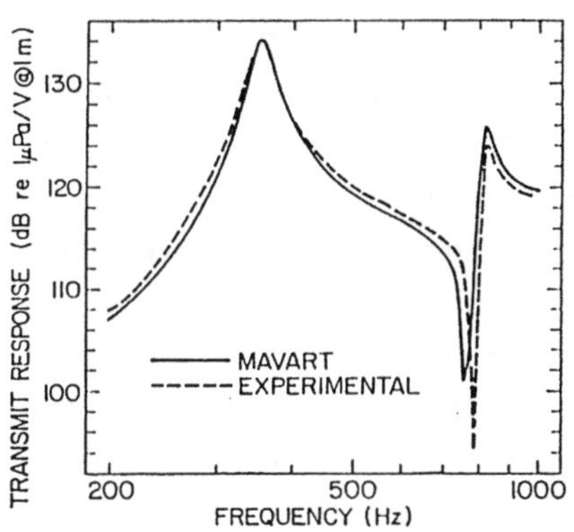

Fig. 6.8 Transmitting voltage response for the projector shown in Figs. 6.6 and 6.7. The solid curve is computed by MAVART finite-element analysis and the dashed curve is the measured response

Fig. 6.9 The 'Jellyfish' array partially deployed from its canister. Only four of the eight elements are seen in this photo

Figure 6.11 is a photograph of a **B**-size projector. Although much larger than the **A**-size, it has thick, deeply curved shells to give the higher resonance frequency. A number of these transducers have been purchased by the US Naval Underwater Systems Center, and have been driven to a source level of 214 dB at their Seneca Lake test facility.

The **B**-size projectors have been used in two, 2-element arrays with half-wavelength spacing. The first, known as the 'Bicycle' array, is a horizontal array designed for towing at speeds up to 10 knots, and is shown in Fig. 6.12. It can be driven either in-phase for broadside steering or out-of-phase for bidirectional end-fire steering. The second is a vertical array that is deployed in a drifting mode to minimize doppler frequency spreading due to vertical motion of the source. The maximum source level available from these arrays is 217 dB.

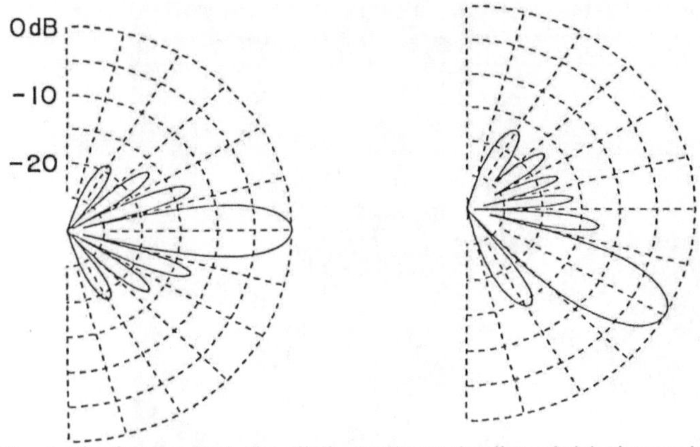

Fig. 6.10 Computed directivity patterns for the eight-element "Jellyfish" array at 355 Hz. Shown are the patterns for constant voltage drive with broadside steering and 30° downward steering of the beam

Fig. 6.11 A 610 Hz, 10 kW ring-shell projector

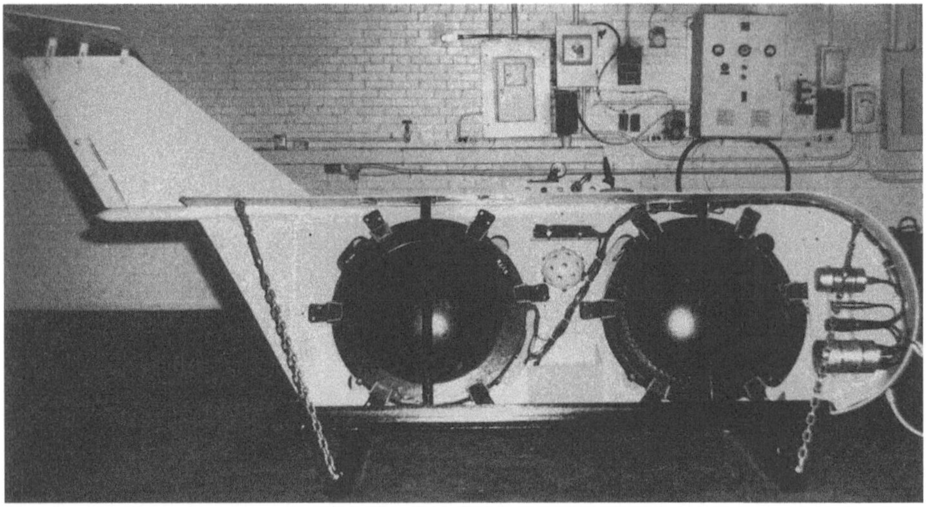

Fig. 6.12 The 'Bicycle' array, containing two of the 610 Hz projectors seen in Fig. 6.11

6.6 ARRAY INTERACTIONS

Flextensional transducers are generally small in comparison with a wavelength at their fundamental resonance and can be considered as point sources when modeling interaction effects in an array. With the dominant effect of water loading on the resonance, interactions are simple to compute, requiring only seven easily measured parameters to describe an element (6.9). These are the resonance frequencies in air and in water, the electromechanical coupling factor at the resonance, the dielectric loss factor, the free capacitance, the electro-acoustic efficiency, and the -3dB bandwidth.

Interaction effects for the eight-element Jellyfish array have been computed for vertical steering angles within ± 45° of broadside. Typical results are shown in Fig. 6.13, which is a three-dimensional plot of element conductances versus steering angle and frequency, with a constant voltage drive. We see that interaction effects are not severe in the Jellyfish array for this range of steering angles; there is an upward shift of the resonance for near-broadside steering. However, when the element spacing is decreased, or if the steering angle approaches end-fire, interactions can be very severe, sometimes leading to negative conductances if the drive amplitudes and phases are not carefully controlled. The situation is even worse in planar or volume arrays.

A later session at this Workshop will address interaction problems in more detail.

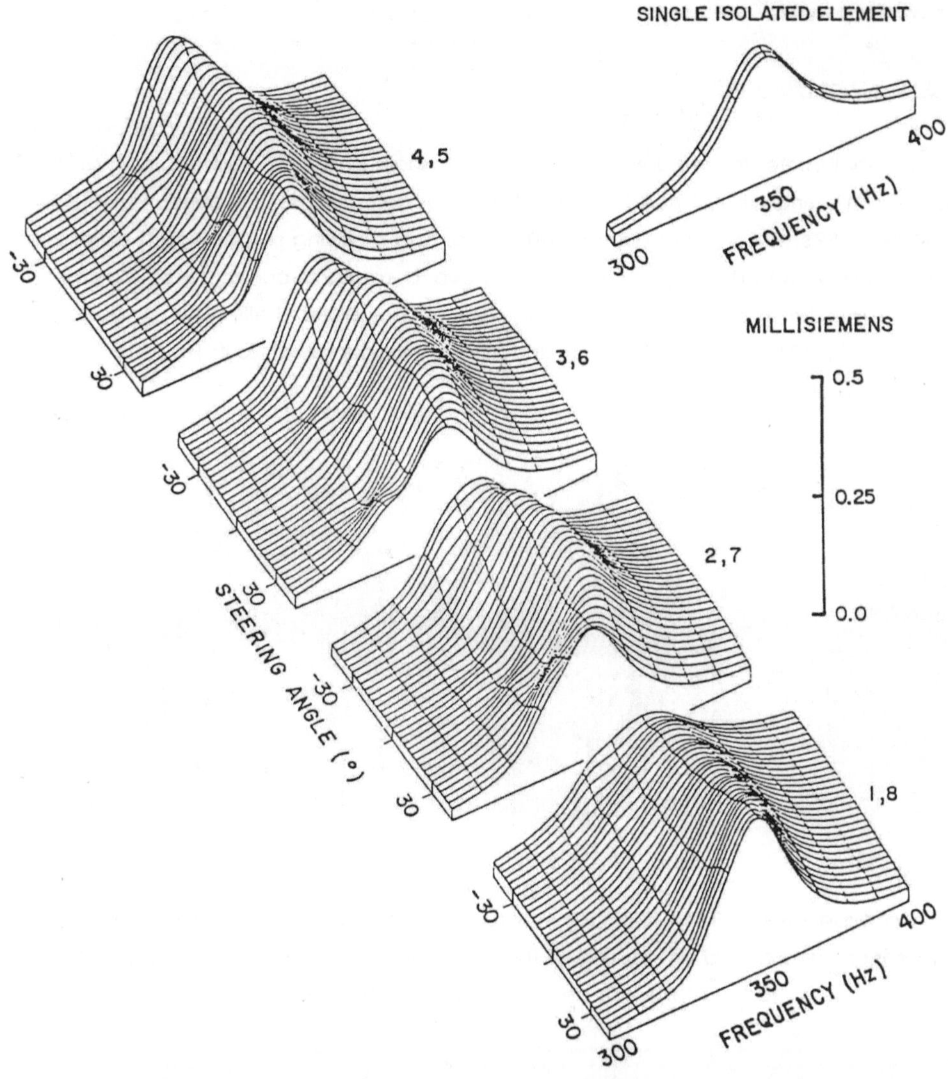

Fig. 6.13 Element conductances vs. frequency and steering angle for elements 1 to 4 of the eight-element 'Jellyfish' array under constant voltage drive. The sign of the steering angle is reversed to display the values for elements 5 to 8

6.7 RECENT DEVELOPMENTS

6.7.1 Mark 3 Ring-Shell Projector

The Mark 3 ring-shell projector features interchangeable shells and external bolts, so that the metal wedges in the ring do not need to be drilled to accommodate the bolts. Figure 6.14 shows two Mark 3 RSP's with shells of different curvature, and a driving ring on which either set of shells could be mounted. The shallow shells have a radius-of-curvature of 42 cm, a thickness of 2.4 mm, and give a resonance of 970 Hz. The deep shells have a radius-of-curvature of 21 cm, a thickness of 3.6 mm, and give a resonance of 2300 Hz. The overall diameter of these projectors is 27.5 cm.

Fig. 6.14 Two experimental Mark 3 ring-shell projectors having interchangeable shells of different dimensions and a third driving ring of the same dimensions as those in the projectors. It is 26.5 cm in diameter and contains 76% ceramic and 24% steel in the piezo-active part

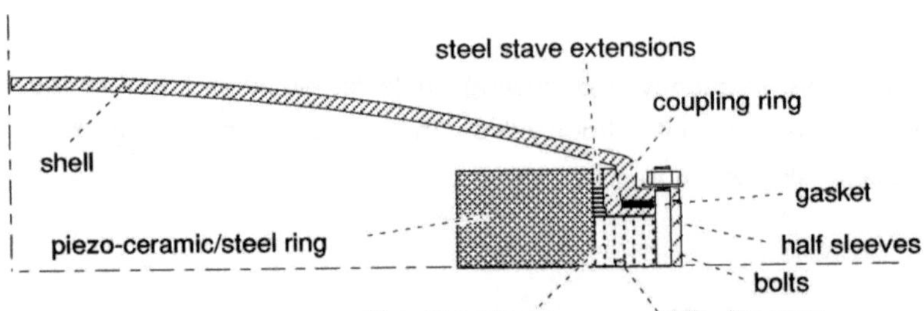

Fig. 6.15 Radial cross-section of the Mark 3 ring-shell projector resonant at 970 Hz

A cross-sectional view of the Mark 3 RSP with shallow shells is given in Fig. 6.15. The composite ceramic/steel ring is made in a similar way to that of Figs. 6.6 and 6.7, but the steel wedges have outward extensions that contact two coupling rings. These mate with the shell rims along a conical surface, so that close machining tolerances are required between the shell rims and the coupling rings. The shells are forced onto the coupling rings in an hydraulic press and the waterproof seal is formed by the compressed rubber gasket. No depth compensation system was installed on the experimental units shown in Fig. 6.14.

Calibration measurements were made on the two projectors shown in Fig. 6.14, both as shown and with the shells interchanged. The two driving rings differ in that one is 24% steel, having 48 ceramic plates and 24 steel wedges; the other is 36% steel, with 40 ceramic plates and 20 steel wedges. Since the densities for steel and ceramic are nearly equal, weight and volume percent are almost the same. Transmitting voltage responses for the two rings fitted with both sets of shells are plotted in Figs. 6.16(a) and 6.16(b). The responses are very similar, with about 1.5 dB higher response from the units that have the greater amount of ceramic in the ring, which shows that consistent coupling of the shells to the rings can be accomplished by this technique.

6.7.2 Robust Ring-Shell Projector

For some military sonar applications, the transducer must be able to withstand severe explosive shocks. A finite-element study has been conducted on a concave-shell RSP that could provide this capability. Depth compensation should not be necessary; if it can survive the shock, it will have a good depth capability.

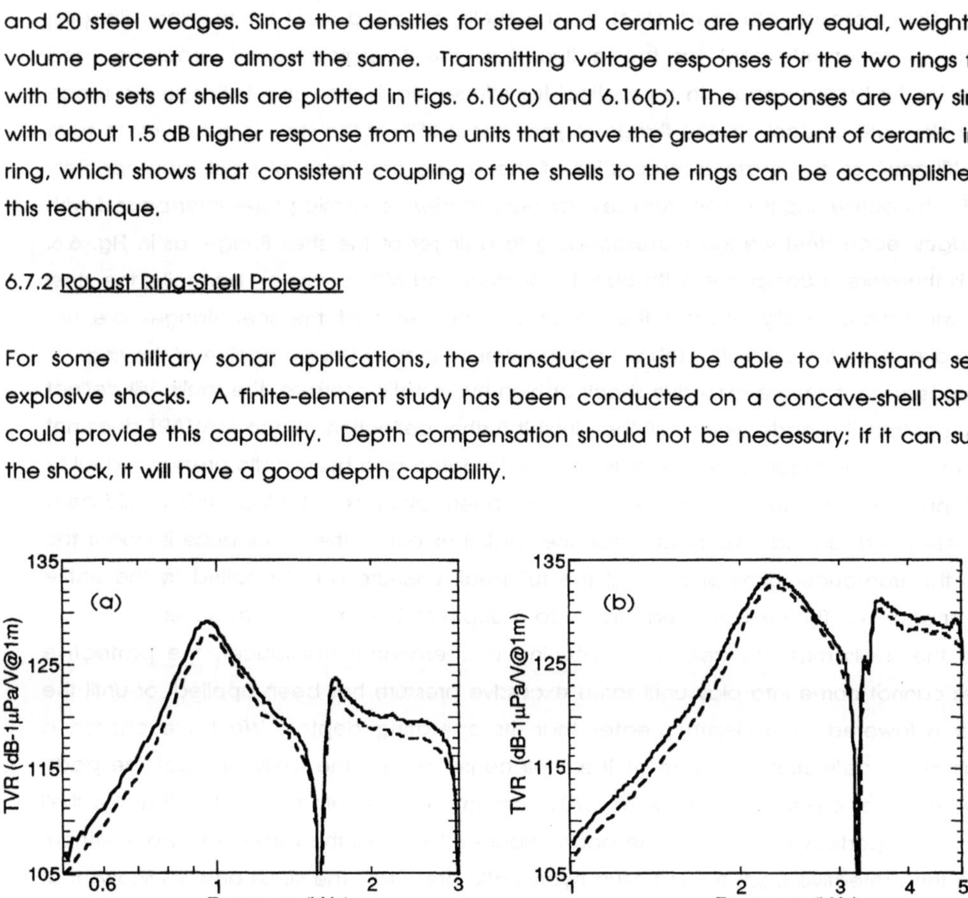

Fig. 6.16 Measured transmitting voltage responses of Mark 3 ring-shell projectors, with two different driving rings and two different sets of shells: (a) shallow, thin shells, and (b) deep, thick shells. The solid curve is for the driving ring with 76% ceramic and the dashed curve is for the ring with 64% ceramic. The curves show unsmoothed experimental data

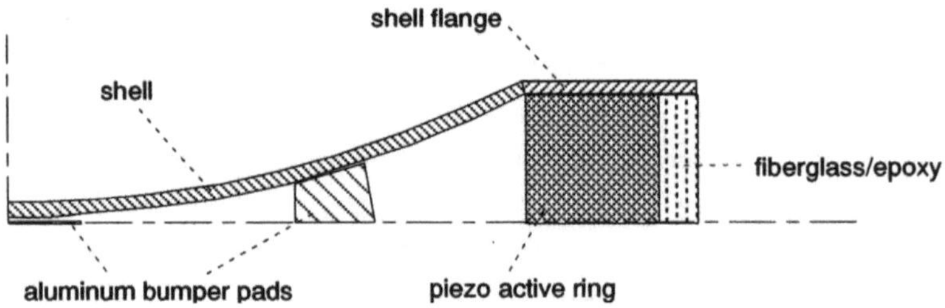

Fig. 6.17 Cross-sectional view of the robust RSP as modeled with MAVART

Figure 6.17 is a radial cross-sectional diagram of the robust transducer, as modeled with MAVART. Five materials are used: steel for the shells, modified steel for the shell flanges, composite piezoceramic-steel for the active ring, the fiberglass-epoxy wrapping, and aluminum for the bumper pads. In a practical transducer, each shell and its flange are made from a single plate of steel. In the flange region, about 35% of the steel is cut away to form hooded "fingers", so the material properties of the flange are those of steel, appropriately modified. The active ring is made from several pairs of piezo-ceramic plates interspersed with steel wedges, each steel wedge corresponding to a finger of the shell flange, as in Fig. 6.6. The ring is therefore a composite, with about 35% steel and 65% ceramic. Normally, the steel wedges will extend axially beyond the ceramic plates, so that the shell flanges are not attached directly to the ceramic. This will reduce stresses in the ceramic portion of the ring.

On exposure to a severe explosive shock or high hydrostatic pressure, the shells will deflect until they contact the pads, which will then inhibit further deflection. Since MAVART does not support transient fluid loading, the shock load was simulated as a hydrostatic pressure equal to the peak pressure in the incident shock, which has been taken as 16.5 MPa. This should be a more severe condition than the blast, since the spatial extent of the shock pulse is about the same as the transducer dimensions, and the full peak pressure is not applied to the entire transducer at once. This pressure is equivalent to a depth of 1630 m (163 atmospheres).

Since the shells must be free to vibrate in an operational transducer, the protective measures cannot come into play until some excessive pressure has been applied, or until the projector is lowered to a depth greater than its operating depth. We have chosen a maximum static deflection of 1 mm at the shell center before the shells contact the pads, which will occur at a pressure of 5.8 MPa (570 m depth). For the remaining 10.7 MPa, the shell deflections are partially blocked by the pads. Figure 6.18 shows the deflections both without and with the protective pads for 16.5 MPa hydrostatic pressure. The stress analysis shows that the use of a pad at the center of the shells only, will reduce the peak stresses by about 10%. The ring pad provides additional support in a critical area, and peak stresses in Fig. 6.18.b are about 35% lower than in Fig. 6.18.a. High strength materials must still be used in the transducer if it is to have any chance of surviving a severe explosive shock.

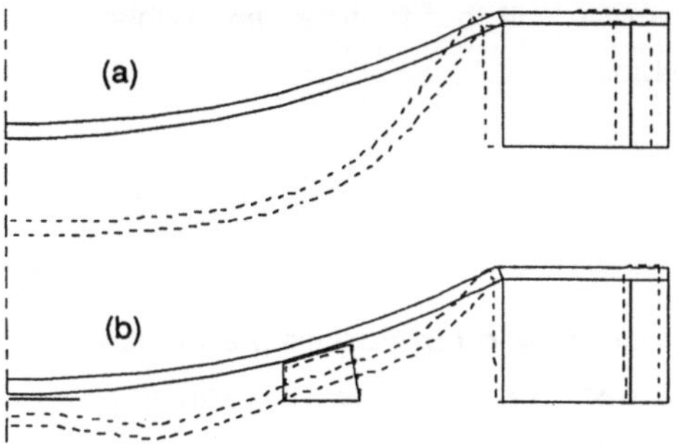

Fig. 6.18 Relative static deflections computed by MAVART analysis of the robust ring-shell projector (a) without and (b) with the aluminum bumper pads, for a pressure of 16.5 MPa. In (b) the shells contact the pads after a pressure of 5.8 MPa is applied

The pads were not included in the acoustic analysis, as the shells remain free of the pads under normal operating conditions. The peak dynamic deflection is 0.1 mm and, since the shells contact the pads at 570 m depth, the maximum operating depth is about 500 m. The predicted source level of this projector at the 970 Hz resonance is 207 dB, and the -3dB bandwidth is 290 Hz.

Sparton of Canada has assembled an experimental transducer incorporating most of the design features mentioned. It has basically the same dimensions as the transducer modeled here but with a shorter active ring, so that stand-off ring extensions were necessary between the active ring and the shell flanges. This is equivalent to having shorter ceramic plates and extending the metal wedges axially. Severe explosive shock tests have been conducted on this transducer with promising results; no serious structural damage to the unit occurred, but the sealant around the shell flanges was ruptured, allowing water ingress. After removing the water and resealing, the transducer operated the same as before the tests. It appears that identification of a tougher, more adherent sealant material will solve this problem. Joint DREA/Sparton patent action on the robust ring-shell projector is underway.

6.8. CONCLUSION

The ring-shell flextensional transducer has been shown to be a versatile high power, low frequency underwater sound source. An internal, passive pressure compensation system can be used to achieve greater operating depths at lower frequencies than uncompensated flextensionals of similar size and weight.

The DREA finite-element code MAVART has proven to be a powerful tool for the design of axi-symmetric transducers.

Two new ring-shell designs have been described: a versatile unit that features removable, interchangeable shells, and a robust, shock-hardened design.

6.9. ACKNOWLEDGEMENTS

The author would like to thank his present and former co-workers who have contributed much to the development of ring-shell projectors: B.A. Armstrong, J.B. Lee, B.L. Fanning, K.A. Edmonds, G.L.-G. Fraser, G.A. Brownell, and D.J. Lewis.

REFERENCES

6.1 F.R. Abbott, 'Compliant Housings to Enhance Low-Frequency Performance of Electrostrictive Transducers, ' NEL Report 574 (1954).

6.2 F.R. Abbott, 'Broad Band Electroacoustic Transducer,' U. S. Patent #2,895,062, 14 July 1959.

6.3 R.H. Pearson, 'Pressure Sensitive Hydrophone,' U. S. Patent #3,660,809, 2 May, 1972.

6.4 G.W. McMahon, and B.A. Armstrong, 'A Pressure-Compensated Ring-Shell Projector,' IOA Conf. Proc. Transducers for Sonar Applications, Birmingham, U. K., 1980.

6.5 G.W. McMahon, and B.A. Armstrong, 'Underwater Transducer with Depth Compensation,' Canadian Patent #1,171,950, 31 July 1984.

6.6 G.W. McMahon, and B.A. Armstrong, 'Underwater Transducer with Depth Compensation,' U. S. Patent #4,524,693, 25 June 1985.

6.7 B.A. Armstrong, and G.W. McMahon, 'Discussion of the finite-element modelling and performance of ring-shell projectors,' IEE Proc. **131**, part F, 275-279 (1984).

6.8 G.W. McMahon, and B. A. Armstrong, 'A 10-kW Ring-Shell Projector,' in *Progress in Underwater Acoustics*, edited by H. M. Merklinger, Plenum Publishing Corp., 647-650 (1987).

6.9 G.W. McMahon, 'Interactions in Ring-Shell Projector Arrays,' DREA Tech. Memo 84/J, (1984).

7. Transducers for Great Depths

P. Dufourcq, J. Adda, M. Letiche, and E. Sernit

Thomson Sintra Activités Sous Marines, Route des dolines,
Parc d'activité de Valbonne, B.P. 38, F-06561 Valbonne Cedex, France

After an overview of the problems of deeply submerged transducers, this paper describes the classical pressure compensation methods. Historically, early research in this area was concerned with free-flooded transducers. For ten years, new solutions for deeply submerged tonpilz transducers have been developed, the most important being the air-compensated tonpilz (with an inner or outer deformable bellows) and the composite filter tonpilz. We present here each of these solutions and their field of application in sonar arrays. More recently, a new system has been developed for a great depth low frequency flextensional transducer for which results are presented. In conclusion, we discuss prospects for future developments.

7.1 INTRODUCTION

Transducers for applications at great depths are more and more used to improve sonar system performance, first for fixed equipment such as localisation systems, but also for active sonars for submarines, even if the active mode remains a secondary function. But the most important need is for Variable Depth Sonar Systems (VDS) and Airborne Sonar Systems in which the operational depth must be increasedto work with the best bathythermal conditions. These basic needs have led, for about fifteen years, to new developments of successful products. We may notice two kinds of existing deep submergence transducers:

- free-flooded transducers which are waterbacked, and thus nearly unlimited in depth of immersion. This kind of transducer presents some problems in acoustical radiation because of the water backing interaction. Even if free-flooded ring transducers have been developed for low frequency applications (using the resonance of the inner cavity), it seems that some problems (for example, high sensitivity to environment) do not allow their use for VDS application,
- air-backed compensated transducers such as tonpilz, flextensional transducers or ceramic rings, which need a special compensation mechanism or optimized geometries. The static pressure on the outer surface must be compensated by an equivalent internal pressure or a compliant element able to undergo a local deformation due to the static pressure effects.

This paper presents these concepts and associated transducers. First, we describe the free-flooded technologies and main associated concepts. Then, we analyse the two great depth tonpilz solutions. Then, we present the new deeply submerged class IV flextensional transducer developed by Thomson Sintra ASM Company.

7.2 FREE-FLOODED TRANSDUCER

From the 60's, the interest has developed in free-flooded transducers. The first studies (7.1-7.5) have shown that this kind of acoustical source could operate at great depth, be light, have a large bandwidth and develop high power. Since then, many analytical, finite element and technological studies (7.6-7.10) have permitted the development of modeling tools and the clarification of working applications of these sources, particularly in the medium and low frequency fields.

The free-flooded ring transducer described in Fig 7.1 is an open, thin walled, cylindrical ceramic completely immersed, with two main resonant modes: the cavity resonance F_c and the radial resonant frequency F_R.

7.2.1 Typical Features

7.2.1.1 Immersion

The free-flooded transducer requires no pressure compensation system. It represents the ideal solution for very great depth transducer since no part of the transducer is subject to differential pressure. The piezoelectric ceramic is only under hydrostatic stress and the electroacoustical stability of the transducer with respect to the pressure will depend basically on the choice of the ceramic.

The piezoelectric ceramic may be either radially polarized, made in one piece ($\varnothing_e \leq$ 150 mm) or tangentially polarized (prestressed stave ring ceramic for $\varnothing >$ 150 mm and high power applications).

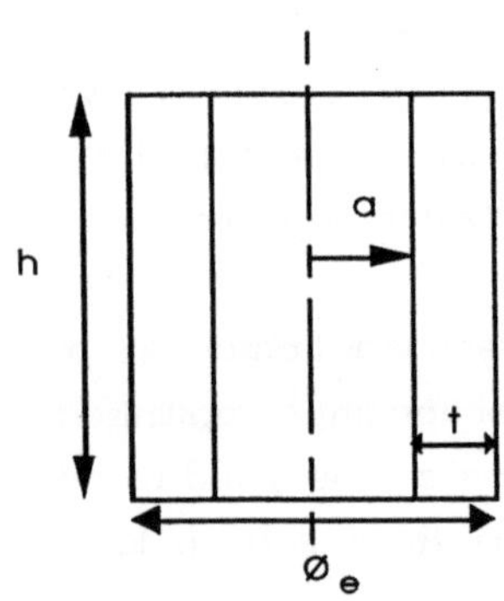

Fig. 7.1 Free-flooded ring transducer

7.2.1.2 Acoustical Characteristics

The radial frequency F_R depends, roughly, on only the mean radius and the sound velocity in the ceramic; the cavity frequency F_c is a function off h, a, of the compliance of the shell and of the water column.

Figure 7.2 shows an example of the variations of F_R and F_c versus h/a for a = 240 mm (finite element results). When h/a increases, F_c decreases, but the coupling between F_R and F_c becomes less efficient with an increase in the quality factor Q_{cavity}. Free-flooded ring transducers for very low frequency (< 1 kHz), large bandwidth and high power applications will be heavy and large.

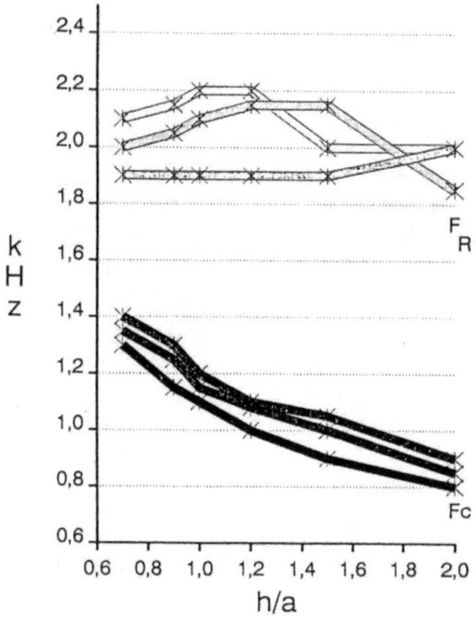

Fig. 7.2 Example of the variations of F_R and F_c versus h/a for a = 240 mm (finite element results)

7.2.1.3 Directivity

A single free-flooded ring has an advantage of being a directional source, although it is small with respect to the wavelength λ (dipole directivity in elevation for h = λ/4). The directivity may be increased when the rings are mounted in a coaxial array.

As an example, Fig. 7.3 shows the experimental directivity pattern in elevation for an array of two rings the overall length of which is 0.47 λ. The elevation beamwidth is 56° compared to the equivalent beamwidth of 106° that would result from a line source of the same length.

It is important to note that the influence of the environment (box with electronics, mechanical mounting of rings, cables) must be minimized to keep good symmetrical characteristics of directivity for free-flooded arrays.

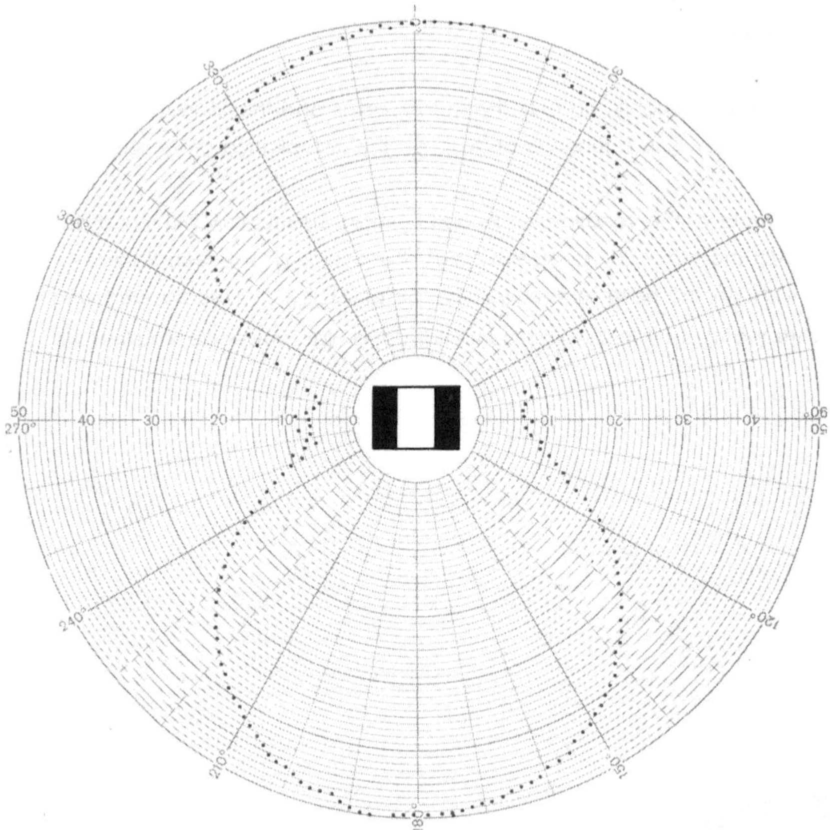

Fig. 7.3 Directivity pattern of two coaxial rings, l = 0.47 λ

7.2.2 Conclusion

For more than twenty years, Thomson Sintra ASM has been involved in theoretical and experimental studies on free-flooded ring transducers (single source and arrays). These studies and the experience with development of such operational transducers point out that:

- for great depth, medium frequency (≥ 2 kHz), large bandwith (≥ one octave), high power applications, an array of free-flooded rings provides a suitable solution, with a correct size and weight,
- for great depth, low or very low frequency (between 500 Hz and 1500 Hz) applications, and if size and weight must remain within acceptable limits with respect to operational conditions, the free-flooded ring sources will have small bandwidth ($Q > 5$) and moderate power. In this last case, the transducer can be a composite ring (ceramic/metal) with a low coupling coefficient. For these low frequency applications, flextensional transducers are a much better solution (Cf § 7.4).

7.3 TONPILZ FOR GREAT DEPTHS

The well-known tonpilz transducer is the most frequently used in underwater arrays for transmitting or receiving applications where directivity and beamforming are necessary. It is composed of a piezoelectric driver which transmits a vibration to a headmass exposed to the water (Fig. 7.4), generating an acoustic wave. On the other end of the driver is a tailmass which has a great mechanical impedance and so permits maximizing vibrations of the headmass.

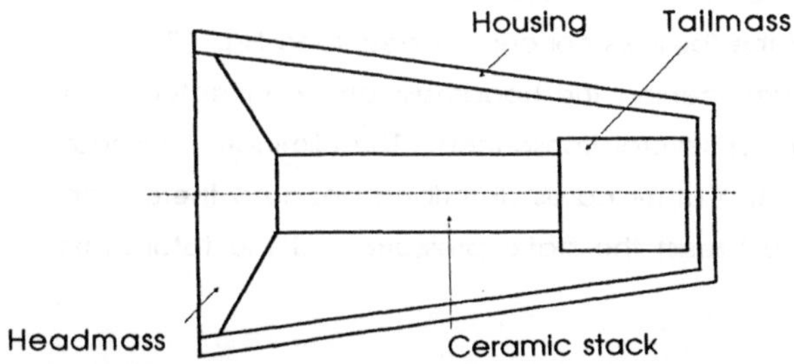

Fig. 7.4 Tonpilz

A rigid envelope permits air-backed conditions. This kind of technology is quickly depth limited, because of the static pressure effects on the box and on the headmass. The main problem is due to the stress amplification on the ceramic stack because of the ratio of cross sections of the headmass and ceramic driver.

For acoustical requirements, the headmass must be uncoupled from the box by a system which must be compliant enough to permit a strain due to the static pressure on the headmass. That is the one solution to solve the great depth problem. Another solution consists in putting the air contained in the housing at the same pressure as the exterior.

These two approaches induced several technological solutions that are presented here.

7.3.1 Air Compensated Tonpilz

The objective is to realize a pressure compensated transducer, the inner pressure being the same as external pressure. The simplest solution consists in injecting gas from bottles, for example. But the large mass of such systems led to the development of other more accurate techniques such as the addition of compliant bellows around the transducer. One example is described in Fig. 7.5.

The upper half of the Fig. 7.5 shows the position of the bellows on the transducer at atmospheric pressure. Due to the static pressure, at great depth, the water compresses the bellows inside around the ceramic stack and the masses as shown in the lower part. So there is no longer the stress amplification due to the large section of the headmass compared to the small section of the ceramic stack. The box itself is subject to the same internal and external static pressure and, in consequence, the design is simple and the mass is kept small (thin wall). The membrane must be pre-inflated to increase the depth performance. The most important problems which can appear are, for great power applications, high temperature inside the transducer because of the difficulty of poor thermal conduction through the rubbery membrane.

This transducer is commonly used in VDS for operational depth of several hundreds of meters. Another example is shown in Fig. 7.6. In this case, the membrane is not in contact with the headmass which permits elimination of all the interactions between them. Note that the form of the tailmass around the ceramic stack which permits an increase of this mass in a smaller volume. Another simplified design of the "bellows concept" is described Fig. 7.7.

In this case, the membrane is only located behind the headmass and connected to a deformable bellows which may be common for several transducers. The elimination of heat around the ceramic stack is better than for the other cases described because there is no rubber around it. Nevertheless, the box must resist the static pressure and the total mass increases.

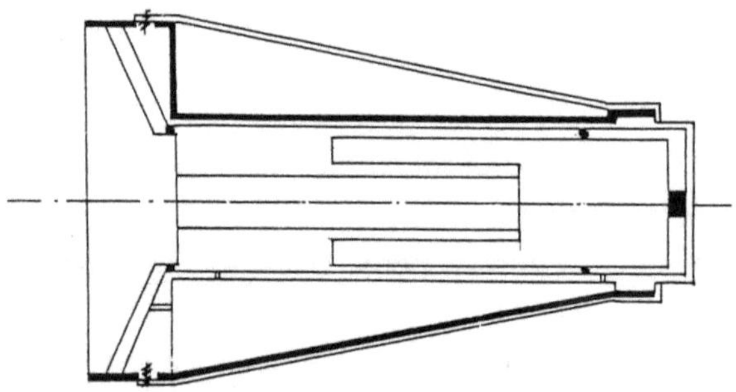

Fig. 7.5 Gas compensated tonpilz with bellows

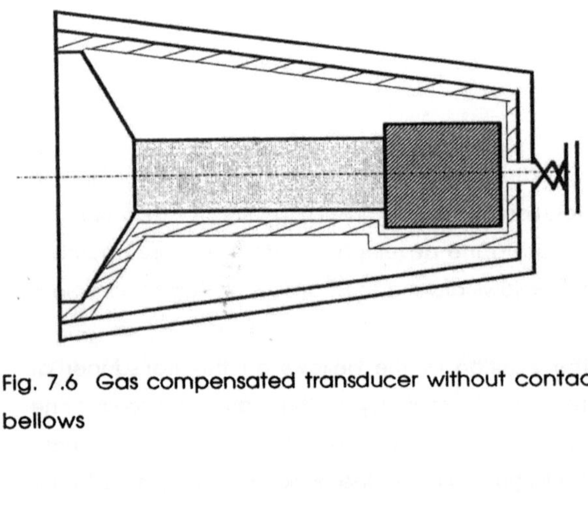

Fig. 7.6 Gas compensated transducer without contact between the headmass and the inner bellows

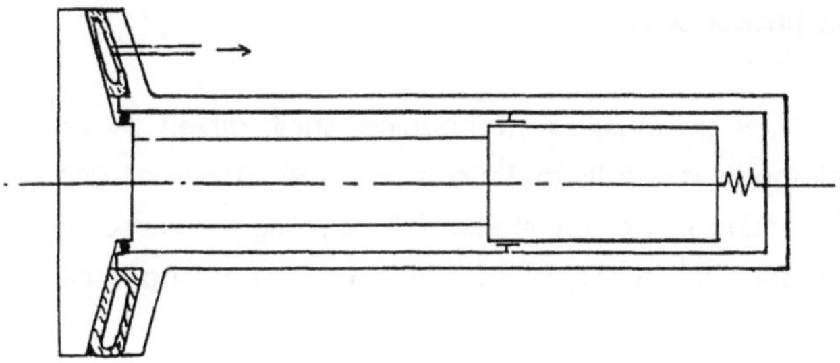

Fig. 7.7 Gas compensated transducer with bellows located behind the headmass

All these technologies have been successfully employed for applications at depths greater than 500 meters and in several cases greater than 1000 m. The design, including choice of materials, are generally optimized by finite element methods which permits prediction of both static and dynamic response.

7.3.2 Composite Filter

Another solution (7.13) introduces a compliant element between the headmass and the body of the transducer (Fig. 7.8). The function of the compliant ring is to decouple dynamically the body and the headmass and to support static pressure loads on the head face.

The main difficulty with such a simple concept is finding the best inhomogeneous material to be used as the compliant ring (high longitudinal compliance, low axial compliance). Development of understanding and knowledge about GRP materials has helped to realize this kind of transducer now used in several VDS of Thomson Sintra. The main advantages are the simplicity, the easy elimination of heat, and a good efficiency. But the body must also resist hydrostatic pressure so that the mass may be large. The design of such transducers must be optimized, specially concerning the ring filter, taking into account the longitudinal wave propagation along the ring. Another advantage is that there is no deformation around the transducer and thus there is no variation of baffling conditions for an array (That is not the case for the bellows solution). Thomson Sintra ASM has manufactured several thousands of transducers with this technology for ten years.

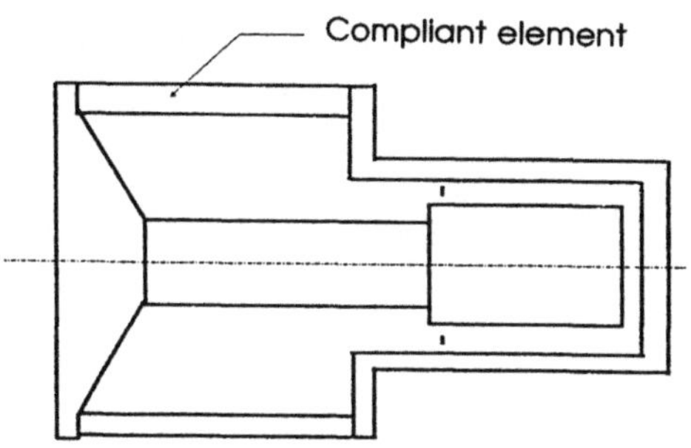

Fig. 7.8 Composite filter tonpilz

7.4 GREAT DEPTH CLASS IV FLEXTENSIONAL TRANSDUCER

The class IV flextensional concept consists of an elliptical mechanical shell, driven by an internal stack of piezoelectric ceramic elements along its major axis (Fig. 7.9). The shell and the piezoelectric stack are coupled at the boundaries of the stack. This coupling is ensured by inserting the stack into the shell and prestressing the system so that during the operating cycle the two parts remain in contact.

The longitudinal dynamic strains in the ceramic bar cause radial flexure in the oval shell. Therefore, small displacements along the major axis produce large transverse displacements of the central surface of the shell. Although the displacement volumes from motions along the major and minor axis are of opposite phase, the net volume displacement is large, resulting in high acoustic power output.

However, the effect of hydrostatic pressure is to squash the ellipse in its minor axis and elongate the major axis, and so subjects the ceramic stack to tensile forces. Therefore, in most cases, the compromise between operating power and operating depth is a significant limitation in the design of this type of transducer.

7.4.1 The Dynamic Filter Device

In order to obviate tensile stress, the ceramic stack is prestressed, which can be done during the assembling of the transducer. In this case, the shell major axis is expanded and an oversize ceramic stack is inserted. The shell is then relaxed to compress the ceramic stack. It is also possible to use a prestressed bolt that goes through the ceramic stack and which is fastened to the shell. These two systems do not allow great depth performance. The main problem which appears, if the working depth is too great, is that the deformation of the shell induces a tensile stress on the ceramic stack that cancels the pre-stress. It appears that the maximum depth capability of a thick shell transducer is typically 200 to 300 m (7.17,7.18).

To avoid these difficulties, Thomson Sintra ASM has developed an original Class IV flextensional transducer. The technical novelty of this transducer is that the dynamic filter is integrated into the piezoceramic stacks. The aim of this apparatus is to make the piezoceramic prestress and the effect of the hydrostatic pressure independent.

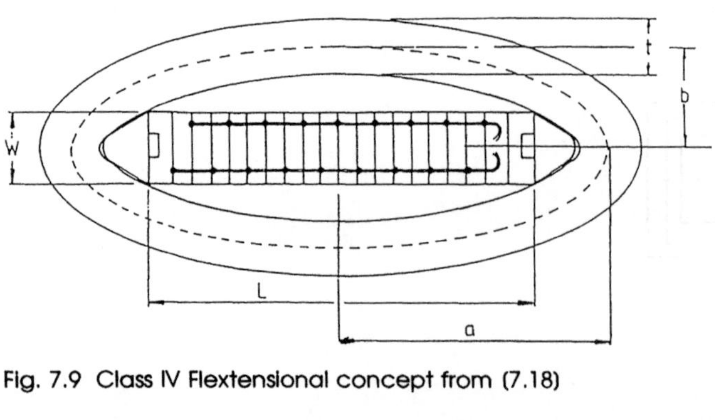

Fig. 7.9 Class IV Flextensional concept from (7.18)

Under static conditions, the filter has a low stiffness which permits to the ceramic stack to follow the shell deformation without losing its own prestress. On the contrary, under dynamic conditions, the stiffness of the filter becomes high which provides a good coupling between the driver and the shell.

Therefore, inserting this filter between the ceramic stack and the shell permits simplification of the assembly (the first deformation of the shell is no longer necessary).

7.4.2 Great Depth Flextensional Transducer Performance

A Class IV flextensional transducer using the Thomson Sintra ASM filter device has been manufactured and tested in an anechoic pressure tank from 0 to 50 bars.

The acoustic performance characteristics of this 1.7 kHz resonance frequency transducer as a function of the hydrostatic pressure are shown in Fig. 7.10.

The stability of the transmitting response versus pressure is quite good because the frequency shift is less than 60 Hz and the decrease in the transmitting response is quite low (≅ 2 dB) between 0 and 50 bars. The impedance measurements support the previous results and confirm the immersion stability of this transducer. The next step will be to improve this device to reach greater depths.

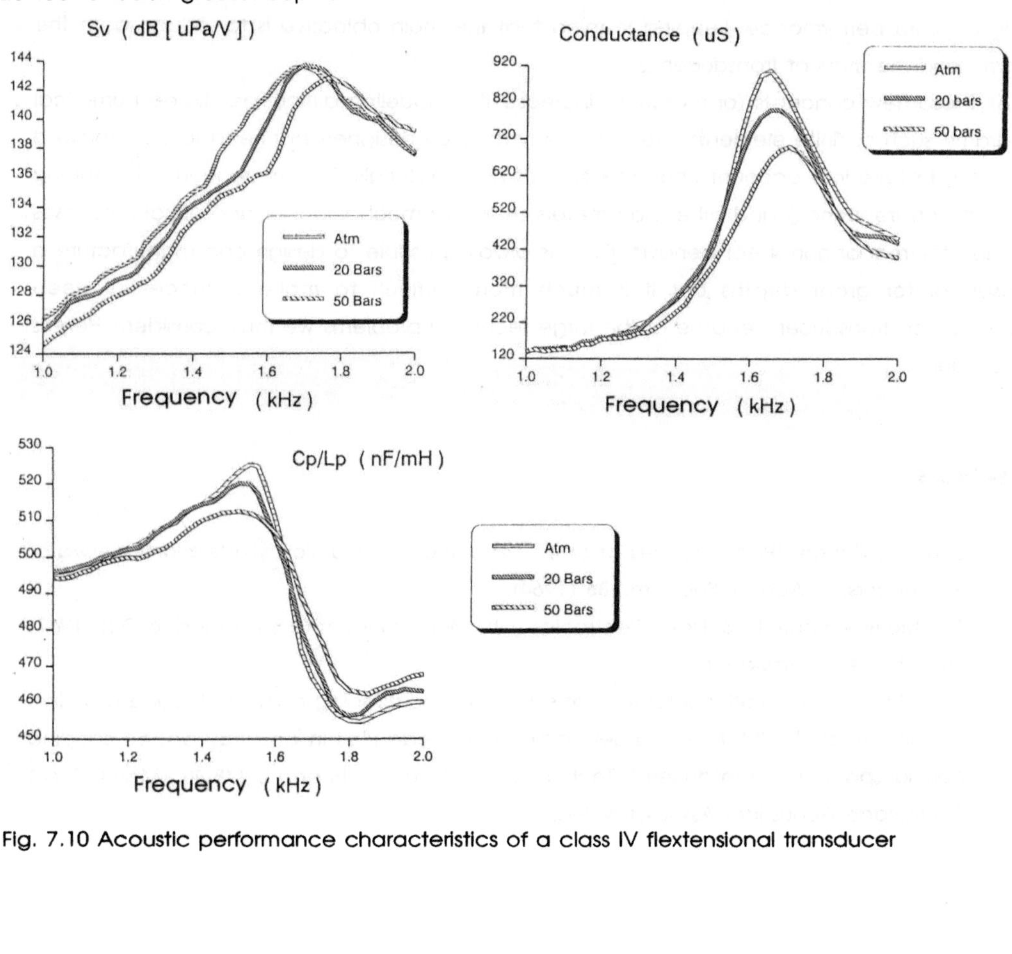

Fig. 7.10 Acoustic performance characteristics of a class IV flextensional transducer

7.5. CONCLUSION

This paper has described some of the principal means used to solve the problem of hydrostatic pressure on transducers for deep submergence. With the support of the French Navy and, considering the necessity to increase the operational depth of sonar arrays, Thomson Sintra ASM has developed for twenty years new solutions for deep submergence tonpilz and free-flooded rings.

More recent is the dynamic filter used in the flextensional transducer. Positive results give us hope that great improvements for the coming years in very low frequency, deep submergence, high power transducers can be expected. We may begin to consider that low frequencies and great depths are not necessarily antagonistic concepts as we thought in the past. The necessity of a large compliance to achieve low frequencies is not yet actually limited to shallow depth operations, because of the use of mechanical filters. Future developments in this field will have to take into account the continuous improvements of the knowledge of new materials such as GRP and viscoelastic materials, the first ones for their mechanical characteristics, the second ones for the variation of their properties with frequency. We will also have to consider and to do research on new drivers such as magnetostrictive materials because of their potentialities for low frequency applications and static pressure performance, keeping in mind that the main objective is to reduce both the volume and the mass of transducers.

All these new concepts (or materials) increase the modeling difficulties. Large numerical programs, such as finite elements, are commonly used by designers but need to be improved everyday to take into account characteristics of new materials (inhomogeneous, depending generally on frequency) and other parameters (not only mechanical or acoustical responses but also thermal or non-linear behaviour). It is always possible to design and manufacture a transducer for great depths but it is much more difficult to make a successful deep submergence transducer because of the large technical problems we must consider. Here is the challenge !

REFERENCES

7.1 G.W. McMahon, 'Performances of open Ferroelectric ceramic cylinders in underwater transducers,' J. Acoust. Soc. Am., **36** (1964).

7.2 A.S. Merriweatherl, U. S. Navy Electronics Lab. Memorandum (unpublished, 6 Oct. 1961) cited by G.W. McMahon.

7.3 M.C. Junger, 'A variational solution of solid and free-flooding cylindrical sound radiators of finite length,' - 'Mutual and Self radiation impedances in an array of free-flooding coaxial space ring transducers,'. Technical reports, U 177-48 and U 178-48; (March 1964) Cambridge Acoustical Associates, Inc.

7.4 W.T. Chin, "Self radiation impedances of a finite free-flooding cylindrical radiator with JUNGER'S End correction," U.S.L. Technical Memorandum N° 960-75-64, August 1964 U.S. Navy Underwater Sound Laboratory, New London CT.

7.5 D.T. Porter, "Method for computing the electrical and acoustical behavior of free-flooded cylindrical transducer array," J. Acoust. Soc. Am., **44** (1968).

7.6 J.P.D. Wilkinson, and M.J. Da Costa, "Underwater Behavior of free-flooded ceramic ring transducers Journal of Engineering for industry," Transactions of the ASME, August 1971.

7.7 R.R Smith, J.T. Hunt, and D. Barach, "Finite element analysis of acoustical radiating structure with application to sonar transducer," J. Acoust. Soc. Am., **55** (1974).

7.8 M.R. Knittel, C.S. Nichols, R.R. Smith, and D. Barach, "Comments on "Finite element analysis of acoustically radiating structure with application to sonar transducers," J. Acoust. Soc. Am., **56** (1974).

7.9 R. Bossut, "Modélisation de transducteurs piézoélectriques annulaires immergés par la méthode des éléments finis," Thèse de doctorat 3ème cycle - Université de Valenciennes, in French (1985).

7.10 B. Dubus, "Analyse des limitations de puissance des transducteurs piézoélectriques," Thèse Doctorat en Physique, Université des Sciences et Techniques de Lille, in French (1989).

7.11 B. Tocquet, "Piezoelectric transducers and acoustic antennas which can be immersed to a great depth," US Patent N° 4151437 (1977).

7.12 B. Tocquet, C. Pohlenz, and D. Boucher,"Transducteurs piézoélectriques et antennes de sonar pouvant être immergés à grande profondeur," Brevet FR N° 8412833 (1984).

7.13 M. Lagier, "Transducteur électroacoustique pour immersion profonde," Brevet FR N° 7437073 (1974).

7.14 H. Louit, and R. Gagno, "Transducteur électro-acoustique de puissance conçu pour les immersions grande profondeur," Brevet FR n° 8101440 (1981).

7.15 A.W. Elston, and G.A. Vincent, Pressure compensated transducer US Patent 3, 241, 099 (1966).

7.16 G.R. Douglas, J.H. Thompson, and R.H. Whitaker, "Pressure compensated transducer," US Patent 3, 263, 208 (1963).

7.17 J.R. Oswin, and A. Turner, "Design limitations of aluminium shell, Class IV flextensional transducers," Proc. Inst. Acoustics **5**, part 3, (1984).

7.18 J.R. Oswin, and J. Dunn, "Frequency, Power and depth performance of Class IV flextensional transducers," in *Power Sonic and Ultrasonic Transducers Design* (Edited by B. Hamonic and J.N. Decarpigny) Springer-Verlag Berlin Heidelberg (1988).

7.19 P. Dufourcq, "Low Frequency Flextensional Transducer modelling," Proc. of the First French Conference on Acoustics, Jour. de Physique, FASC. 2, **1**, C2-611-615 (1990).

8. Physics of Array Element Interaction Phenomena

R.T. Richards, J.B. Blottman III, and B. McTaggart

Naval Underwater Systems Center, New London Laboratory,
New London, Connecticut 06320, USA

In a sonar array, the interaction between sources affects the radiation loading on each individual source, the power it radiates and the radiation pattern of the array. This interaction is characterized by the mutual radiation impedance which is an essential parameter determining array behavior. Low-frequency transducers in a volumetric array with small size requirements are subject to much larger mutual interaction and scattering than in conventional arrays. Recent deep water tests by NUSC have shown significant deviations between experimental results and the classical theoretical treatment. This paper presents a review of array interaction phenomena, classical theoretical developments, experimental measurements and array performance effects. A case study of a recent low-frequency volumetric array test is discussed.

8.1 INTRODUCTION

In an array of sonar transducers, the mutual coupling between sources affects the radiation loading on each individual source, the power it radiates and the beam pattern of the array. Understanding the physics of array interaction phenomena is imperative if one is to design an efficient and well-behaved acoustic array.

Traditional array interaction physics generally assume small ka, where k is the wavenumber and a is the radius of the radiating surface. This assumption is not valid in the case of low frequency acoustic arrays. Transducer elements used to populate these arrays are large multimode radiators with ka close to 1. These radiators are physically large in size and, in an array, require close packing (small kd, where d is the distance between the acoustic centers of the projectors) to minimize the volume of the array. These array requirements accentuate the mutual coupling.

8.2 ARRAY INTERACTION PHYSICS

8.2.1 Mutual radiation impedance

Acoustic projectors convert electric input power to sound pressure. This conversion is highly dependent upon frequency, array geometry and nearby surfaces (for example, baffles or array structure). In terms of the classical Thevenin equivalent circuit description, (see Fig. 8.1) each transducer in the array may be characterized by the equation:

$$F = NE - Z^T V \tag{8.1}$$

relating force F, voltage E and velocity V. N is the electromechanical transformation ratio, Z^T is the internal impedance and $NE = F^T$ is the Thevenin driving force. The force F arises due to the sound pressure at the radiating transducer face, and can be expressed in terms of a radiation impedance Z^R:

$$F = Z^R V \tag{8.2}$$

Since the total force on the transducers face can be found by simply integrating the pressure, P, over the surface area, S, of the transducer, the radiation impedance can be expressed as:

$$Z^R = \frac{1}{V} \iint P \, dS \tag{8.3}$$

An inherent assumption in removing the velocity, V, from within the integral is that the velocity be constant across the face of the transducer.

The pressure is composed of contributions from all the transducers in the array:

$$P_j = \sum_{i=1}^{N} P_i \tag{8.4}$$

The radiation impedance of the jth transducer is:

$$Z_j^R = \frac{1}{V} \iint_{S_j} \sum_{i=1}^{N} P_i \, dS \tag{8.5}$$

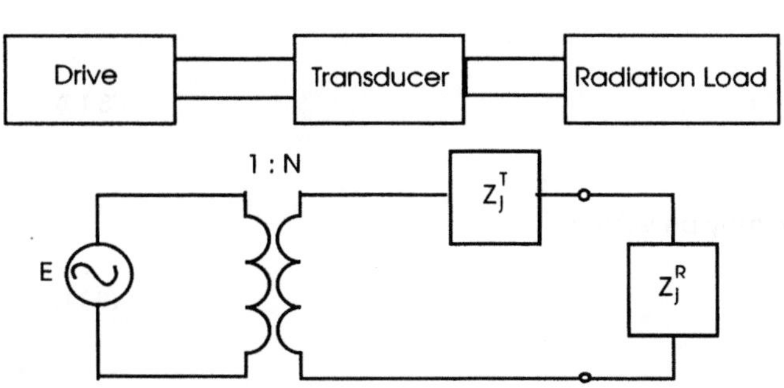

Fig. 8.1 Transducer equivalent circuit

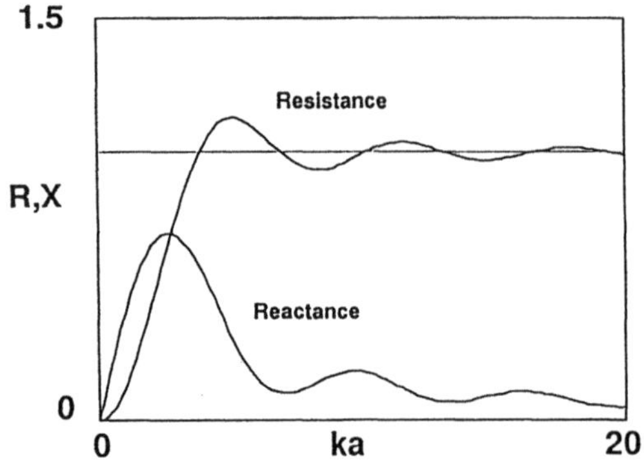

Fig. 8.2 Self radiation impedance of a piston

The i = j term is the self-radiation impedance, i.e., the radiation loading of a single element in free-field:

$$Z_{jj}^{R} = \frac{1}{V_j}\iint_{Sj} P_j \, dS \tag{8.6}$$

Figure 8.2 depicts the self-radiation impedance of a circular piston from Rayleigh (1877).

The remaining terms, $i \neq j$, in the summation are the mutual radiation impedances. Specifically, the mutual impedance of the jth transducer arising from the interaction with the ith transducer is given by:

$$Z_{ij}^{R} = \frac{1}{V_i}\iint_{Sj} P_i \, dS \tag{8.7}$$

Combining these expressions, the coupled array equation for the jth transducer becomes:

$$N\,E_j = V_j\left(Z^T + \sum_{i=1}^{N} Z_{ij}^{R}\frac{V_i}{V_j}\right) \tag{8.8}$$

or,

$$F_j^T = \sum_{i=1}^{N} Z_{ij} V_i \tag{8.9}$$

where,

$$Z_{ij}^{R} = Z_{ij} - \delta_{ij} Z^T \tag{8.10}$$

and δ_{ij} is the Kronecker delta.

In matrix notation, the array equation may be written:

$$\left[F_j^T\right] = \left[Z_{ij}\right]\left[V_i\right] \tag{8.11}$$

It may be noted here, that near mechanical resonance, when the internal impedance is at a minimum, the mutual-radiation impedance can play a dominant role in determining the radiating surface velocity.

8.2.2 Theoretical Development

The calculation of mutual-radiation impedance consists of solving the wave equation for the sound pressure produced by one transducer and then integrating that pressure over the radiating surface of another transducer. The earliest theoretical attempt to do this was by Karnovskii (1941).

Equations for the mutual impedance between circular pistons on a rigid plane were derived by Pritchard in 1951, 1960. Figure 8.3 shows a plot of Pritchard's equation. For small ka, Pritchard's equation can be approximated by the first term in his summation:

$$Z_{ij} \approx \frac{1}{2}(ka)^2 i \frac{e^{ikd}}{kd} \rho c A \tag{8.12}$$

where a is the piston radius and d is the separation distance between pistons. For small ka, the self radiation resistance may be approximated by

$$R_{ii} \approx \frac{1}{2}(ka)^2 \rho c A \tag{8.13}$$

Therefore, the mutual impedance may be expressed in terms of the self resistance:

$$Z_{ij} \approx R_{ii}\, i \frac{e^{ikd}}{kd} \tag{8.14}$$

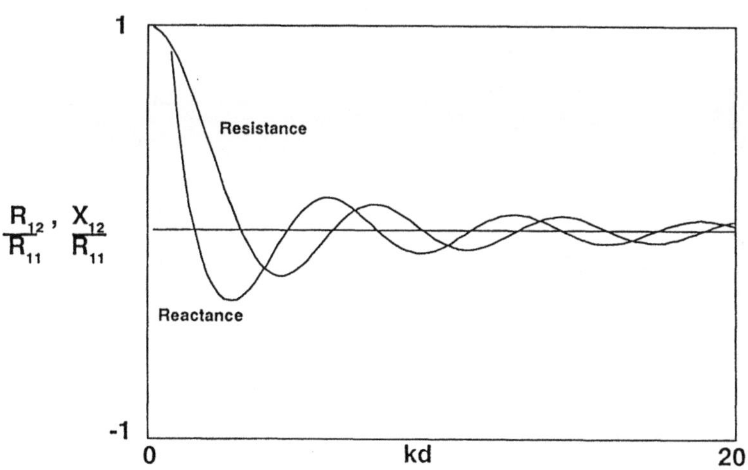

Fig. 8.3 Mutual radiation impedance of two circular pistons

The magnitude of Z_{ij} is approximated by

$$|Z_{ij}| \approx \frac{R_{ii}}{kd} \tag{8.15}$$

and decreases with increasing kd, as should be expected.

Circular pistons on spherical baffles may be approximated by Pritchard's planar rule: for $a/\lambda > 3$, there is little difference in the self impedance terms; for small kd, the mutual impedance terms are similar, and for large kd, the mutual impedance term is negligible. Square or rectangular pistons may also be approximated by Pritchard's rule in the case of large kd. For small kd, or when the length and width of the rectangular piston differ greatly, Arase (1964) derived a more exact analytical expression. Karnovskii (1956) evaluated the complex mutual impedance for pulsing spheres of small ka.

8.2.3 Experimental Development

The earliest reported measurements of radiation resistance appear to be that of Pordes and Sherman (1959) at the US Naval Underwater Sound Laboratory (now NUSC). These measurements were carried forward by Stumpf (1960, 1964) and his students at Ohio University. Figures 8.4 and 8.5 show experimental measurements for radiation resistance and reactance, respectively, for two square pistons at an air-water interface taken by Richards while at Ohio University. Figure 8.4 compares these experimental results with Sherman's (1965) interacting dipole model. Figure 8.5, reported by Stumpf, Richards and Crum (1968) is one of the earliest successful measurements of the radiation reactance term; Butler (1968) references Stumpf and Crum (1966) at the first.

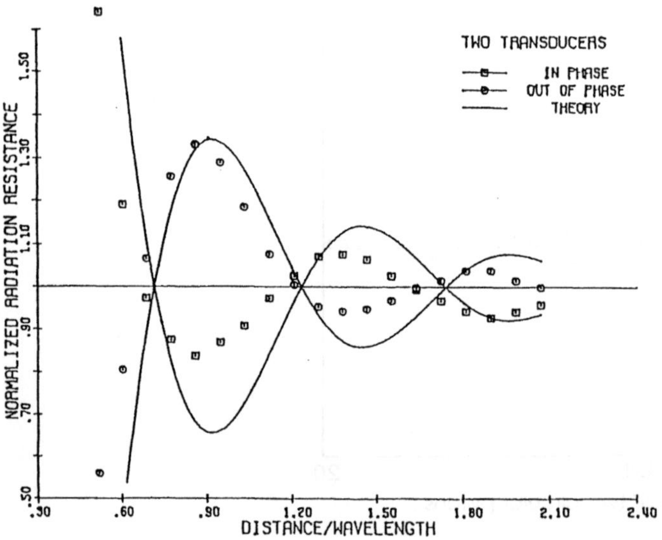

Fig. 8.4 Radiation resistance for two transducers at an air-water interface

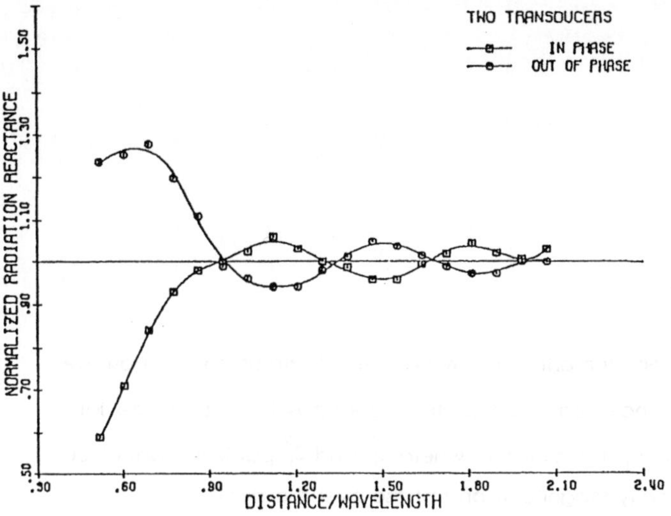

Fig. 8.5 Radiation reactance for two transducers at an air-water interface

8.3 ACOUSTIC INTERACTION EFFECTS

Array interaction creates uneven loading on the elements in the array resulting in significant variations in the volume velocity of each array element. These interactions reduce the acoustic output power of the array and in some severe cases a transducer element may actually absorb acoustic power. This condition is evident as a negative real part of the electrical impedance. The negative electrical resistance can lead to significant problems for the amplifiers used to power the transducer elements. The acoustic power taken in by the element is absorbed back into the driving amplifier and may create an overload, and electrical failure of the amplifier. This phenomenon was experienced in the early 1960s LORAD program. (Martin 1985)

Radiation loading may also induce a mechanical resonance excitation. At resonance, excessive velocity is induced in the element. This may result in a stress failure of the transducer driver as was experienced by the ARTEMIS array in the early 1960s. (Hanish 1985)

The effects of mutual coupling may be examined with the simple circular planar array depicted in Fig. 8.6. If all the elements are driven with the same amplitude and phase, then the face velocities will all have the same amplitude and phase, regardless of d/l or Z^T since each element in the array sees the same radiation impedance. If the elements have different radiation impedances, for example, in a line array, then the coupled array equation for the jth element may be written as:

$$V_j = \frac{F_j^T}{Z^T + Z_j^R} \tag{8.16}$$

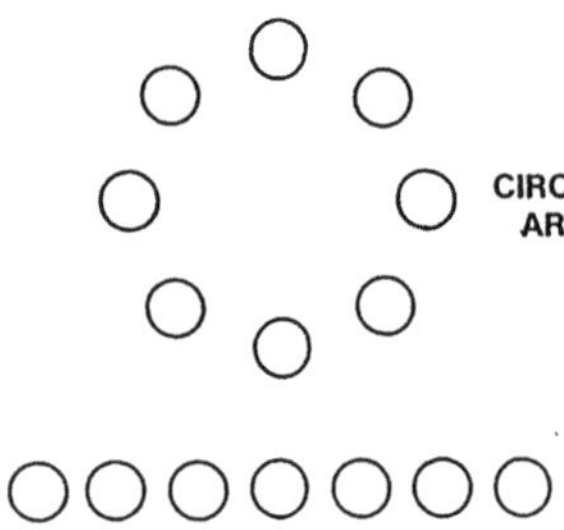

Fig. 8.6 Mutual coupling: planar array

If $Z^T >> Z_j^R$, then changes in Z_j^R between elements will only have small effects on V_j. However, if $Z^T << Z_j^R$, or Z_j^T and Z_j^R partially cancel each other, then changes in Z_j^R produce large changes in V_j. Since Z_j^R depend on V_j, the situation where Z_j^T and Z_j^R partially cancel can produce very large differences in velocity magnitude and phase over the array.

The matrix form of the array equation may be stated as a velocity vector equal to the inverse of the Z-matrix time the driving force vector. If the off-diagonal terms of the Z-matrix are negligible compared to the main-diagonal terms, then the face velocities will follow the driving force vector in magnitude and phase. It can be shown that if the main-diagonal elements of the Z-matrix are much greater than the off-diagonal elements, then the same will be true for the Z-inverse matrix. The largest off-diagonal terms in the Z-matrix tend to be the Z_{ij} coefficients between adjacent transducers. Recall that in the case of a piston, the magnitude of the Z_{ij} is approximately R_{ii} / kd. The ratio of the largest off-diagonal term to the main-diagonal terms is defined as the velocity control factor (VCF).

$$VCF = \frac{R_{ii} / kd}{\left| Z^T + Z_{ii}^R \right|} \tag{8.17}$$

For small values of the VCF, the main-diagonal terms predominate and the array will be well behaved. However, if kd is small, and the Thevenin and self-radiation impedances partially cancel, then the VCF is large and the array may be poorly behaved.

Porter (1965) studied the array coupling effect as a function of ka and Thevenin internal reactance, assuming no internal resistance. Figure 8.7 depicts which combination of ka and X^T produced negative radiation resistance in one or more elements, that is, they were absorbing acoustic power from the rest of the array. It may be observed that for cases where ka > 1.3 negative radiation resistance does not occur, however, for smaller ka, the radiation resistance became negative over a range of normalized Z^T/X_{11} of -3 to -1.

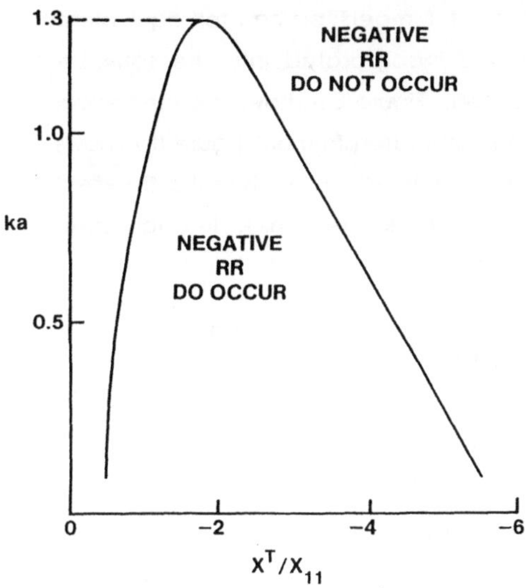

Fig. 8.7 Parametric study: negative radiation resistance. Occurrence of negative radiation resistance (RR) in 80 element array

8.4 INTERACTION IN LOW FREQUENCY - HIGH POWER ARRAYS

8.4.1 Characteristics

In low frequency active arrays, acoustic interaction effects become more prevalent and more severe. Large flextensional transducers are frequently preferred over tonpliz pistons. Low frequency arrays require close packing to reduce real estate, and volumetric configurations rather than baffled planes or spheres. As was pointed out by Sherman, strong interaction effects are felt by the transducers of these sizes since they are not large enough with respect to a wavelength to average out the fluctuations in the pressure field on the surface of the array.

8.4.2 Case Study

Preliminary volumetric array design usually begins with a non-interacting point source model (e.g., Richards 1970). The total pressure field is computed by summing the signal over all elements at each field point. Interaction effects become more important as the individual transducers become more closely packed. In this case study, Pritchard's formulae for circular pistons in a rigid baffle were used to include mutual interactions. The ring-shell transducers used were actually unbaffled. However, since they behave much like symmetrical, double-sided piston vibrators, treating them as single-sided pistons in a baffle proved adequate. (Butler 1989)

The individual ring-shell transducers were modeled as lumped parameter equivalent circuits. The model was developed by Butler (1988) and incorporated into the software program FIRST (Flextensional Integrated Ring Shell Transducer). Figure 8.8 shows the theoretical model versus measured data for a ring-shell in-air with no tuning transformer. Figure 8.9 shows the theoretical admittance loop versus measurement of three ring-shells each in the free-field at test depth. Figure 8.10 shows the corresponding transmit voltage response. In each case, the fit to experimental data is respectable.

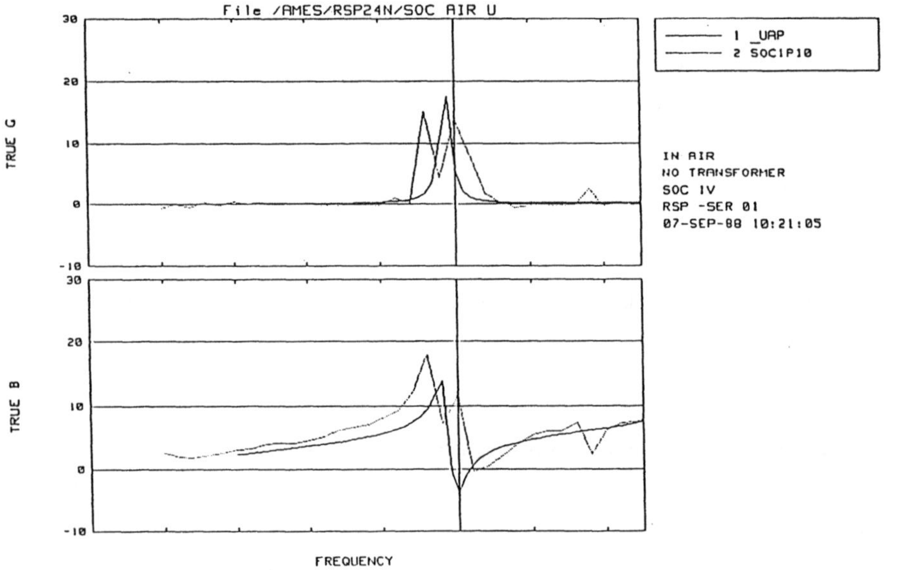

Fig. 8.8 The ring-shell transducer, untuned, in air

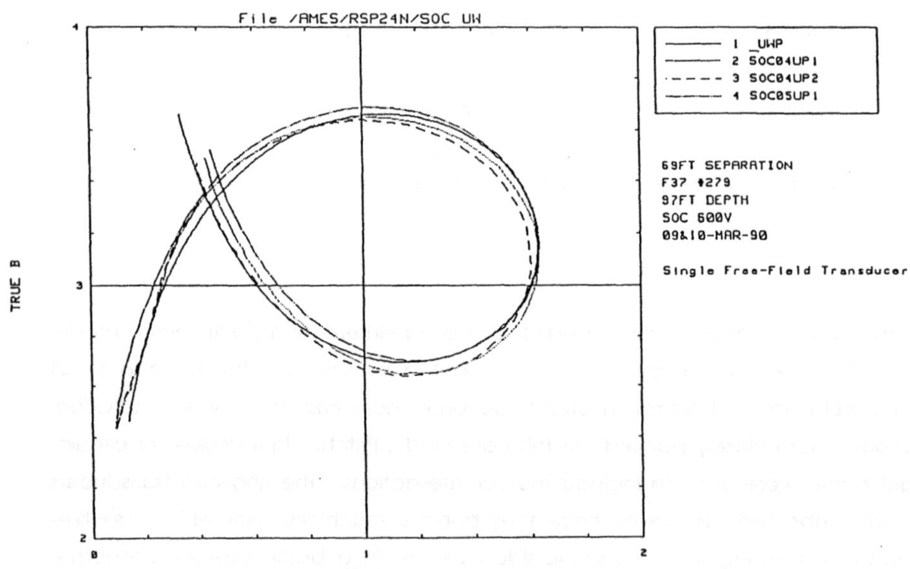

Fig. 8.9 The ring-shell transducer, untuned, in water

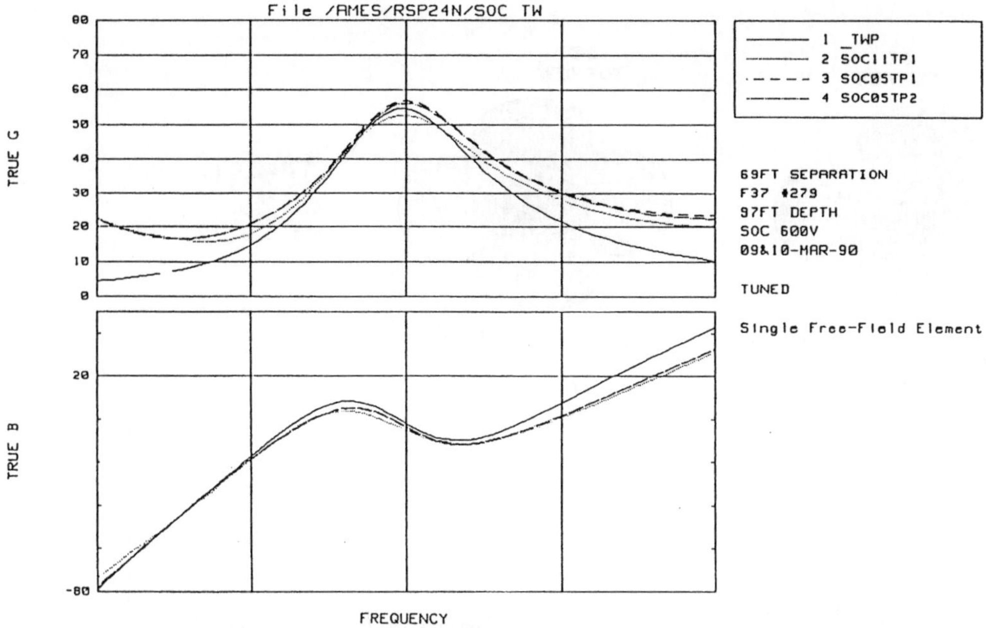

Fig. 8.10 The ring-shell transducer, tuned in water

The case study array consisted of two three-by-four planes for a total of 24 ring-shell transducers. The predicted beam patterns from both the point-source and the FIRST models are shown in Fig. 8.11 for the unsteered array. The customary two-dimensional beam patterns are shown at the bottom of the figure for theory and experimental data. The corresponding three-dimensional beam patterns are shown at the top of the figure and are depicted as looking directly down on the array. Figure 8.12 shows the three-dimensional beam patterns for the point-source and FIRST models in perspective views for broadside steering. These same patterns are repeated in Fig. 8.13 but the view is from above the array to more closely depict the two-dimensional measured data. The 2-dimensional slice through the FIRST patterns shows a remarkable match to experimental measurements. Endfire steering is shown in Fig. 8.14. Again, the comparison to measured data is quite good for the FIRST model, which includes interaction, but not nearly as good for the non-interacting point-source model. Since the array was not perfectly symmetrical, endfire patterns were not exactly the same for the left and right ends of the array. This comparison is shown in Fig. 8.15. Figure 8.16 reports steering at 30 degrees from endfire and Fig. 8.17 reports 60 degrees from endfire. In both cases, the model interaction effects is a considerable improvement over the non-interacting model, but in neither case is the comparison with measured data as dramatic as it was for the broadside and endfire cases. Figure 8.18 shows the three-dimensional perspective views of the 30 and 60 degree steering.

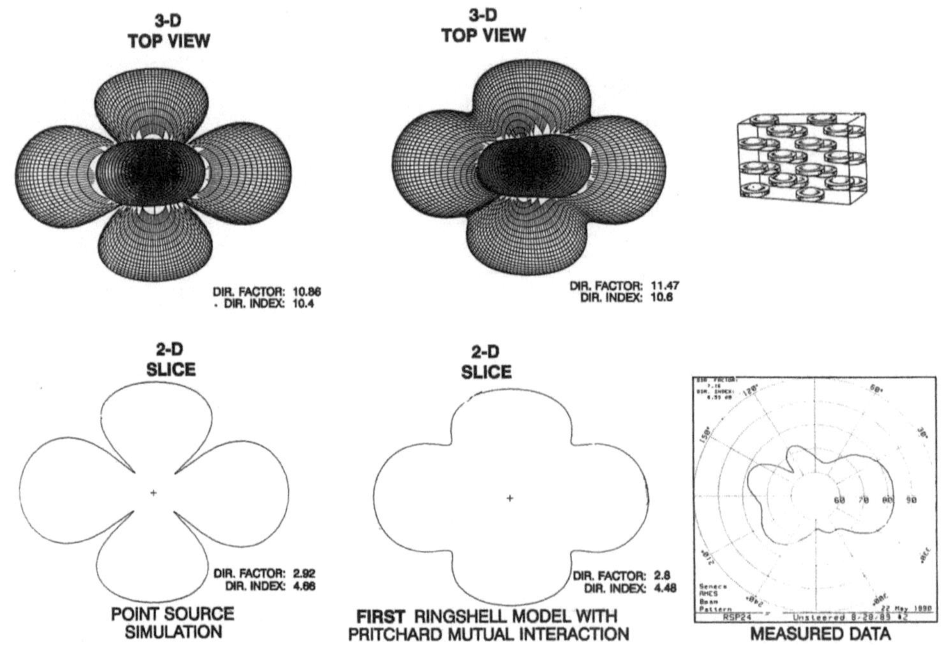

Fig. 8.11 FIRST ring-shell model with Pritchard mutual interaction, point source simulation, unsteered

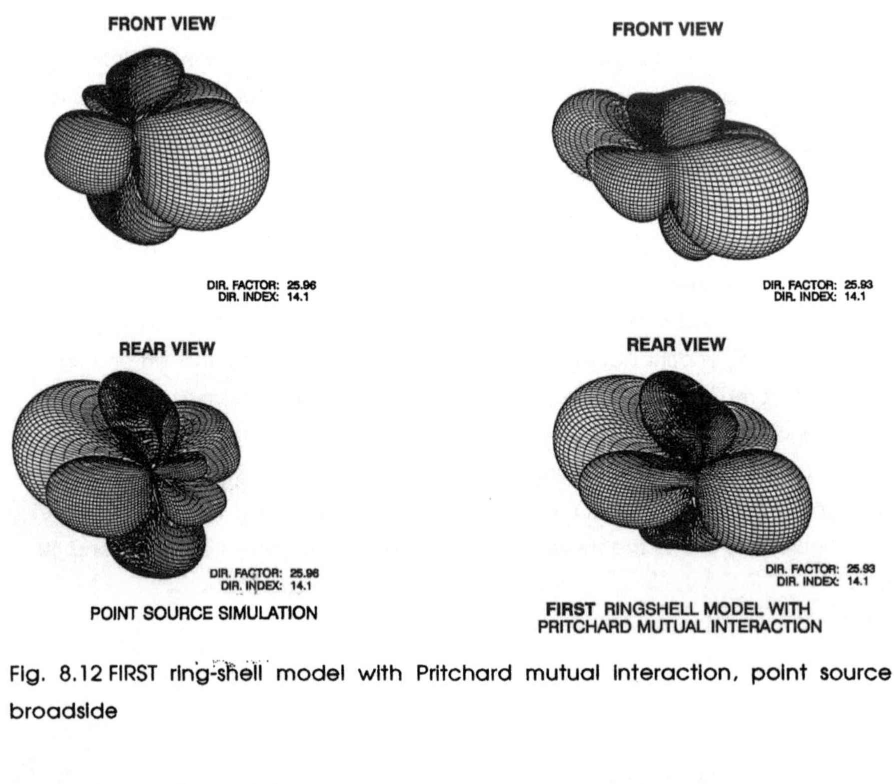

Fig. 8.12 FIRST ring-shell model with Pritchard mutual interaction, point source simulation, broadside

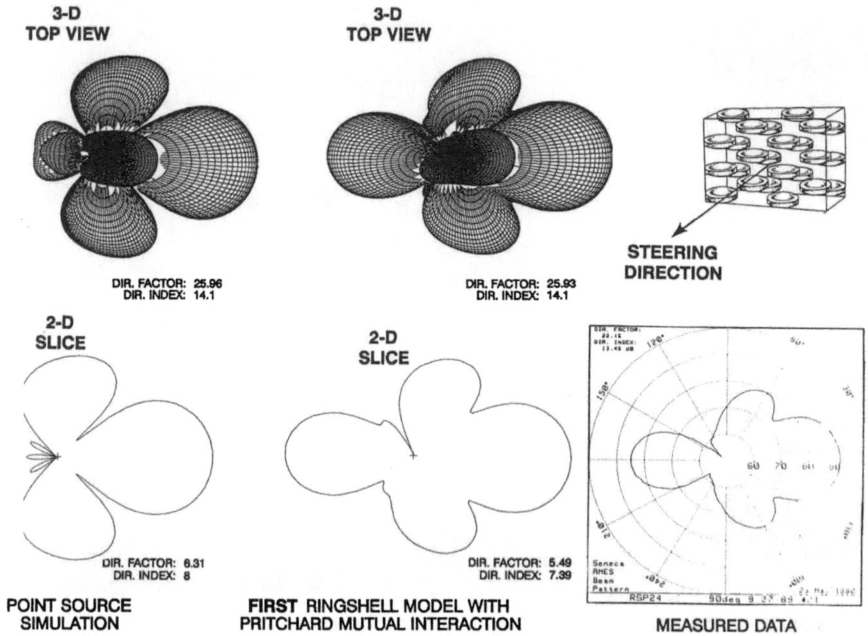

Fig. 8.13 FIRST ring-shell model with Pritchard mutual interaction, point source simulation, broadside

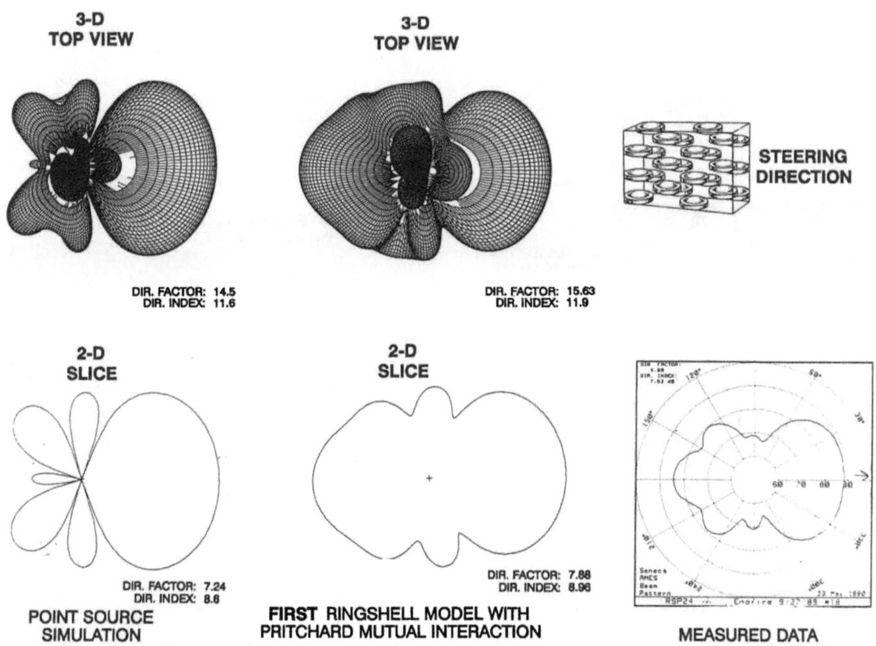

Fig. 8.14 FIRST ring-shell model with Pritchard mutual interaction, point source simulation, endfire

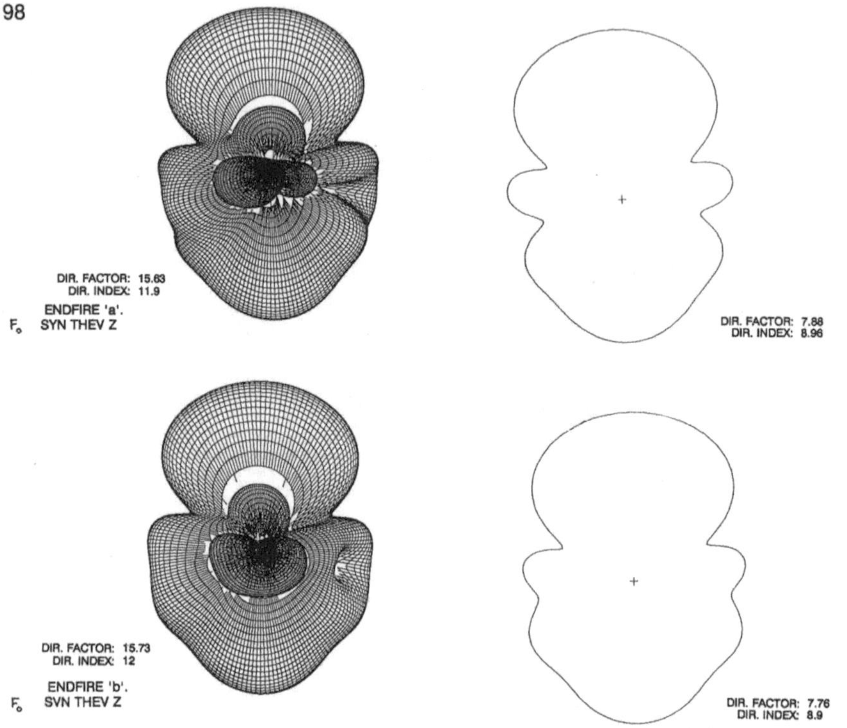

Fig. 8.15 Comparison of endfire patterns for left and right ends of the array

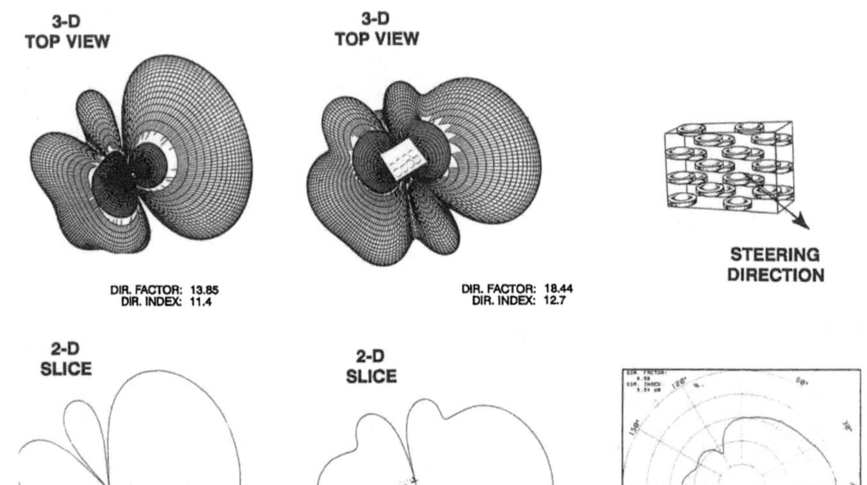

Fig. 8.16 FIRST ring-shell model with Pritchard mutual interaction, point source simulation, steered 30° from endfire

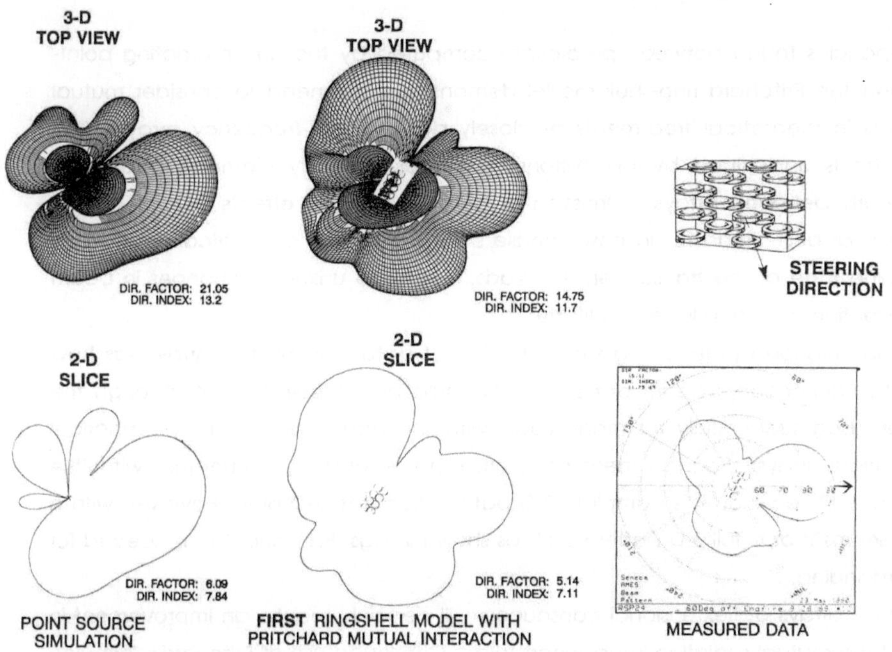

Fig. 8.17 FIRST ring-shell model with Pritchard mutual interaction, point source simulation, steered 60° from endfire

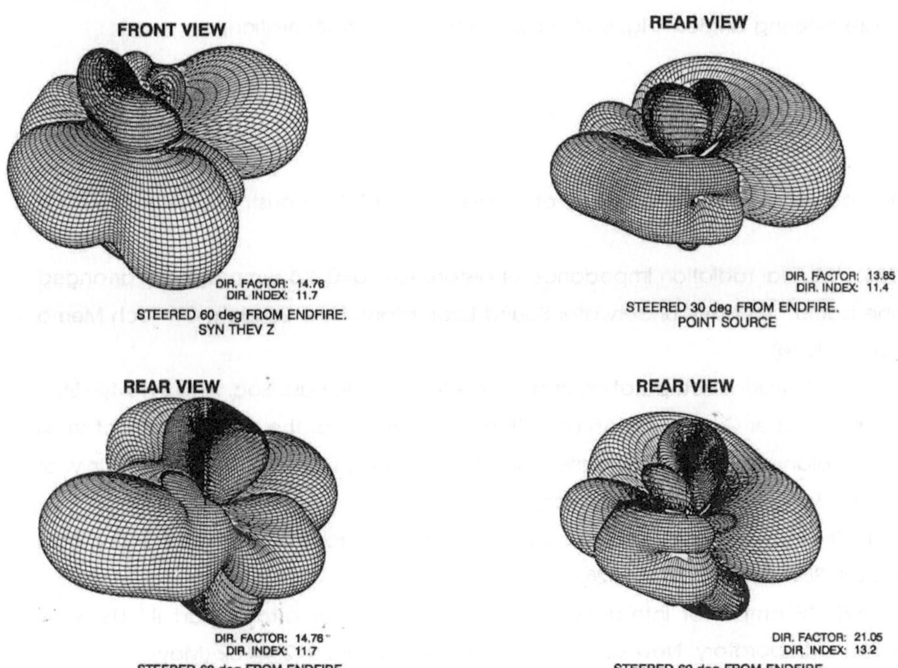

Fig. 8.18 Three-dimensonal perspective views of the 30 and 60 degree steering

8.5 CONCLUSION

The large discrepancies found between predictions computed by the non-interacting point-source model and the Pritchard ring-shell model demonstrate the need to consider mutual interaction effects in theoretical treatments of closely packed low frequency arrays. The importance of this is underlined by the history of failures of array elements and array electronics in poorly behaved arrays. Omission of array interaction effects in such closely packed arrays will, at best, lead to unknown phase shifts between the electrical drive signal and the responding face of the transducer; this leads, in turn, to unknown changes in beam width, steering direction, and side lobe amplitude.

One further observation made during the reduction of data for this test array was how misleading the two-dimensional beam patterns plots could be. These 2-D cuts through the array have been used successfully for many years with line arrays, planar arrays, spherical arrays and cylindrical arrays. Each of these array types has enough symmetry that with little practice one can easily envision the complete 3-D pattern from the 2-D plot. However, with a true 3-D array it seems that a full 3-D pattern such as shown in Figs. 8.11 and 8.17 is needed for complete understanding.

Future work with arrays of flextensional transducers will certainly require an improvement in the estimation of the mutual radiation impedance term. Continued use of Pritchard's formulae for circular pistons in a rigid baffle surely must breakdown in cases of extreme coupling. This seems to be happening with the test case at the upper end of the designed frequency band at and near endfire steering angles. Fig. 8.16 shows some of this deterioration.

REFERENCES

Akay A. (**1990**). 'Radiation from a finite array of simple sources,' J. Acoust. Soc. Am. **87**, 1449-1454

Alperin H.A. (**1956**). 'Mutual radiation impedance of pistons (of ka=0.40) symmetrically arranged in a stiff plane baffle,' US Navy Underwater Sound Laboratory, New London, CT, Tech Memo No. 1150-64-56 (13 June)

Anderson V.C. (**1977**). 'Efficient computation of array patterns,' J. Acoust. Soc. Am. **61**, 744-755

Andrulis M.A.W., and Gruber GJ (**1968**). 'An analytical investigation of the performance of small sound sources in planar arrays,' Underwater Sound Laboratory, Dept. of Mech. Eng., Univ. of Texas, Austin, TX, Tech Report No. 1 (March)

Arase E.M. (**1964**). 'Mutual radiation impedance of square and rectangular pistons in a rigid baffle,' J. Acoust. Soc. Am. **36**, 1521-1525

Atwood J.F. (**1962**). 'Example of interaction effects in transducer arrays: part II,' US Navy Underwater Sound Laboratory, New London, CT, Tech Memo No. 912044-62 (May)

Atwood J.F., and Sherman C.H. (**1962**). 'Example of interaction effects in transducer arrays,' US Navy Underwater Sound Laboratory, New London, CT, Tech Memo No. 912-20-62 (March)

Awatani J. (**1957**). 'Note on acoustic radiation pressure,' J. Acoust. Soc. Am. **29**, 392

Bank G., and Wright J.R. (**1990**). 'Radiation impedance calculations for a rectangular piston,' J Audio Eng Soc **38**, 350-353

Belyakov I.I., and Smaryshev M.D. (**1972**). 'Radiation impedance and gain of a one-dimensional array of rings on an infinite rigid cylinder,' Sov Phys - Acoust **18**, 154

Bergassolli A., and Brodut M. (**1971**). 'Impedance acoustique d'un piston rectangulaire sur cran plan,' Acustica **25**, 61-63

Berkowitz R.L., and Baier R.V. (**1971**). 'Self and mutual acoustic radiation impedances for pistons in a plane infinite rigid baffle,' Naval Research Laboratory, Washington, DC, Report 7008 (18 Jan)

Berman A., and Clay C.S. (**1957**). 'Theory of time-averaged-product arrays,' J. Acoust. Soc. Am. **29**, 805-812

Biot M.A., and Tolstoy I. (**1957**). 'Formulation of wave propagation in infinite media by normal coordinates with an application to diffraction,' J. Acoust. Soc. Am. **29**, 381-391

Bouwkamp C.J. (**1946**). 'A contribution to the theory of acoustic radiation,' Phil Res Rep **1**, 262-264

Bouwkamp C.J. (**1954**) Rep Prog Phys **17**, 35

Brown J.L. Jr, and Rowlands R.O. (1959). 'Design of directional arrays,' J. Acoust. Soc. Am. **31**, 1638-1643

Burnett D.S. (1969). "Radiation impedance functions of rectangular pistons and their applications to sound transmission through finite depth apertures,' PhD Dissertation, University of California, Berkeley.

Burnett D.S., and Soroka W.W. (**1972**). 'Tables of rectangular piston radiation impedance functions with application to sound transmission loss through deep apertures,' J. Acoust. Soc. Am. **51**, 1618-1623

Butler J.L. (**1968**). 'The mutual impedance between two small pistons with and without baffles,' Parke Mathematical Laboratories, Inc., Carlisle, MA, Technical Memorandum No. 2

Butler J.L., and Sherman C.H. (**1969**). 'Acoustic radiation from partially coherent line sources,' J. Acoust. Soc. Am. **47**, 1290-1296

Butler J.L. (**1988**). 'An electro-acoustic model for flextensional ring shell projector,' Image Acoustic, Inc. (31 March)

Butler J.L. (**1989**). 'Flextensional ring-shell transducer 3-D array model with driver options,' Image Acoustic, Inc. (10 March)

Carson D.L. (1962). 'Diagnosis and cure of erratic velocity distributions in sonar projector arrays,' J. Acoust. Soc. Am. **34**, 1191-1196

Carson D.L., Martin G.E., Benthien G.W., and Hickman J.S. (**1963**). 'Control of element velocity distributions in sonar projector arrays,' Proc Navy Sci Symp 7

Carter A.H., and Williams A.O. Jr (**1951**). 'A new expansion for the velocity potential of a piston source,' J. Acoust. Soc. Am. **23**, 179-184

Chakraborty B. (**1988**). 'Mutual interaction effects between array elements at different wavelengths in a coaxial circular array,' J. Acoust. Soc. Am. **83**, 362-364

Chervenak J. (**1972**). 'An experimental study of acoustic interactions between transducer elements in a model array,' Naval Research Laboratory, Washington, DC, Report 7376 (10 March)

Chetaev D.N. (**1951**). 'The impedance of a rectangular piston vibrating in an opening in a flat baffle,' Prik Mat Mekhan **15**, 439-444

Chin N.T. (**1964**). 'Self radiation impedance of a finite free-flooding cylindrical radiator with Junger's end correction. U. S. Navy Underwater Sound Laboratory, New London1). 'The universal radiator, a new development in radiaptechnique,' Elektrotech 5, 365-372

Flad F.R., Kuzneski J.A., and Packard W.B. (**1964**). 'Experimental determination of mutual radiation impedance between coplanar circular pistons,' J. Acoust. Soc. Am. **36**, 2005(A)

Foldy L.L. (**1949**). 'Theory of passive linear electroacoustic transducers with fixed velocity distribution,' J. Acoust. Soc. Am. **21**, 595

Greenspon J.E., and Sherman C.H. (**1964**). 'Mutual radiation impedance and nearfield pressure for pistons on a cylinder,' J. Acoust. Soc. Am. **36**, 149-153

Hanish S. (**1960**). 'The mechanical self resistance and the mechanical mutual resistance of an unbaffled rigid disk (ka < 1) radiating sound from a single face into an acoustic medium,' Naval Research Laboratory, Washington, DC, Report 5538

Hanish S. (**1985**). 'Historical account of mutual acoustic radiation interaction between elementary transducers of a large underwater sound projector array,' J. Acoust. Soc. Am. **78**(Suppl. 1), S73

Hills R. Jr (**1951**). 'Synthesis of line source directivity patterns,' J. Acoust. Soc. Am. **23**, 143-144(A)

Ingard K.U. (**1950**). 'Scattering and absorption by acoustic resonators,'PhD Dissertation, MIT.

Ingard K.U., and Lamb G.L. (**1957**). 'Effect of a reflecting plane on the power output of sound sources,' J. Acoust. Soc. Am. **29**, 743-744

Jacobsen O. (**1976**). 'Some aspects of the self and mutual radiation impedance concept with respect to loudspeakers,' J Audio Eng Soc **24**, 82-92

Johnson H.M. (**1962**). 'Effects of radiation impedance variation on sonar transducer performance in a large array,' J. Acoust. Soc. Am. **34**, 716(A)

Johnson M.P. (**1988**). 'Velocity control and the mechanical impedance of single degree of freedom electromechanical vibrators,' J. Acoust. Soc. Am. **84**, 1994- 2001

Joseph P.M., and Saseendran Pillai P.R. (**1988**). 'An approach towards improving the efficiency of sonar projector arrays,' IEEE ICASSP 5, 2697-2700

Karim M.M.A., and Stumpf F.B. (**1986**). 'Effect of spherical scatterers on the driving-point impedance and directivity pattern of a sonar transducer at an air-water surface,' J. Acoust. Soc. Am. **79**, 865-866

Karnovskii M.I. (**1941**). 'Mutual acoustic impedances of spherical radiators and resonators,' Doklady Akad Nauk SSSR **32**, 40-43

Karnovskii M.I. (**1948**). 'Collected Works of the Kiev Cinema-Engineering Institute, The calculation of mutual acoustic impedances, vol I,' Kiev Cinema- Engineering Institute, Kiev (pp 95-113)

Karnovskii M.I. (**1956**). 'Calculation of the radiation resistance of several types of distributed radiator systems,' Sov Phys - Acoust **2**, 280-293

Kass D.F., and Sherman C.H. (**1959**). 'Radiation Patterns for an Array on a Sphere,' US Naval Underwater Sound Laboratory, New London, CT, Tech Memo No. 1 150-21-59 (7 April)

Kendig P.M. (**1967**). 'Advanced Transducer Developments,' In, Albers VM (ed) Proceedings of the NATO-Sponsored Institutes, Copenhagen, Underwater Acoustics, vol 2. Plenum Press, New York

Klapman S.J. (**1940**). 'Interaction impedance of a system of circular pistons,' J. Acoust. Soc. Am. **11**, 289

Korochentsev V.I., Pyatov A.P., and Subbotin A.G. (**1985**). 'Analysis of the interaction of transducers in an array,' Sov Phys - Acoust. **31**, 363-365

Lasalle D., Steichen W., and Vanderborck G. (**1987**). 'A 3D finite element - plane wave decomposition coupling method to compute mutual radiation impedances in plane arrays,' Ultrason Int Conf Proc -, 437-442

Laird D.T., and Cohen H. (**1952**). 'Directivity patterns for acoustic radiation from a source on a rigid cylinder,' J. Acoust. Soc. Am. **24**, 46-49

Levine H. (**1983**). 'On the radiation impedance of a rectangular piston,' J Sound Vib **89**, 447-455

Locanthi B.N. (**1952**). 'Application of electric circuit analogies to loudspeaker design problems,' Tran I R E PGA-6, 15-36 (Addendum: PGA-7, p 46; oral paper at 1951 IRE meeting)

Lowenstein C.D. (**1959**). 'Synthesis of directive arrays,' Proc Inter Congr Acoust **3**, 680-682

Lowenthal S., and Tournois P. (**1963**). 'Radiation impedance of membranes and plates, their acoustic coupling with the propagating medium,' J. Acoust. Soc. Am. **35**, 1423-1428

Lyon R.H. (**1957**). 'On the low-frequency radiation load of a bass-reflex speaker,' J. Acoust. Soc. Am. **29**, 654

Madella G.B. (**1950**). 'Acoustic coupling of the diaphragms in Duo-Cone loudspeakers,' Alta Frequenza **19**, 267-276

Maidanik G., and Westervelt P.J. (**1957**). 'Acoustical radiation pressure due to incident plane progressive waves on spherical objects,' J. Acoust. Soc. Am. **29**, 936-940

Mangulis V. (**1962**). 'Infinite array of circular pistons in a rigid plane baffle,' J. Acoust. Soc. Am. **34**, 1558-1563

Mangulis V. (**1963**). 'On the radiation of sound from a piston in a nonrigid baffle,' J. Acoust. Soc. Am. **35**, 115-116

Mangulis V. (**1964**). 'Relation between the radiation impedance, pressure in the far field, and baffle impedance,' J. Acoust. Soc. Am. **36**, 211-212

Mangulis V. (**1964**). 'Kramers-Kronig or dispersion relations in acoustics,' J. Acoust. Soc. Am. **36**, 211-212

Mangulis V. (**1966**). 'Near-field pressure for an infinite array of strips,' IEEE Trans Sonics Ultrason **SU-13**, 49-53

Martin G.E., Hickman J.S., and Byrnes F.X. (**1959**). 'Radiation impedances and power output limitations of array elements,' Proc Inter Congr Acoust **3**, 678-679

Martin G.E. (**1985**). 'Analysis of large arrays: brief theory and some techniques used in 1954-1985,' J. Acoust. Soc. Am. **78**(Suppl. 1), S73

Mawardi O.K. (**1951**). 'On the generalization of the concept of impedance in acoustics,' J. Acoust. Soc. Am. **23**, 571-576

McKinney C.M., and Owen W.R. (**1957**). 'Wedge-shaped acoustic horns for underwater sound applications,' J. Acoust. Soc. Am. **29**, 940-947

McMahon G.W. (**1984**). 'Interactions in ring-shell projector arrays,' Defence Research Establishment Atlantic, Dartmouth, Nova Scotia, Tech Memo 84/J

Mechel F.P. (**1988**). 'Notes on the radiation impedance, especially of piston-like radiators,' J Sound Vib **123**, 537-572

Miles J.W. (**1946**). 'The analysis of plane discontinuities in cylindrical tubes: Part I,' J. Acoust. Soc. Am. **17**, 259-271

Miles J.W. (**1946**). 'The analysis of plane discontinuities in cylindrical tubes: Part II,' J. Acoust. Soc. Am. **17**, 272-284

Molloy C.T. (**1948**). 'Calculation of the directivity index for various types of radiators,' J. Acoust. Soc. Am. **20**, 387-405

Nomura Y., and Aida Y. (**1951**). 'On the radiation impedance of a rectangular plate with an infinitely large fixed baffle,' Sci Rep Res Inst Tohoku Univ Ser. B, 1/2, 337-347

Pachner J. (**1951**). 'On the acoustical radiation of an emitter vibrating in an infinite wall,' J. Acoust. Soc. Am. **23**, 185-198

Pachner J. (**1951**). 'On the acoustical radiation of an emitter vibrating freely or in a wall of finite dimensions,' J. Acoust. Soc. Am. **23**, 198-208

Pordes F., and Sherman C.H. (**1959**). 'Measurement of variation of radiation resistance with separation of pairs of underwater transducers,' Proc Inter Congr Acoust **3**, 675-677 (Elsevier, Amsterdam)

Porter D.T. (**1962**). 'Near field sound pressure on square and rectangular pistons,' U. S. Navy Underwater Sound Laboratory, New London, CT, Rept. 557 (AD285963)

Porter D.T. (**1964**). 'Self and mutual radiation impedance and beam patterns for flexural disks in a rigid plane,' J. Acoust. Soc. Am. **36**, 1154-1161

Porter D.T. (**1965**). 'Effect of Thevenin equivalent internal impedance on velocity control and acoustic power for planar broadside arrays for different driving level limitations,' U. S. Navy Underwater Sound Laboratory, New London, CT, Rept. 428 (AD615773)

Porter D.T. (**1965**). 'Broadband velocity control,' US Navy Underwater Sound Laboratory, New London, CT, Technical Memorandum No. 960-47-65

Porter D.T. (**1966**). 'Examples of the effects of array size, spacing, efficiency and driving force uniformity upon array behavior,' US Navy Underwater Sound Laboratory, New London, CT, Tech memo No. 960-64-66 (June)

Porter D.T. (**1966**). 'Computation of transducer velocities for very large transmitting arrays,' US Navy Underwater Sound Laboratory, New London, CT, Report No. 741 (March)

Porter D.T. (**1966**). 'Behavior of small arrays and the effects of undriven elements surrounding them,' US Navy Underwater Sound Laboratory, New London, CT, Report No. 746 (July)

Porter D.T. (**1967**). 'Two FORTRAN programs for computing electroacoustical behavior of transmitting sonar arrays,' US Navy Underwater Sound Laboratory, New London, CT, Technical Report 791 (15 June)

Porter D.T. (**1971**). 'An improved version of the NUSC train of programs for transmitting sonar array prediction,' Naval Underwater Systems Center, New London, CT, Technical Report 4099 (13 August)

Porter D.T. (**1972**). 'December 1971 Status of the NUSC train of computer programs for transmitting sonar array prediction,' Naval Underwater Systems Center, New London, CT, Technical Memo TD12-10-72 (11 January)

Porter D.T. (**1974**). 'January 1974 status of the NUSC train of computer programs for transmitting array prediction,' Naval Underwater Systems Center, New London, CT, Technical Memo TD12-91-74 (12 April)

Porter D.T. (**1985**). 'Effects of nonrigid mounting surfaces on array behavior,' J. Acoust. Soc. Am. **78**(Suppl. 1), S73

Porter D.T. (**1988**). 'NUSC train of computer programs for transmitting array prediction,' Naval Underwater Systems Center, New London, CT, Technical Document 8159 (26 January)

Porter D.T. (**1989**). 'Math model for a transducer array including scattering,' NUSC Ser 92131/18, dtd 23 Jan 89.

Pritchard R.L. (**1951**). 'Discussion of papers by Pachner and by Stenzel on radiation from a circular emitter,' J. Acoust. Soc. Am. **23**, 591

Pritchard R.L. (**1951**). Acoustics Research Laboratory, Harvard University, Cambridge, MA, Tech. Memo. No. 21, Appendix C (Jan 15)

Pritchard R.L. (**1953**). 'Optimum directivity patterns for linear point arrays,' J. Acoust. Soc. Am. **25**, 879-891

Pritchard R.L. (**1954**). 'Maximum directivity index of a linear point array,' J. Acoust. Soc. Am. **26**, 1034-1039

Pritchard R.L. (**1960**). 'Mutual acoustic impedance between radiators in an infinite rigid plane,' J. Acoust. Soc. Am. **32**, 730-737

Pritchard R.L. (**1951**). 'Mutual acoustic impedance between two circular disks,' J. Acoust. Soc. Am. **23**, 143(A)

Pyett J.S. (**1954**). 'Acoustic admittance and impedance as vector quantities,' J. Acoust. Soc. Am. **26**, 870

Queen W.C. (**1969**). 'The directivity of sonar receiving arrays,' J. Acoust. Soc. Am. **47**, 711-720

Reese J.M., and Thompson W. Jr (**1981**). 'Acoustic coupling between two finite-sized spheres; n=2 mode,' J. Acoust. Soc. Am. **69**, 1587-1590

Rhian E. (**1954**). 'An exact method for determining the directivity index of a general three-dimensional array,' J. Acoust. Soc. Am. **26**, 704-706

Richards R.T. (**1970**). 'A general array simulation program for predicting directivity patterns of acoustic arrays,' General Dynamics/Electronics Division, Rochester, NY, Report No. AC- 115-70

Robey D.H. (**1955**). 'On the radiation impedance of an array of finite cylinders,' J. Acoust. Soc. Am. **27**, 706-710 (errata on p 1227)

Robey D.H. (**1955**). 'On the radiation impedance of the liquid-filled squirting cylinder,' J. Acoust. Soc. Am. **27**, 711-714

Rodgers P.H., and Van Buren A.L. (**1978**). 'New approach to a constant beamwidth transducer,' J. Acoust. Soc. Am. **64**, 38-43

Rudgers A.J. (**1986**). 'Application of a Neumann-series method to two problems in acoustic radiation theory that are formulated in terms of Green's functions,' J. Acoust. Soc. Am. **79**, 1211-1222

Rudgers A.J. (**1987**). 'Radiation impedances of interacting acoustic sources expressed in terms of the individual source Green's functions,' J. Acoust. Soc. Am. **82**, 709-710

Rusby J.S.M. (**1960**). 'Measurements of the total acoustic radiation impedance of rigid pistons in an array,' Nature **186**, 144-146

Sawade S. (**1951**). 'Note on radiator-array technique,' Elektrotech **Z 72**, 720

Schenck H.A. (**1968**). 'Improved integral formulation for acoustic radiation problems,' J. Acoust. Soc. Am. **44**, 41-58

Schoch A. (**1941**). Akust **Z 6**, 318

Sherman C.H. (**1958**). 'Mutual radiation impedance between pistons on spheres and cylinders,' US Navy Underwater Sound Laboratory, New London, CT,Research Report No. 405

Sherman C.H. (**1958**). 'Interaction radiation impedance between rectangular pistons on an infinite cylinder,' US Naval Underwater Sound Laboratory, New London, CT, Tech Memo No. 150-29-58 (13 June)

Sherman C.H. (**1958**). 'Acoustic Radiation from spherical arrays,' US Naval Underwater Sound Laboratory, New London, CT, Tech Memo No. 1150-5-58 (20 Jan)

Sherman C.H. (**1959**). 'Mutual radiation impedance of sources on a sphere,' J. Acoust. Soc. Am. **31**, 947-952

Sherman C.H. (**1961**). 'Mutual radiation impedance of fixed velocity distribution transducers,' U. S. Navy Underwater Sound Laboratory, New London, CT, Tech. Memo. 912-1-74-61

Sherman C.H. (**1962**). 'Example of interaction effects in transducer arrays: part III,' U. S. Navy Underwater Sound Laboratory, New London, CT, Tech. Memo. 912-82-62

Sherman C.H. (**1963**). 'Effect of the nearfield on the cavitation limit of transducers,' J. Acoust. Soc. Am. **35**, 1409-1412

Sherman C.H. (**1965**). 'Theoretical model for mutual radiation resistance of small transducers at an air-water surface,' J. Acoust. Soc. Am. **37**, 532-533

Sherman C.H. (**1966**). 'Analysis of acoustic interactions in transducer arrays,' IEEE Trans Sonics Ultrason **SU-13**, 9-15

Sherman C.H., and Kass D.F. (**1958**). 'Radiation impedances for an array on a sphere,' US Naval Underwater Sound Laboratory, New London, CT, Tech Memo No. 1150-70-58 (25 Nov)

Sherman C.H., Kass D.F. (**1959**). 'Radiation impedances, radiation patterns and efficiency for a large array on a sphere,' U. S. Navy Underwater Sound Laboratory, New London, CT, Report No. 429 (17 July)

Shindo T., Yoshioka T., and Fukuyama K. (**1990**). 'Calculation of sound radiation from an unbaffled, rectangular-cross-section horn loudspeaker using combined analytical and boundary-element methods,' J Audio Eng Soc **38**, 340-349

Skorheim R.D. (**1957**). 'The radiation impedance of a colinear array of finite cylindrical radiators,' MS Dissertation, US Naval Postgraduate School.

Skudrzyk E.J. (**1954**). 'Die Grundlagen der Akustik,' Springer-Verlag, Vienna

Skudrzyk E.J. (**1971**). 'The foundation of acoustics,' Springer-Verlag, Vienna, New York

Smaryshev M.D. (**1973**). 'Directivity of underwater acoustic arrays,' Sudostroenie, Leningrad (in Russian)

Smaryshev M.D. (**1979**). 'Directivity of underwater acoustic arrays,' 2nd ed. Sudostroenie, Leningrad (in Russian)

Steinberg B.D. (**1976**). 'Principles of aperture and array system design,' John Wiley, New York

Stenzel H. (**1930**). Ann Physik **(5)7**, 952,965-968

Stenzel H. (**1942**). Ann Physik **(5)41**, 245-260

Stenzel H. (**1949**). Ann Physik **(6)4**, 303-324

Stumpf F.B. (**1964**). 'Interaction radiation resistance for a line array of two and three magnetostrictive-stack transducers at an air-water surface,' J. Acoust. Soc. Am. **36**, 174-176

Stumpf F.B., and Bulger C.R. (**1973**). 'Effect of spherical scatterers on the radiation resistance of a transducer at an air-water surface,' J. Acoust. Soc. Am. **53**, 955- 956

Stumpf F.B., and Crum L.A. (**1966**). 'Interaction radiation resistance and reactance measurements for two small transducers at an air- water surface,' J. Acoust. Soc. Am. **40**, 1554-1555

Stumpf F.B., and Junit A.M. (**1980**). 'Effect of a spherical scatterer on the radiation reactance of a transducer at an air-water surface,' J. Acoust. Soc. Am. **67**, 715- 716

Stumpf F.B., and Lam Y.Y. (**1970**). 'Radiation resistance of a small transducer at a water surface near plane boundaries,' J. Acoust. Soc. Am. **47**, 332-338

Stumpf F.B., and Lukman F.J. (**1960**). 'Radiation resistance of magnetostrictive-stack transducer in presence of second transducer at air-water surface,' J. Acoust. Soc. Am. **32**, 1420-1422

Stumpf F.B., Richards R.T., and Crum L.A. (**1968**). 'Acoustic interaction for arrays of small magnetostrictive-stack sonar transducers at an air-water surface,' Ohio University Research Institute, Athens, Ohio, Final Technical Report (Contract Nonr4419 (01))

Stumpf F.B., and Shukla S.S. (**1977**). 'Effect of a hollow spherical scatter on the radiation resistance of a transducer at an air-water surface,' J. Acoust. Soc. Am. **61**, 221- 222

Stumpf F.B., and Sugar R.J. (**1974**). 'Effect of cylindrical scatterers on the radiation resistance of a transducer at an air-water surface,' J. Acoust. Soc. Am. **55**, 194- 195

Swenson G.W. Jr, and Johnson W.E. (**1952**). 'Radiation impedance of a rigid square piston in an infinite baffle,' J. Acoust. Soc. Am. **24**, 84

Teer K. (**1965**). 'The reactance barrier in the transducing process,' Proc Inter Congr Acoust 5, paper J17

Thompson W. Jr (**1971**). 'The computation of self- and mutual- radiation impedances for annular and elliptic pistons using Bouwkamp's integral,' J Sound Vib **17**, 221-233

Thompson W. Jr (**1976**). 'Radiation from a spherical acoustic source near a scattering sphere,' J. Acoust. Soc. Am. **60**, 781-787

Thompson W. Jr (**1977**). 'Acoustic coupling between two finite-sized spherical sources,' J. Acoust. Soc. Am. **62**, 8-11

Thompson W. Jr, and Reese J.M. (**1983**). 'Acoustic coupling between a pulsating and an oscillating sphere,' J. Acoust. Soc. Am. **74**, 1048- 1050

Thorn J.V., and Booth N.O., Lockwood JC (**1980**). 'Random and Partially random acoustic arrays,' J. Acoust. Soc. Am. **67**, 1277-1286

Toulis W.J. (**1957**). 'Radiation load on arrays of small pistons,' J. Acoust. Soc. Am. **29**, 346-348

Toulis W.J. (**1963**). 'Mutual coupling with dipoles in arrays,' J. Acoust. Soc. Am. **35**, 1062-1063

Wang Wei-M (**1972**). 'Mutual coupling through radiation between transducer elements in an infinite array,' IEEE Trans Sonics Ultrason **SU-19**, 28-31

Wang Wei-M (**1972**). 'Radiation patterns of acoustic transducer elements in an infinite array,' IEEE Trans Sonics Ultrason **SU-19**, 31-33

Waterhouse R.V. (**1961**). 'The effect of rigid and pressure release boundaries on various multipole sources,' Cambridge Acoustical Associates, Cambridge, MA, Inc. Rep. U-117-48

Waterhouse R.V. (**1958**). 'Output of a sound source in a reverberation chamber and other reflecting environments,' J. Acoust. Soc. Am. **30**, 4-13

Waterhouse R.V. (**1963**). 'Radiation impedance of a source near reflectors,' J. Acoust. Soc. Am. **35**, 1144-1151

Waterhouse R.V. (**1965**). 'Mutual impedance of acoustic sources,' Proc Inter Congr Acoust 5th, Paper J-21

Westervelt P.J. (**1951**). 'Acoustic impedance in terms of energy functions,' J. Acoust. Soc. Am. **23**, 347-348

Westervelt P.J. (**1951**). 'The interaction of a finite amplitude acoustic wave with small obstacles and orifices,' PhD Dissertation, MIT. Physics Department, (January)

Williams W., Parke N.G., Moran D.A., and Sherman C.H. (**1964**). 'Acoustic radiation from a finite cylinder,' J. Acoust. Soc. Am. **36**, 2316-2322

Willms W. (**1954**). 'Bemerkungen zur Definition des vektoriellen Schall-leitwertes, des akustischen Widerstandes und der akustischen Elemente,' Acustica **4**, 133-136

Willms W. (**1954**). 'Zum Begriff der Schallimpedanz,' Acustica **4**, 427-432

Wolff I., and Malter L. (**1929**). Phys Rev **33**, 1061

Woollett R.S. (**1959**). 'Effects of radiation impedance variation on sonar transducer performance in a large array,' U. S. Navy Underwater Sound Laboratory, New London, CT, Report No. 428 (OTS No. PB150291)

Woollett R.S. (**1963**). 'Trends and problems in sonar transducer design,' IEEE Trans Ultrason Eng **UE-10**, 116-124

Yatsenko V.K. (**1977**). 'Method for measuring the mutual impedance of two piezoelectric transducers,' Izd. Rost. Univ., Rostov-on-Don, No. 4

9. Numerical Modeling of Array Interactions

G.W. Benthien

Naval Ocean Systems Center San Diego, CA 92152-5000, USA

Numerical methods are frequently used to calculate array interactions. This paper summarizes some of the advantages that numerical methods have over the more classical analytical methods. Many of the numerical methods for calculating array interactions are based on the Helmholtz integral relations. This paper describes one such method and shows how it can be coupled to a finite element model for the individual transducers in the array. A very useful approximation for the array interaction matrices is also presented. To illustrate the validity of this numerical model, a comparison between numerical and experimental results for a three element flextensional array is also presented.

9.1 INTRODUCTION

The methods for calculating acoustic interactions between transducer elements in an array can be roughly grouped into two classes: analytical methods and numerical methods. The term *analytical methods* refers to those methods which lead to closed-form expressions for the acoustic interactions in which the dependence on parameters such as frequency, size, and separation is explicit. The term *numerical methods* refers to those methods which attempt to solve the basic differential equations, integral equations, or variational principle by direct approximation. Each of these classes has certain advantages over the other. Analytical methods offer the potential of greater insight since the dependence on the relevant parameters is explicit. They also frequently lead to more efficient computational algorithms. In some cases, however, the resulting analytical expressions are so complicated that their potential advantages in computational speed and better insight are not realized. Numerical methods also provide certain advantages. For example:

- numerical methods are not limited to a few idealized geometries, but can handle quite general configurations. Analytical methods, on the other hand, are usually limited to simple geometries such as planes, spheres, infinite cylinders, etc.,
- numerical methods can handle more general velocity distributions on the array elements. In fact, with most numerical methods, it is not necessary to specify in advance what the velocity distribution on each element will be,
- most numerical methods for calculating array interactions can be easily coupled to numerical structural models (e.g., finite element models) for the transducer elements.

In addition to the above advantages, it should be pointed out that the tremendous increase in computational power over the last few years has made numerical methods much more feasible than they were in the past.

This paper will concentrate on a particular numerical method called CHIEF (Combined Helmholtz Integral Equation Formulation) which is based on the Hemholtz integral relations. It will show how this method can be used to calculate array interactions and how it can be coupled to a finite element model for the individual transducer elements. It will also present a useful approximation to the array interaction matrices which is a generalization of an approximation introduced by Pritchard (9.1). To illustrate the validity of this numerical method it will present both numerical and experimental results for an array of flextensional transducers.

9.2 FINITE ELEMENT EQUATIONS

The finite element method is now being used extensively in transducer element analysis. For example, in flextensional transducers the finite element method is used to model the flexing of the radiating shell and in longitudinal vibrators it is used to model the flexing of the radiating head. Therefore, any general numerical method for calculating array interactions must be capable of being coupled with a finite element model for the transducer element.

In the finite element method, the structural displacement is represented as a linear combination of vector interpolation functions, i.e.,

$$u(x) = \sum_{m=1}^{M} U_m \phi_m(x). \tag{9.1}$$

Usually the coefficients U_m represent displacement degrees-of-freedom at points in the body called nodes. The vector interpolation functions $\phi_m(x)$ allow the displacement degrees-of-freedom to be interpolated at points in between the nodes. Typically, a finite element program will give the user a choice of various types of interpolation functions (elements). A finite element is said to be isoparametric if the same interpolation functions $\phi_m(x)$ used to represent displacements are also used to represent the boundary surface of the body.

Combination of the representation (9.1) with a variational formulation of the equations of linear elasticity produces a set of linear equations

$$M\ddot{U} + KU = F^{rad} + F^{d} \tag{9.2}$$

where U is a vector the components of whichare the coefficients U_m shown in equation (9.1), M is the mass matrix, K is the stiffness matrix, F^{rad} is the acoustic load vector, and F^d is the load vector representing the forces driving the structure. The vector F^d might represent the force produced by a ceramic driver on a flextensional shell or a radiating head mass. The dots represent time derivatives. If each of the vector functions of time is replaced by its Fourier transform, e.g.,

$$U(t) = \frac{1}{2\pi}\int_{-\infty}^{+\infty} U(\omega)\, e^{i\omega t}\, d\omega, \qquad (9.3)$$

then equation (9.2) becomes

$$(-\omega^2 M + K)\, U = F^{rad} + F^{d}. \qquad (9.4)$$

where U, F^{rad}, and F^{d} are now complex functions of frequency.

The acoustic interactions couple with the finite element method through the vector F^{rad}. The components of F^{rad} are given by

$$F_m^{rad} = -\int_S p\phi_m \cdot n\, dS \qquad (9.5)$$

where p is the acoustic pressure, S is the radiating surface of the structure, and n is the unit normal to S extending into the fluid.

It will sometimes be convenient to represent the finite element equations in terms of normal modes. The matrix E, whose columns are the normal modes of the structure, has the properties

$$E^T M E = I, \qquad (9.6a)$$
$$E^T K E = \Omega \qquad (9.6b)$$

where I is the identity matrix and $\Omega = \mathrm{diag}(\omega_1^2, \ldots \omega_n^2)$ is a diagonal matrix built with *in vacuo* resonance frequencies. Making use of the relations (9.6), the finite element equations (9.4) can be written

$$(-\omega^2 I + \Omega)\,\hat{U} = \hat{F}^{rad} + \hat{F}^{d}. \qquad (9.7)$$

where :

$U = E\hat{U}$, $\hat{F}^{rad} = E^T F^{rad}$, and $\hat{F}^{d} = E^T F^{d}$.

Thus, the normal modes allow the finite element equations to be diagonalized or decoupled. Equation (9.7) is a very convenient form when the problem requires the calculation of the response at a large number of frequencies or when the response can be approximately represented in terms of a small number of normal modes.

9.3 CHIEF EQUATIONS

Many of the numerical techniques for solving the acoustic equations are based on the Helmholtz integral relations. One such method called CHIEF (Combined Helmholtz Integral Equation Formulation) (9.2, 9.3) which has been used for many years at the Naval Ocean Systems Center is based on the Helmholtz integral relations

$$\int_S p(\sigma)\frac{\partial G(x,\sigma)}{\partial n_\sigma}\,dS(\sigma) + i\omega\rho\int_S v(\sigma)\,G(x,\sigma)\,dS(\sigma) = \frac{1}{2}p(x) \qquad x \in S \tag{9.8}$$

$$= 0 \qquad x \text{ interior to } S \tag{9.9}$$

where:

$$G(x,\sigma) = \frac{e^{-ik|x-\sigma|}}{2\pi\,|x-\sigma|}\,.$$

If v is prescribed on S, then equation (9.8) becomes an integral equation of the second kind for p on S. This integral equation is well suited to numerical solution for most wavenumbers k, but its solution is not unique whenever k is an eigenvalue of the interior Dirichlet problem. The CHIEF approach to overcoming this difficulty is to form an overdetermined system by supplementing the integral equation (9.8) with the integral equation (9.9) evaluated at selected interior points.

The combination of equations (9.8) and (9.9) can be solved numerically by approximating p and v on S by piecewise constant functions, i.e.,

$$p(\sigma) \doteq p_n \quad \text{and} \quad v(\sigma) \doteq v_n \quad \text{for } \sigma \in S_n \tag{9.10}$$

where $S_1,\ldots,S_n$, are nonoverlapping subregions of S the union of which is S. Collocation of equation (9.8) at N_s surface points σ_n and equation (9.9) at N'_s interior points x_n leads to the matrix relation

$$A\,P = B\,V \tag{9.11}$$

where:

$$A_{mn} = \frac{1}{2}\delta_{mn}\int_{Sn}\frac{\partial G(\sigma_m,\sigma)}{\partial n_\sigma}\,dS(\sigma) \qquad m = 1,\ldots,N_s$$

$$= -\int_{Sn}\frac{\partial G(x_m,\sigma)}{\partial n_\sigma}\,dS(\sigma) \qquad m = N_s+1,\ldots,N_s+N_s'$$

$$B_{mn} = i\omega\rho\int_{Sn}\partial G(\sigma_m,\sigma)\,dS(\sigma) \qquad m = 1,\ldots,N_s$$

$$= i\omega\rho\int_{Sn}\partial G(x_m,\sigma)\,dS(\sigma) \qquad m = N_s+1,\ldots,N_s+N_s'$$

and:

$$P = \begin{pmatrix} p_1 \\ \vdots \\ p_n \end{pmatrix}, \quad V = \begin{pmatrix} v_1 \\ \vdots \\ v_n \end{pmatrix}$$

For specified V, equation (9.11) is an overdetermined system of equations for P. This overdetermined system is solved in the least-square sense using the Householder reduction technique (9.4). In this technique a unitary matrix Q is constructed so that

$$Q\,A = \begin{pmatrix} U \\ 0 \end{pmatrix} \tag{9.12}$$

where U is an $N_s \times N_s$ upper triangular matrix. The least-square solution of equation (9.11) is given by

$$P = (U^{-1}, 0)\; Q\; B\; V. \tag{9.13}$$

If an impedance matrix Z is defined by

$$Z = (U^{-1}, 0)\; Q\; B, \tag{9.14}$$

then

$$P = Z\; V. \tag{9.15}$$

Equation (9.15) will serve as the basis for the acoustic interaction relations.

9.4 CHIEF ARRAY EQUATIONS

In the CHIEF approach it is not required that the boundary surface S encloses a connected volume. For an array the surface S can be taken to be the union of the boundary surfaces of the array elements. The CHIEF impedance relation (9.15) for an array can be partitioned as follows:

$$\begin{pmatrix} P(1) \\ P(2) \\ \vdots \\ P(n) \end{pmatrix} = \begin{pmatrix} Z_{11} & Z_{12} & \cdots & Z_{1N} \\ Z_{21} & Z_{22} & \cdots & Z_{2N} \\ \vdots & \vdots & & \vdots \\ Z_{N1} & Z_{N2} & \cdots & Z_{NN} \end{pmatrix} \begin{pmatrix} V(1) \\ V(2) \\ \vdots \\ V(n) \end{pmatrix} \tag{9.16}$$

where P(n) is the surface pressure vector and V(n) is the surface normal velocity vector for array element n. It follows from equation (9.16) that

$$P(m) = \sum_{n=0}^{N} Z_{mn} V(n) \quad m = 1,2,\ldots,N. \tag{9.17}$$

The matrix Z_{mn} is the mutual radiation impedance between elements m and n .

In a paper on the mutual radiation impedance between circular pistons, Pritchard (9.1) introduced an approximate mutual radiation impedance given by

$$\bar{Z}_{mn} = i\,\mathrm{Real}(\,\bar{Z}_{self}\,)\,\frac{e^{-i\,kd_{mn}}}{k\,d_{mn}} \tag{9.18}$$

where $\bar{Z}_{self}$ is the self radiation impedance and d_{mn} is the center-to-center separation between the m-th and n-th circular pistons. The relation (9.18) can be generalized to the mutual radiation impedance matrices Z_{mn} as follows:

$$Z_{mn} = i\,\mathrm{Real}(\,Z_{self}\,)\,\frac{e^{-i\,kd_{mn}}}{k\,d_{mn}} \tag{9.19}$$

where Z_{self} is the self radiation impedance matrix of a single element in the free-field. Z_{self} can be computed using CHIEF. The approximation (9.19) has proven to be amazingly accurate for most array configurations that occur in practice. Results for a three element flextensional array are presented in section 9.7 which show how the Pritchard approximation compares with the full CHIEF computation.

9.5 FORCE AND VELOCITY COUPLING

In this section the coupling equations relating finite element forces and displacements on the surface to CHIEF pressures and velocities will be developed. It will be assumed that the finite elements on the surface coincide with the CHIEF subdivisions. If the piecewise constant pressure approximation is used in equation (9.5) for the radiation force components, the following result is obtained

$$F_m^{rad} = -\int_S p\phi_m \cdot n\,dS = -\sum_{n=1}^{N_s} (p_n S_n)\frac{1}{S_n}\int_{S_n} \phi_m \cdot n\,dS. \tag{9.20}$$

Equation (9.20) can be written in the matrix form

$$F^{rad} = -\,C\;D\;P \tag{9.21}$$

where

$$C_{mn} = \frac{1}{S_n}\int_{S_n} \phi_m \ . \ n \, dS. \tag{9.22}$$

and

$$D = \text{diag}\ (S_1, \ldots, S_{N_S}). \tag{9.23}$$

The matrix C will be referred to as the coupling matrix.

Since the CHIEF and finite element interpolation functions are different, it is impossible to enforce exact normal velocity continuity across S. Instead, normal velocity continuity is enforced in the average sense

$$v_n = \frac{1}{S_n}\int_{S_n} (i\,\omega\, u\ .\ n)\, dS \tag{9.24}$$

where v_n is the CHIEF normal velocity associated with the n-th subregion S_n, and u is the finite element displacement function on S. Combination of equation (9.24) with the displacement expansion (1) yields

$$v_n = i\omega \sum_{m=1}^{N_s} U_m \frac{1}{S_n}\int_{S_n} \phi_m \ . \ n \, dS \tag{9.25}$$

or equivalently,

$$V = i\,\omega\, C^T U. \tag{9.26}$$

Equations (9.21) and (9.26) are the desired force and velocity coupling equations.

9.6 ARRAY INTERACTION EQUATIONS

There are basically two ways that the equations of the previous section can be combined in order to solve the array interaction problem. The CHIEF relations can be substituted into the finite element equations or the finite element relations can be substituted into the CHIEF equations. Let us consider the first approach. Writing the coupling relation (9.21) for the m-th array element and making use of the interaction equation (9.17), there results

$$F^{rad}(m) = -C\,D\,P(m) = \sum_{n=1}^{N} C\,D\,Z_{mn}\ V(n) \qquad m = 1, 2, \ldots, N. \tag{9.27}$$

Substitution of the velocity coupling relation (26) into equation (9.27) yields

$$F^{rad}(m) = i\omega \sum_{n=1}^{N} C D Z_{mn} C^T U(n) \qquad m = 1, 2, \ldots, N. \tag{9.28}$$

Substitution of equation (9.28) into the finite element equation (9.4) yields

$$(-\omega^2 M + K)\, U(m) = i\omega \sum_{n=1}^{N} C D Z_{mn} C^T\, U(n) + F^d(m) \qquad m = 1, 2, \ldots, N. \tag{9.29}$$

Given the driving forces $F^d(m)$, the set of equations (9.29) can be solved for the displacement vectors U(1),....,U(N).

Let us now consider the second approach. If the finite element equation (9.4) for the m-th array element is solved for U(m), there results

$$U(m) = (-\omega^2 M + K)^{-1} F^{rad}(m) + (-\omega^2 M + K)^{-1} F^d(m). \tag{9.30}$$

Substitution of equation (9.30) into the velocity coupling relation (9.26) yields

$$V(m) = i\omega C^T U(m) = i\omega C^T (-\omega^2 M + K)^{-1} \left[F^{rad}(m) + F^d(m)\right]. \tag{9.31}$$

Combination of equation (9.31) with equation (9.27) gives

$$\begin{aligned} V(m) &= i\omega C^T (-\omega^2 M + K)^{-1} \sum_{n=1}^{N} C D Z_{mn}\, V(n) \cdot \\ &+ i\omega C^T (-\omega^2 M + K)^{-1} \left[F^d(m)\right] \quad m = 1, 2, \ldots, N. \end{aligned} \tag{9.32}$$

Given the driving forces F^d (m), the system of equations (9.32) can be solved for the normal velocity vectors V(1),...., V(N). This second method is the one we use at the Naval Ocean Systems Center since it leads to a smaller system of equations. The inverse $(-\omega^2 M + K)^{-1}$ that appears in equation (9.32) is calculated using normal modes, i.e.,

$$(-\omega^2 M + K)^{-1} = E\,(-\omega^2 I + \Omega)^{-1} E^T. \tag{9.33}$$

Of course the inverse does not exist if ω is exactly equal to one of the *in vacuo* eigenfrequencies, but these resonances are so sharp that this poses no problem in practice. The problem can be eliminated entirely by adding a small amount of damping to the structure. It should also be pointed out that the acoustic interaction matrices are smooth enough functions of frequency that frequency interpolation can be employed in their calculation.

Rapid changes in the response are due more to the response of the structure than to any rapid change in the acoustic interactions.

9.7 COMPUTATIONAL AND EXPERIMENTAL ARRAY RESULTS

In this section I will show a comparison of computed and experimental results for a three element array of flextensional transducers as shown in Figure 9.1. The flextensional shell was modeled using the finite element program MARTSAM. The ceramic driver was modeled using a one-dimensional analytic model. The array interactions were modeled using the CHIEF program developed at the Naval Ocean Systems Center. The radiating part of the transducer elements consisted of an E-glass shell having a major to minor diameter ratio of 2.92, a thickness to major diameter ratio of .0406, and a height to major diameter ratio of 1. The ratio of interelement array spacing to major diameter is 2.06. The parameters of the ceramic and shell were adjusted slightly to match the in-air resonances of a typical element. The array elements were driven with individual amplifiers, and the amplifiers were included in the model as Thevenin equivalents. The experimental results used as a comparison were obtained at Lake Seneca in New York State.

Figures 9.2-9.6 show comparisons of model and experimental results for the three element array. In each of these figures the solid curve is the model result and the dotted line is the experimental result. Figure 9.2 shows the source level for the array versus ka (a=major diameter) and figures 9.3-9.6 show the real and imaginary parts of input impedance for center and end elements versus ka. Note that the model correctly predicts that the center element will take in power in the vicinity of ka= 0.85 (figure 9.5). In general, the agreement between experimental and model results is reasonably good and the model predicts all of the main features.

Figures 9.7-9.10 show a comparison of CHIEF interaction impedances for the dominant flexural mode of the flextensional transducers, i.e., the modal force is computed using the CHIEF interaction matrices and a velocity distribution corresponding to the dominant flexural mode. Figures 9.7 and 9.8 show the real and imaginary parts of the interaction impedance between an end element and the center element. Figures 9.9 and 9.10 show the real and imaginary parts of the interaction impedance between the two end elements. In each case the Pritchard approximation agrees quite well with the CHIEF results.

Figure 9.11 shows the array source level computed both by the CHIEF model and with the Pritchard approximation. Notice that the two curves are virtually indistinguishable. Figure 9.12 shows a similar result for the real part of input impedance for the center element.

The above results show that the Pritchard approximation is quite good for this three element array configuration. Our experience at the Naval Ocean Systems Center is that the Pritchard approximation has worked well in almost every array configuration that we have tried. In fact it is very difficult to find a situation in which the Pritchard approximation doesn't produce adequate results.

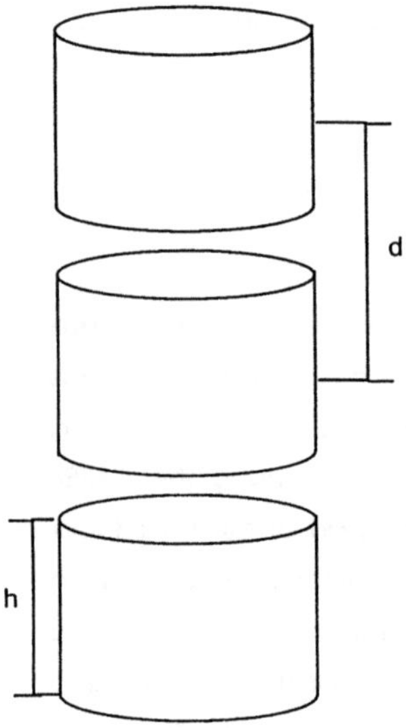

Fig. 9.1 Three element flextensional array

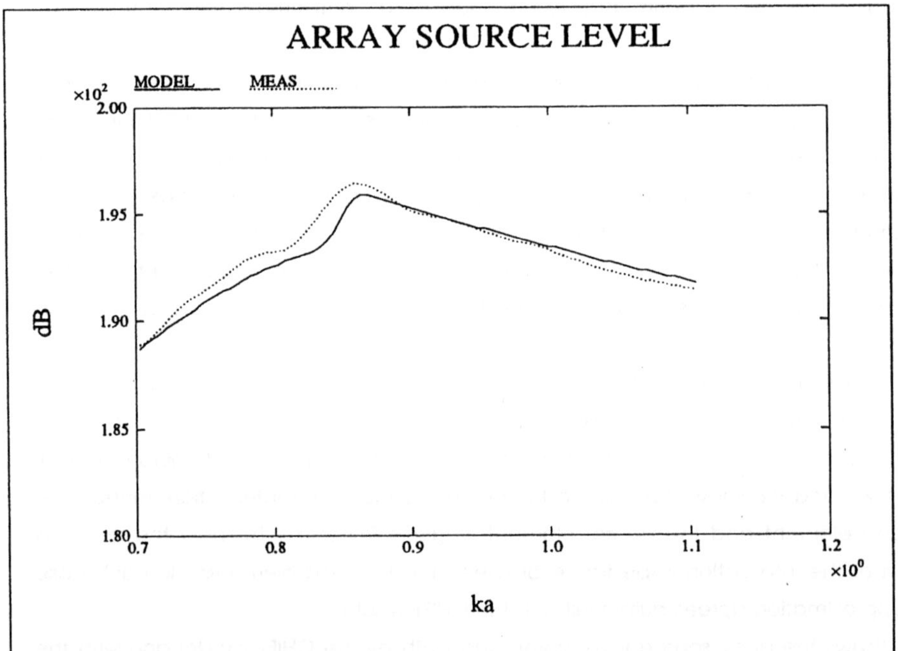

Fig. 9.2 Source Level---model vs. measured

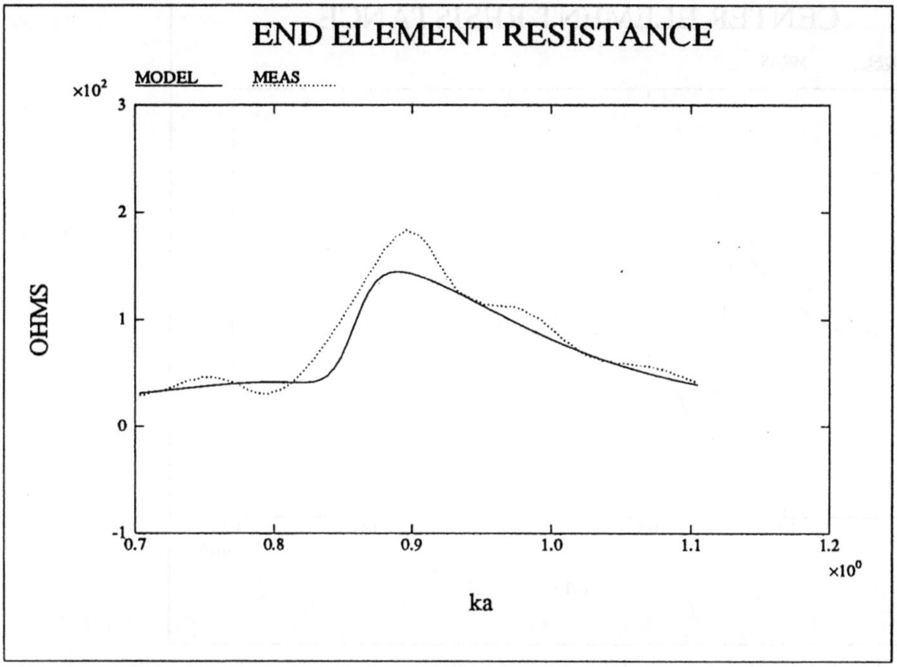

Fig 9.3 End element resistance---model vs. measured

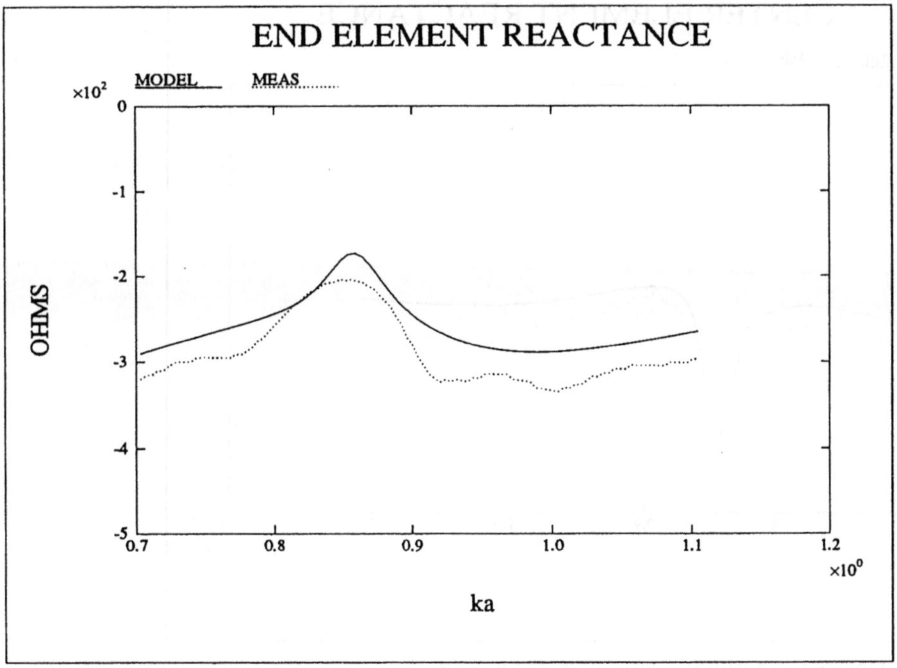

Fig. 9.4 End element reactance---model vs. measured

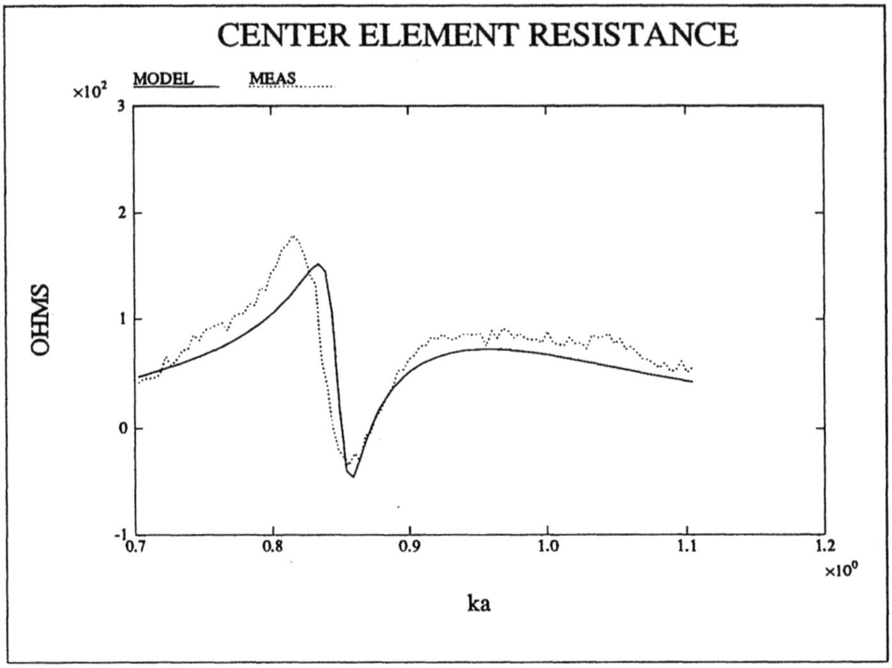

Fig. 9.5 Center element resistance---model vs. measured

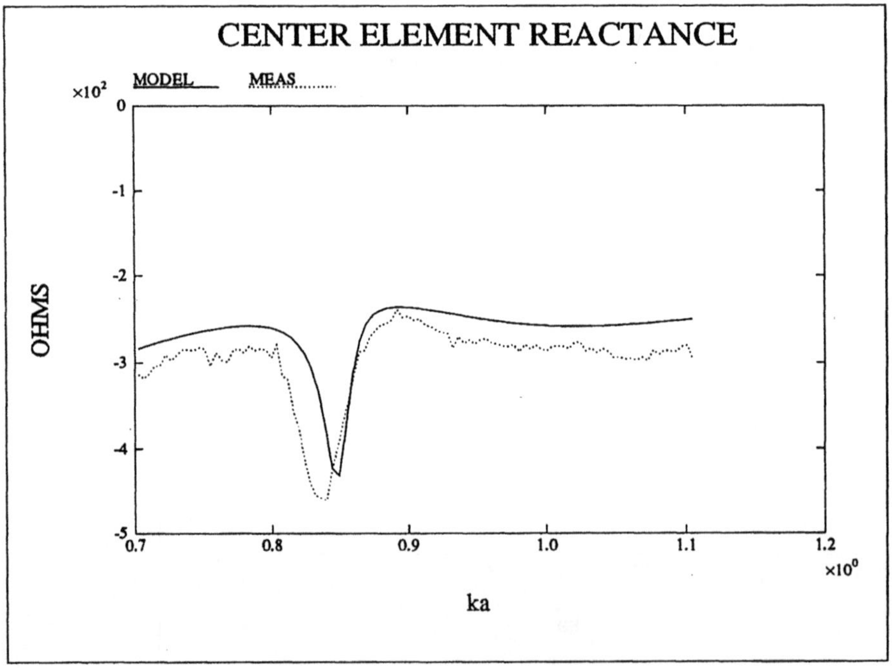

Fig. 9.6 Center element reactance---model vs. measured

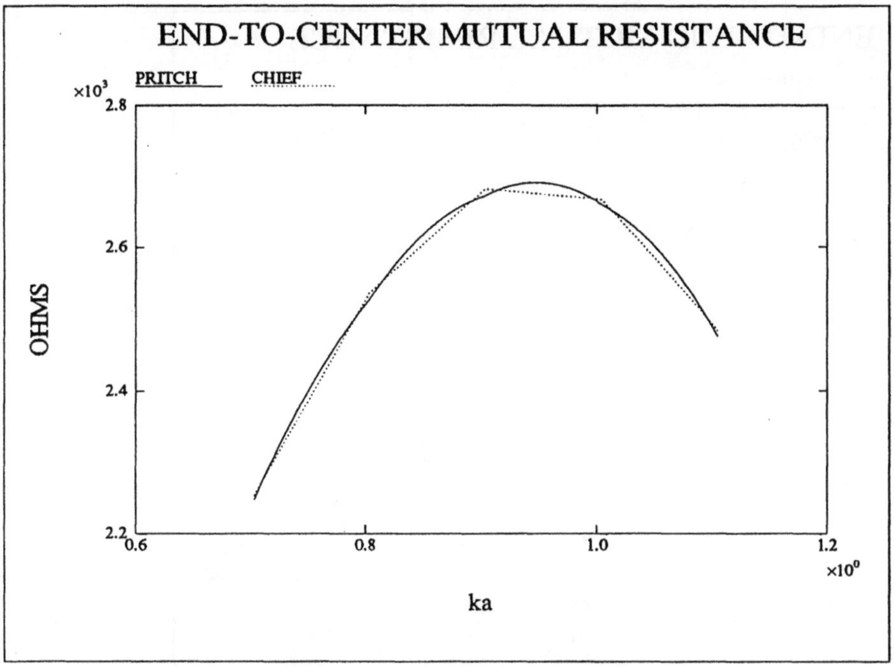

Fig. 9.7 Pritchard approximation, end-to-center mutual resistance

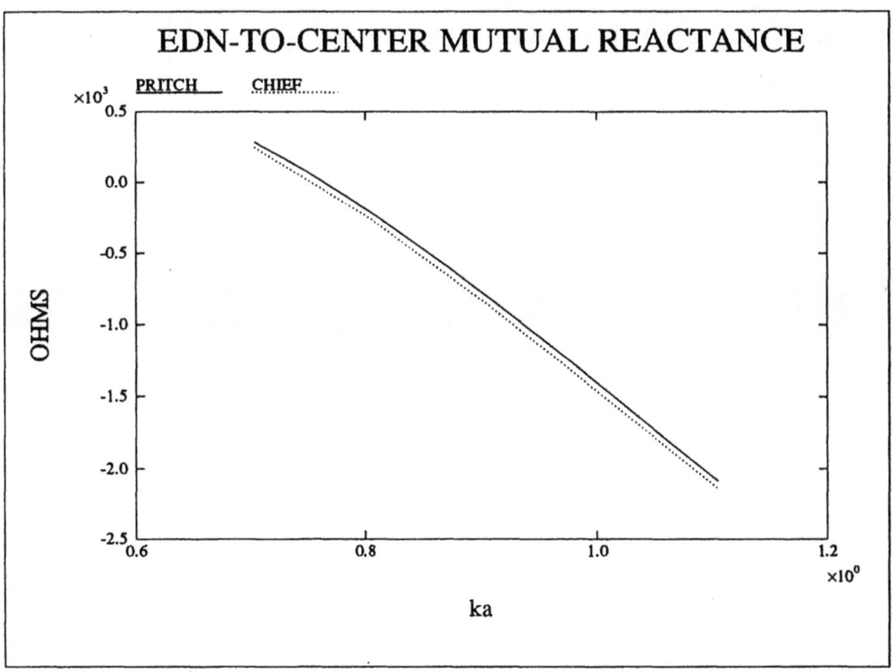

Fig. 9.8 Pritchard approximation, end-to-center mutual reactance

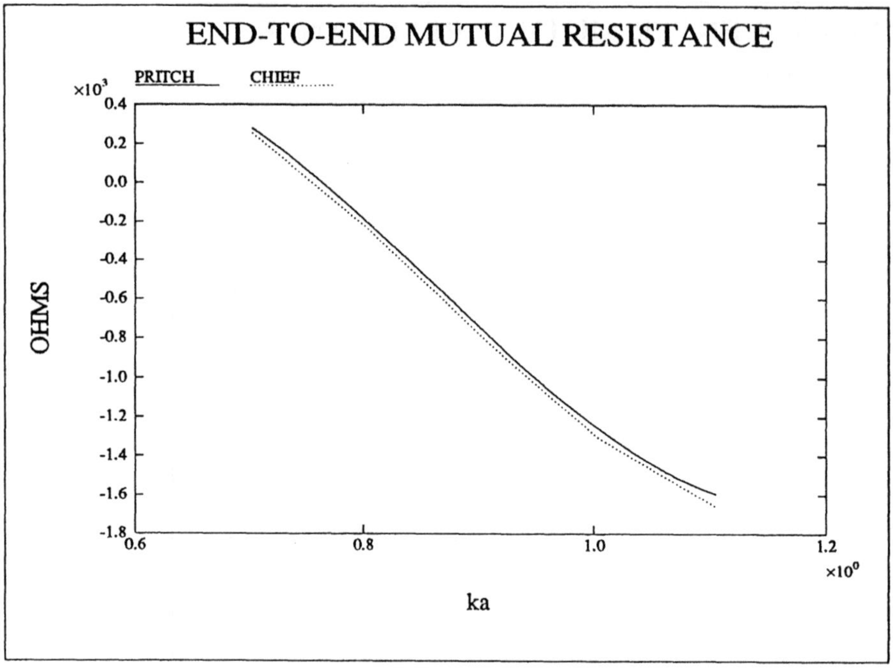

Fig. 9.9 Pritchard approximation, end-to-end mutual resistance

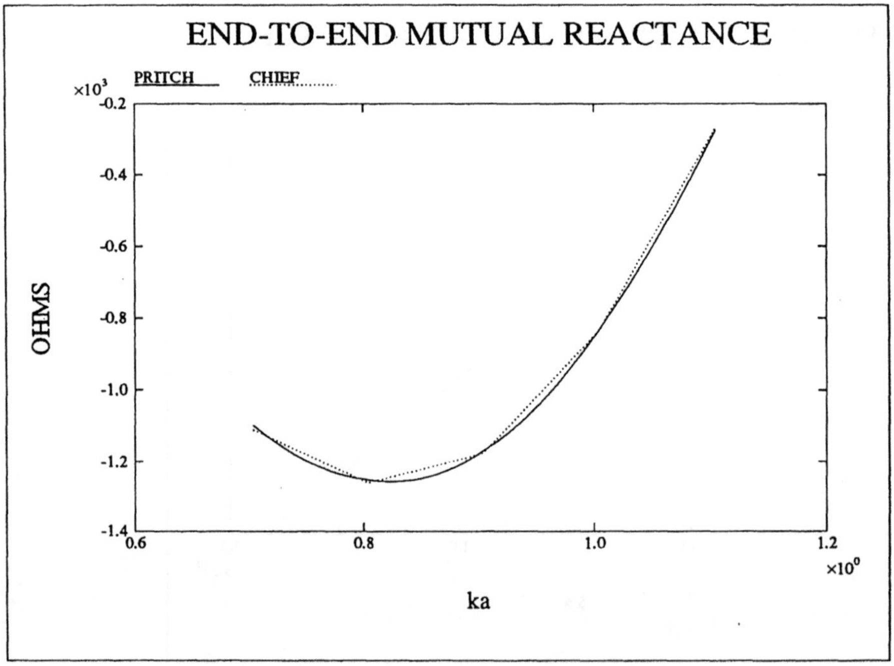

Fig. 9.10 Pritchard approximation, end-to-end mutual reactance

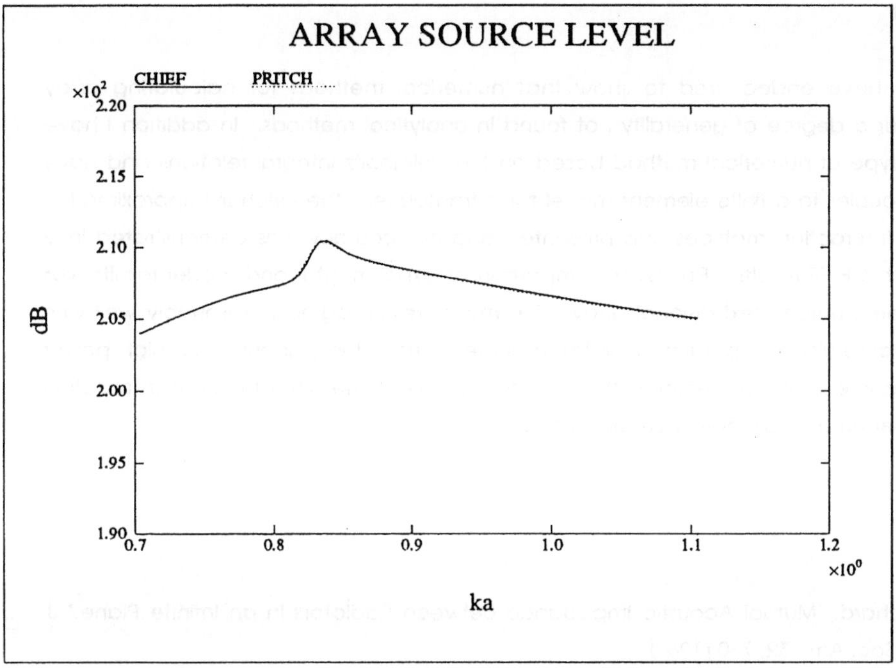

Fig. 9.11 Source level---Pritchard vs. CHIEF

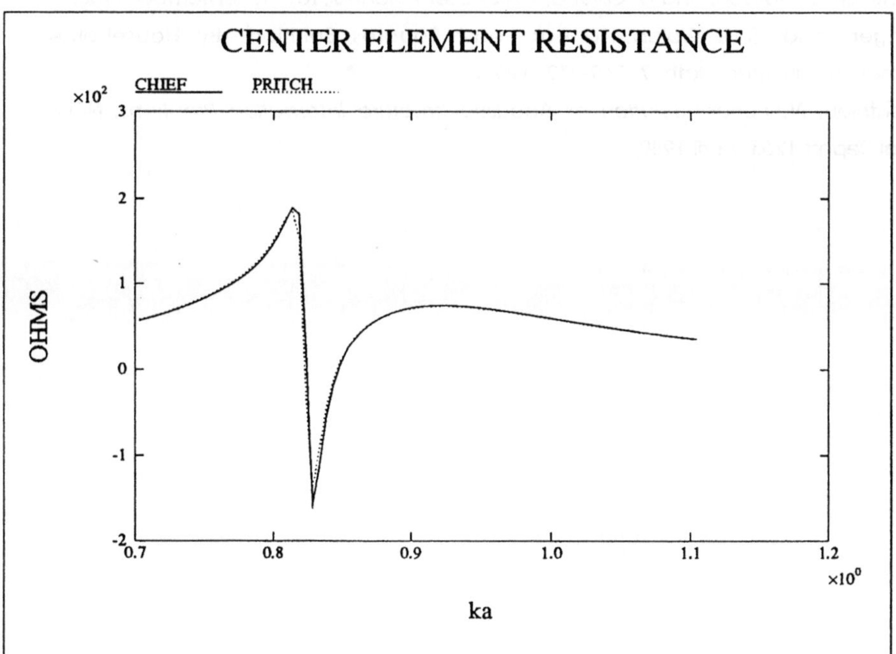

Fig. 9.12 Center element resistance-Pritchard vs. CHIEF

9.8 CONCLUSION

In this paper I have endeavored to show that numerical methods for calculating array interactions offer a degree of generality not found in analytical methods. In addition I have discussed one type of numerical method based on the Helmholtz integral relations and have shown how it couples to a finite element model for a transducer. The Pritchard approximation for the mutual interaction matrices was presented and its accuracy was demonstrated in a comparison with CHIEF results. Finally, a comparison of experimental and model results was given for a three element flextensional array. The model results agreed reasonably well with the experimental results and predicted all the main features in the response. As high power computers become more accessible, the use of numerical methods in array interaction problems will become even more commonplace.

REFERENCES

9.1 R.L. Pritchard, "Mutual Acoustic Impedance Between Radiators in an Infinite Plane," J. Acoust. Soc. Am., **32**, 730 (1960).

9.2 H.A. Schenk, "Improved Integral Formulation for Acoustic Radiation Problems," J. Acoust. Soc. Am., **44**, 41-58 (1968).

9.3 G.W. Benthien, D. Barach, and D. Gillette, *CHIEF User's Manual*, rev. 1, September 1988.

9.4 P. Businger, and G.H. Golub, "Linear Least Squares Solutions By Householder Transformations," Numer. Math. **7**, 269-279 (1965).

9.5 G.W. Benthien, *Numerical Solution of Acoustic-Structure Interaction Problems*, NOSC Technical Report 1263, April 1989.

10. Magnetostrictive Materials

D. McMasters

Edge Technologies, Inc. Ames, IA 50010, USA

The large magnetostrictive strain, λ, possessed by Terfenol-D, $Tb_xDy_{1-x}Fe_y$, has been sustained from the research laboratory to commercial production. Transducer drive elements produced by two methods, free-standing zone melt (FSZM) and modified Bridgman (MB), are available in various sizes and shapes. These processes are briefly described. The FSZM single crystal material has about 20% more strain than does the grain oriented MB material. The effects of heat treatment and stoichiometry on the magnetostriction of these drive elements are reported. The recent price reduction is the result of an extensive evaluation of the raw materials, Tb and Dy obtained from several sources. The purity and therefore the cost of these raw materials are the variables and the product was not acceptable if performance was sacrificed. Improved processing techniques and large volumes will lead to further price reductions. Recently, emphasis has been placed on performance consistency, both end-to-end and lot-to-lot, and the results of these efforts to produce matched sets of drivers are reported. The potential uses of these giant magnetostrictive materials remain large. This is evident from the numerous first and second stage prototype devices that have been successfully completed. The commercialization of a device that uses a large volume of Terfenol-D drivers will have a major impact on the current applications activities and we are seeing evidence of this becoming a reality in the near future, especially in the areas of sonar, vibration isolation, high force actuators and linear motors.

10.1 INTRODUCTION

Terfenol-D ($Tb_{0.3}Dy_{0.7}Fe_{1.9-1.95}$) is a highly magnetostrictive alloy that has been found to be ideal for use in many transducer applications (10.1-10.4). The alloy $Tb_xDy_{1-x}Fe_y$ has been developed and brought to commercialization. The addition of Dy (dysprosium) to the original Terfenol ($TbFe_2$ intermetallic compound) compensates for the magnetic anisotropy and allows the material to be driven at practical magnetic field levels (10.5). Earlier studies dealt with the basic properties of the materials and processing techniques that led to the most efficient transducer drive elements (10.5-10.9). In recent years these developmental studies have been more device/application oriented (10.10-10.14), which makes these 'giant' magnetostrictive drive elements more user ready. Most of these programs are in the prototype stages and small quantities of various sizes, shapes, stoichiometries - customized to meet the design criteria - are the usual request.

Since the report of the magnetostriction 'jumps' in twinned single crystals (10.8), no major improvements in the performance of these drivers have been made. However, the evaluation of cheaper starting materials, terbium and dysprosium, has allowed a price reduction without sacrificing strain. The current drive elements are therefore more cost effective on an energy density basis. Improved processing techniques have led to a more consistent product and the progress toward the reduction of crystalline defects in both the FSZM and MB materials that will result in higher strain values is reported.

10.2 EXPERIMENT

The magnetostrictive drive elements of $Tb_xDy_{1-x}Fe_y$ (Terfenol-D) used in these studies were prepared by the free standing zone melt (FSZM) and the modified Bridgman (MB) methods. These processes are described in detail in the patents issued (10.15, 10.16) and techniques to improve the performance of these drivers are reported in a subsequent patent (10.17) The goal of both methods is to produce defect-free single crystal elements with the <111> crystallographic direction along the drive axis. The natural solidification direction is <112> and this is obtained in both the FSZM and MB materials. The dendritic growth produces twins which constitute the major crystalline defect that restricts the magnetostriction.

The procedures used to characterize most of these magnetostrictive rods have been described elsewhere (10.8,10.9,10.11). Independent magnetostriction (λ) versus applied field (H) characterization curves were obtained using a lever arm dead weight prestress apparatus. The magnetic field was applied by a solenoid coil and the strain (λ) was measured by either a strain gauge or an LVDT.

Of significance to present and future price reductions was the evaluation of the raw materials, Tb and Dy, from four suppliers. The performance of both FSZM and MB drive elements prepared from these different starting materials was compared to that of the current materials. Maximum strain, slope, hysteresis, consistency and long term stability (in progress) were compared.

10.3 RESULTS

Figure 10.1 shows the typical response of twinned single crystalline $Tb_{.3}Dy_{.7}Fe_{1.9}$ material prepared by the FSZM method (10.8). In this report by A.E. Clark et al. (10.8) it is predicted that this 'jump' in magnetostriction could be increased from about 1000 ppm in these twinned materials to about 2000 ppm in untwinned single crystals. Therefore, it is worthwhile to continue efforts to develop twin free drive elements. The effect of composition on these magnetostriction jumping alloys is reported by J.D. Verhoeven et al. (10.9). It was found that for maximum performance in these (Terfenol-D) $Tb_xDy_{1-x}Fe_y$ drivers, the value of x should be greater than 0.3 and y should lie between 1.92 and 1.95.

The magnetostrictive properties of the grain oriented MB material have been compared to those of the single crystal, bicrystal and polycrystal FSZM material. The MB material was comparable to the 3 and 4 grain FSZM material and the magnetostriction was less than the FSZM single crystal rods by about 20%.

An extensive characterization of the MB materials was recently completed (10.11). Figure 10.2 shows the strain versus field curves for five of the large diameter drive elements. These curves represent the lot-to-lot consistency that is being achieved in these grain-oriented Bridgman materials. Similar curves were obtained for higher prestress conditions and the agreement between the samples at high field levels was even better.

Since the Bridgman crystal growth method is inherently susceptible to macrosegregation the end-to-end consistency was also determined for this study. The MB rods were 3.175 cm in diameter and 18 cm in length. Six 2.5 cm sections were machined from this rod and strain versus field curves were compared. The strain was measured using an LVDT for these relatively low field (1.4 kOe) and low prestress (9 MPa) conditions. Figure 10.3 shows the maximum variance of the six curves which were section numbered one through six from the bottom to top of the rod. The curves for the other four sections lie between the curves shown in Fig. 10.3 which represents the worst consistency of a series of three rods so characterized. The three rod analysis allowed a simultaneous lot-to-lot consistency check and the overall consistency is shown in Fig. 10.4.

The price of these Terfenol-D transducer drive elements is raw-materials cost sensitive. The price of the raw materials, terbium and dysprosium, is purity and volume controlled. For now it is a matter of determining the impurities and levels that can be tolerated without sacrificing strain. Terbium and dysprosium from four suppliers were converted into $Tb_{.3}Dy_{.7}Fe_{1.9-1.95}$ drive elements. The magnetostrictive properties of these drivers were compared and the long term stability is being evaluated.

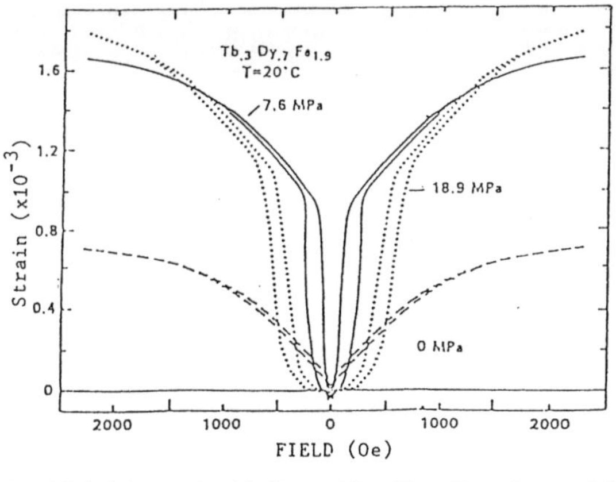

Fig. 10.1 Magnetostriction of the $Tb_{0.3}Dy_{0.7}Fe_{1.9}$ at 0, 7.6 and 18.9 MPa

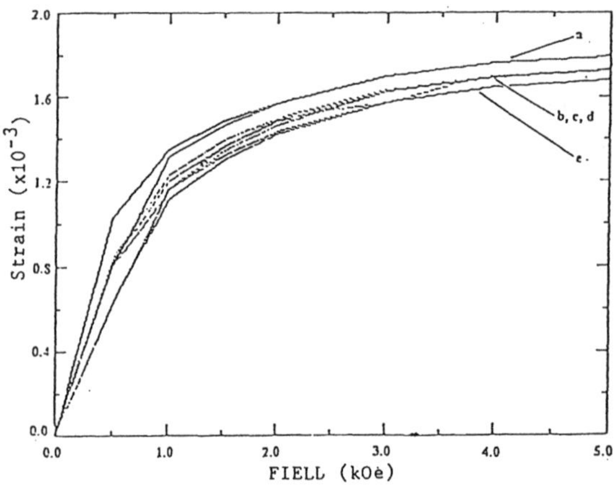

Fig. 10.2 Strain vs. field for five 3.5 cm diameter x 10 cm long Bridgman-grow Terfenol-D rods at a pre-stress of 15.5 MPa. Rods b, c and d are nearly equivalent

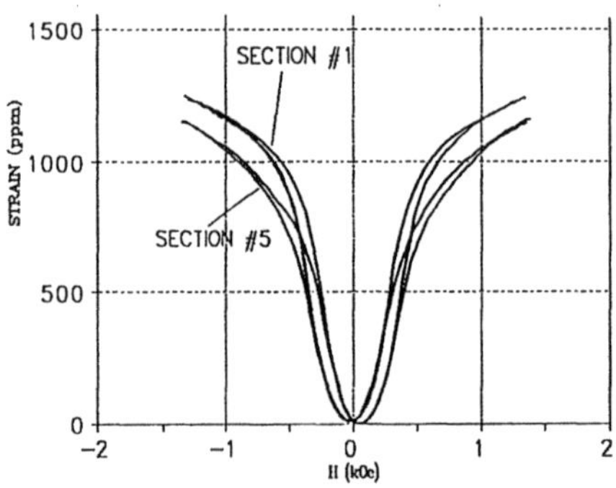

Fig. 10.3 Maximum variation in strain of six sections from Rod c at a prestress of 9 MPa. Sections 32 mm diameter x 25 mm long

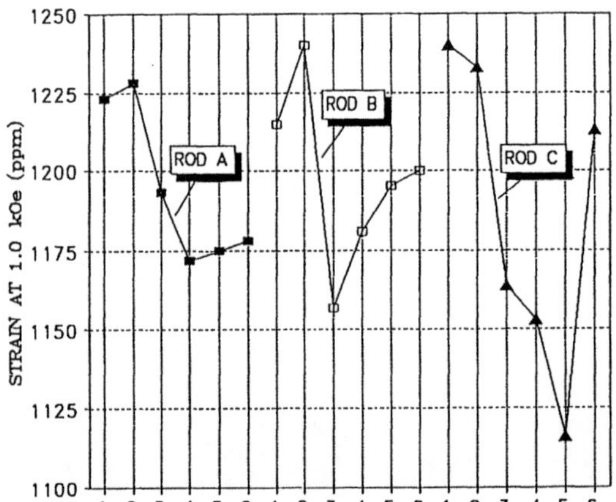

Fig. 10.4 Strain values for 18 sections, 32 mm diameter x 25 mm long, from 3 rods (lots) at 1.0 kOe and 9 MPa prestress

The major interstitial impurities (ppm by weight.) are listed in Table 10.1 for these metals. Supplier #1 is the current material being used in the production of these drive elements. In the evaluation the metals were rated on a scale of one to ten (ten being the highest) on the basis of performance (magnetostrictive properties), consistency, long term stability, short term price, volume discount price and overall. As shown in Table 10.1, supplier #4 has been eliminated. Although their initial performance was close to current standards, the rods are slowly disintegrating. Two lots of terbium from supplier #3 yielded good to excellent performance, but lacked consistency and are still being tested for stability. This terbium could be a back-up source. Supplier #2 has the highest long term potential and is serving as a second source for these raw materials.

Table 10.1 Interstitial impurity contents in ppm by weight for raw materials from four suppliers. Evaluation ratings from 0 to 10.

Element	Supplier #1		Supplier #2		Supplier #3		Supplier #4	
	Dy	Tb	Dy	Tb	Tb	Tb	Dy	Tb
O	320	<400	140	<100	1900	2000	1000	1250
N	13	< 50	<100	<100	400	600	----	----
C	60	<100	160	<100	80	90	<500	120
Performance	10		8		9	8	7	
Consistency	9		9		8	7	7	
Stability	10		9		8	7	0	
Short Term Price	10		10		10	10	?	
Volume Discount Price	7		10		8	8	?	
Overall	9		10		8			

Future price reductions can be estimated on the basis of volume discounts for raw materials and the assumption that applications needs will reach these levels in a reasonable time frame. Performance improvements and more efficient processing are also factors in these projections. Figure 10.5. shows the expected percent reduction for the five year period that includes the recent reduction of about 18 percent in 1990. User's volume of finished Terfenol-D drive elements in kilograms necessary to enact these discounts are also plotted in Fig. 10.5.

Based on the 1989 price lists, an average price reduction of 17.5% was put into effect on March 15, 1990. The projected price reduction of 45% in 1993 is contingent upon the 1000 kg quantity of finished drive elements consumed. A major spike in this curve to the 10,000 kg user's quantity and a 60 percent price reduction is projected for 1995.

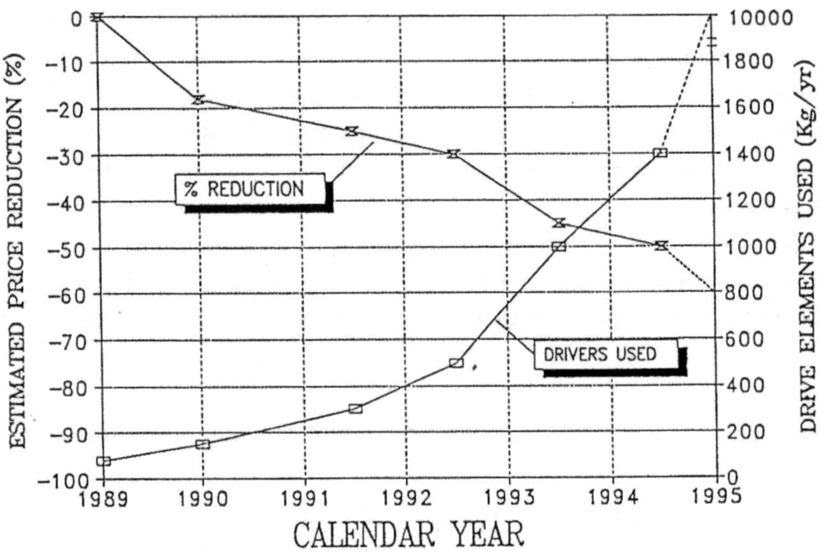

Fig. 10.5 Price projections for five year period based on finished drive element quantity used in Kg/Year

10.4 CONCLUSION

Two reliable raw material sources and competitive volume discount price estimates have been established. The reliability of these magnetostrictive transducer drive elements has increased and verified on the basis of performance and consistency. One or two applications that require large volumes of these materials will result in price reductions for all users. A major breakthrough in performance comparable to the "jump" effect is needed. Our efforts as a drive element supplier will be concentrated in this area. Your efforts as an applications user is needed in order to bring the potential of these giant magnetostrictive materials to reality.

REFERENCES

10.1 J.L. Butler, and S.J. Ciosek, "Rare earth iron octagonal transducer," J. Acoust. Soc. Am., **67**, 1809-1811 (1980).

10.2 F. Claeyssen, D. Boucher, and C. Pohlenz, "Analysis of a magnetostrictive Tonpilz transducer," J. Acoust. Soc. Am., **81**, S89 (1987).

10.3 R. Porzio, W.J. Harrold, and J.R. Sturges, "Self-biased modular magnetostrictive driver and transducer," U.S. Pat. No. 4, 845, 450 (July 4, 1989).

10.4 T. Akuta, "An application of giant magnetostrictive material to high power actuators," Proceedings of the Tenth International Workshop on Rare-earth Magnets and Their Applications, Vol. I, 359-368, (1989).

10.5 A.E. Clark, "Magnetostrictive rare earth - Fe_2 compounds," Chap. 7 in Ferromagnetic Materials, Vol. 1, ed. by E.P. Wohlfarth (North-Holland, 1980), pp. 531-589.

10.6 H.T. Savage, R. Abbundi, A.E. Clark, and O.D .McMasters, "Permeability, magnetomechanical coupling and magnetostriction in grain-oriented rare earth-iron alloys," J. Appl. Phys. **50**, 1674-1676 (1979).

10.7 A.E. Clark, and H.T. Savage, "Magnetostriction of rare earth - Fe_2 compounds under compressive stress," Journal of Magnetism and Magnetic Materials 31-34 (North-Holland, 1983), pp. 849-851.

10.8 A.E. Clark, J.P. Teter, and O.D. McMasters, "Magnetostriction "jumps" in twinned $Tb_{0.3}Dy_{0.7}Fe_{1.9}$," J. Appl. Phys. **63**, 3910-3912 (1988).

10.9 J.D. Verhoeven, J.E. Ostenson, E.D. Gibson, and O.D. McMasters, "The effect of composition and magnetic heat treatment on the magnetostriction of $Tb_xDy_{1-x}Fe_y$ twinned single crystals," J. Appl. Phys., **66**(2), 772-779 (1989).

10.10 F. Claeyssen, D. Boucher, and S. Faure, "Characterization of length expander magnetostrictive rare earth-iron rods under normal use conditions in transducers," J. Acoust. Soc. Am., **83**, S20 (1988).

10.11 A.E. Clark, J.P. Teter, M. Wun-Fogle, M. Moffett, and J. Lindberg, "Magnetomechanical coupling in Bridgman-grown $Tb_{0.3}Dy_{0.7}Fe_{1.9}$ at high drive levels," Proc. 34th Annual Conf. on Magnetism and Magnetic Materials, Boston, MA, Nov., 1989.

10.12 M.B. Moffett, A.E. Clark, M. Wun-Fogle, J.F. Lindberg, J.P. Teter, and E.A. McLaughlin, "Characterization of Terfenol-D for magnetostrictive transducers," submitted to J. Acoust. Soc. Am., Feb. 1990.

10.13 F. Claeyssen, D. Boucher, and S. Faure, "Comparative study of Terfenol-D piezomagnetic constants," Proceedings of the Second International Conference on Giant Magnetostrictive Alloys and Amorphous Alloys for Actuators and Sensors, (Marbella, Spain, Oct. 12-14, 1988).

10.14 K.C. Pitman, A. Branwood, and A.R. Piercy, "Device oriented measurements in Terfenol materials," loc. cit., Ref. 10.13.

10.15 O.D. McMasters, "Method of forming magnetostrictive rods from rare earth-iron alloys," U.S. Patent No. 4, 609, 402, September 2, 1986.

10.16 E.D. Gibson, J.D. Verhoeven, F.A. Schmidt, and O.D. McMasters, "Continuous method for manufacturing grain oriented magnetostrictive bodies," U.S. Patent No. 4,770,704, September 13, 1988.

10.17 J.D. Verhoeven, O.D. McMasters, and E.D. Gibson, "Thermal treatment for increasing magnetostrictive response of rare earth-iron alloy rods," U.S. Patent No. 4, 849, 034, July 18, 1989.

11. Modeling and Characterization of the Magnetostrictive Coupling

F. Claeyssen
CEDRAT Recherche,Ch. du Pré Carré, ZIRST, F-38240 Meylan, France

R. Bossut
ISEN, 41 Bd Vauban, F-59046 Lille CEDEX, France

D. Boucher
GERDSM -DCAN Toulon, DCN, Le Brusc, F-83140 Six Fours Les Plages, France

The physical parameters which are relevant to the design and construction of magnetostrictive transducers are discussed through the analysis of a lumped parameter model (LPM). Numerical modeling of the magnetostrictive coupling, including a plane wave method (PWM) and the finite element method (FEM) which has been implemented in the ATILA code is presented. A process to determine all the magneto-elastic constants for use in the models is described. A complete magneto-elastic matrix of a Terfenol-D alloy is given. A test characterization device is analyzed using this matrix with the three types of model. The good agreement between measurements and modeling is a validation for both the models and the characterization method.

11.1 INTRODUCTION

Because propagation conditions are better and sonar range is increased at low frequency, the trend in transmitting transducers is to lower their working resonance (11.1). Until now, the piezoelectric tonpilz-type transducers have been satisfactory, but with the new desired range of frequencies, their size has become a major problem.

The first way to deal with such a problem has been to study new geometries, which has led to the flextensional concept (11.2). However, even if the desired frequencies have been reached, not all the technological problems are solved (shell aging, etc.). Moreover, the transducers may still be very bulky in the case of very low frequencies.

Another research approach is to retain (temporarily) the conventional design concepts and to use a new active material, better fitted to the goal. This opportunity is offered by the attractive properties of magnetostrictive rare earth-iron alloys, especially terbium dysprosium iron alloy (11.3,11.4), currently called Terfenol-D: this alloy, discovered by A.E. Clark (11.5) displays a 'giant' magnetostriction of about 2000 ppm, a low Young's modulus and a good electro-mechanical coupling factor.

Previous works have already proved that using Terfenol-D instead of ceramics leads to a significant decrease in the resonant frequency (11.6). A first prototype, a tonpilz-type named the Quadripode (11.1, 11.7-11.9), has also proved that using Terfenol-D, high powers (1.6 kW) in a compact size may be achieved. The interest for this research approach having been shown, it appeared useful to try and get a better understanding of the transducers behavior in order to be able to design them more carefully. For that purpose, different types of models (11.9-11.12) have been studied and the characterization of the Terfenol-D properties required by the models has been undertaken (11.9,11.13,11.14).

The purpose of this paper is to deal with the dynamic coupling in transducers. The computation of the static working point of the transducer (magnetic bias and the prestress) is another subject which has already been tackled elsewhere (11.15)). Different models including classical models and the finite element models are presented. A characterization process to determine the complete magneto-elastic matrix of the Terfenol-D is described and illustrated by an example. And last, an application of these models to a characterization device is given.

11.2 CLASSICAL MODELING

Two types of classical models (11.16,11.17) for magnetostrictive transducers exist, the lumped parameter and the distributed parameter models. Both of them are uni-dimensional. They may be derived leading either to analytical formulae or to a transfer matrix treatment, a numerical method currently used for piezoelectric transducers (11.18). However, as it is generally done, the lumped model, which is the simplest one, is used to give a physical understanding of the transducer parameters. So the development in analytical formulae through equivalent circuits is preferred. This type of model may be applied to tonpilz, double-ended ("Janus"), ring and polygonal transducers. The distributed or plane wave model has been applied to free rods (11.19) or rods in Tonpilz transducers (11.20). It is a more accurate model than the previous one because the distribution of masses and elasticities are taken into account. But it leads to a more complex formulation which, for a whole transducer, requires a numerical computation, such as the transfer matrix method.

11.2.1 The Lumped Parameter Model of the Tonpilz

One of the greatest drawbacks of the lumped parameter models is that their dependence on the type of transducer. However, the conclusions that can be drawn from the analysis of a Tonpilz model are rather representative of any transducer. As a consequence, only the main formula for the tonpilz-type transducer will be presented, the complete analysis of different transducer types ("Janus", polygonal, Wiedemann-effect twisting pendulum (11.9)) being detailed elsewhere. Just to introduce these formulae, it can be recalled that they are established using some algebra from the equivalent circuit analysis (Fig. 11.1), the elements of which are defined as follows:

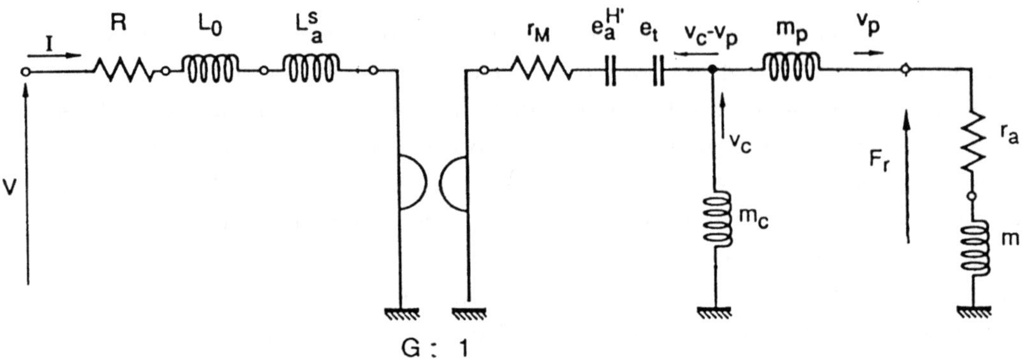

Fig. 11.1 The 'electro-mechano-acoustic' equivalent circuit of a magnetostrictive tonpilz-type transducer in water

R: the circuit resistance, when the active material is clamped (it is the sum of the coil resistance and the eddy current losses resistance),

L_o: the leakage inductance (due to the thickness of the coil),

L_a^S: the clamped inductance of the active material,

r_M: the mechanical resistance of the transducer (due to the mechanical internal losses in the material and to the assembling losses),

e_a^H: the active material elasticity at constant field,

e_t: the stress bolt elasticity,

m_c: the tailmass,

m_p: the headmass,

r_A: the acoustic resistance,

m_A: the acoustic mass,

G: the electro-mechanical gyrator ratio.

This circuit takes into account both loss terms (R, r_M) and terms which represent a storage of inactive energy (L_o, e_t). These formulae will permit to analyze the role of the transducer parameters on its performances and to answer two questions:

- which material should be chosen and at what static working point?
- on which transducer parameter must the designer operate and how?

The model chosen will also permit to see the influence of parameters with known undesirable consequences. In particular, it will be possible to look at the influences of the leakage flux, a parameter which is not encountered in piezoelectric transducers.

11.2.1.1 The Mechanical Resonance Frequency

The angular mechanical resonance frequency at constant current, the antiresonance frequency, is given as following:

$$\omega_l = 1 / (e^H m)^{1/2}, \tag{11.1}$$

where $m = m_c (m_p+m_A) / (m_c+m_p+m_A)$ is the 'effective' mass of the transducer (obtained by the transformation of the two mechanical branches in Fig. 11.1 into one equivalent branch) and $e^H = 1/ (1/e_a^H + 1/e_t)$ is the transducer total elasticity.

In this classical result, it should be noted that as the total elasticity is more or less equal to that of the active material, the resonance frequency is nearly proportional to the square of the material Young's modulus Y^H. It means that the type of Terfenol-D and the choice of the bias point affect.the transducer resonance frequency.

11.2.1.2 Acoustic Power at the Antiresonance Frequency

The acoustic power at the antiresonance frequency is given by

$$P_a = e_{ma}\, \omega_l\, E_e / Q_m \tag{11.2}$$

where e_{ma} is the mechano-acoustic efficiency, Q_m the mechanical quality factor and E_e the elastic energy stored in the transducer which is only slightly different from the product of the active material volume by the elastic energy density in the active material ε_e:

$$\varepsilon_e = 1/2\, Y^H S_3^2 \tag{11.3}$$

These expressions show that the acoustic power is mainly limited by the maximum of material energy density which is defined by the choice of the prestress value and the maximum of magneto-strain available in the chosen conditions of bias point.

Through the mechano-acoustic efficiency, the acoustic power depends also, with a lower importance, on the mechanical quality factor of the active material Q^H. On the other hand, the acoustic power limit at constant current does not depend on the electric loss term R and on the leakage flux.

11.2.1.3 The Efficiency

The electro-acoustic efficiency e_{ea} which is the ratio of the output acoustic power to the input electrical power, is often given as the product of the electro-mechanical efficiency e_{em} and the mechano-acoustic efficiency e_{ma} (11.16). At the antiresonance frequency, these terms have rather simple expressions.

The mechano-acoustic efficiency is the ratio of the modified acoustic resistance r_A' (modified by the transformation of the two mechanical branches of Fig. 11.1 into one equivalent branch) to the sum r, of r_A' and r_M, the mechanical resistance of the transducer. This last term depends on the mechanical quality factor of the active material Q^H and on the care taken in the technological design and in the assembling.

$$e_{ma} = r_A' / r \tag{11.4}$$

The electro-mechanical efficiency is defined by:

$$e_{em} = k_{33}^2 / (k_{33}^2 + 1/ Q_{am} Q_{aE}) \qquad (11.5)$$

where:

k_{33} is the electro-mechanical coupling factor of the active material for the longitudinal mode,

$Q_{am} = 1/\omega_l e_a^H r$ is a mechanical quality factor reckoned with the active material elasticity e_a^H,

$Q_{aE} = \omega_l L_a^T / R$ is an over-voltage factor reckoned with the free active material self inductance L_a^T.

These formulae show that the efficiency only depends on resistive terms and on the material coupling factor. It is of primary importance to note that it does not depend on the effective coupling factor of the transducer, nor on the prestress bolt and the leakage flux. Thus, as k_{33} is high in Terfenol-D (50-70%), the electro-mechanical efficiency is good, although the magnetostrictive transducer effective coupling factor is often only acceptable.

On the other hand as the mechanical quality factor of the active material Q^H is often very low (10-20), the mechano-acoustic efficiency of magnetostrictive transducers is much more sensitive to the acoustic resistance value than the mechano-acoustic efficiency of piezoelectric transducers. That is the reason why, for a given transducer size, the largest emitting surface is highly recommended in the case of magnetostrictive transducers.

11.2.1.4 The Passband Width

The passband width is inversely proportional to the transducer mechanical quality factor Q_m defined by:

$$Q_m = 1 / \omega_l e^H r. \qquad (11.6)$$

As the Terfenol-D mechanical quality factor Q^H is low, the transducer passband width is naturally larger than with piezoelectric ceramics. On the other hand, as the resistance R is not negligible in the case of rare-earth transducers (Q_{aE} is not high compared to one, the relative permeability being low), the efficiency is not constant in the passband.

11.2.1.5 The Effective Coupling Factor

The effective coupling factor k_{eff} of a transducer is a parameter giving the proportion of stored mechanical energy converted into the input stored electrical energy. There is no accurate analytical formulation for this factor in the general case (11.21). However, a good approximation can be proposed:

$$k_{eff} \approx k_M k_{33} k_E \qquad (11.7)$$

where k_M is the 'magnetic coupling factor' the square of which is defined as the ratio of the magnetic energy stored in the active material to the total magnetic energy (equal to the electric stored energy) and k_E is the 'elastic coupling factor' the square of which is defined as the ratio of the elastic energy stored in the active material to the total elastic energy stored in the transducer.

The 'magnetic coupling factor' may be smaller than one for many reasons:

- when the coils are thick, the leakage inductance L_o is not negligible compared to the free self inductance of the active material L_a^T, then:

$$k_M^2 = L_a^T / (L_a^T + L_o) \tag{11.8}$$

- when the reluctance of the return path R_r is is not negligible compared to the clamped reluctance of the active material R_a^S, then:

$$k_M^2 = R_a^S / (R_a^S + R_r) \tag{11.9}$$

- when the eddy currents are large, the Terfenol-D permeability is multiplied by a complex factor $\chi = \chi_R + j\,\chi_I$ which decreases the magnetic energy stored in the active material, then:

$$k_M^2 = \chi_R \tag{11.10}$$

In the general case, all these effects exist and, at the first order, the square of the magnetic coupling factor is equal to the product of all these expressions.

In the same way, the elastic coupling factor may be lower than one for many reasons such as the existence of inactive elasticities in series or in parallel to the active material elasticity or in the case where the strain field in the active material is not uniform (which cannot be studied in the lumped parameters formalism). The most frequent cause for obtaining an elastic coupling factor smaller than one is when the prestress bolt elasticity e_t is not high compared to the constant-induction active material elasticity e_a^B then:

$$k_E^2 = e_t / (e_t + e_a^B) \tag{11.11}$$

The problem is more severe in magnetostrictive transducers than in piezoelectric transducers, especially with high prestress, because the Terfenol-D Young's modulus Y^B is small compared to that of steel.

Due to k_M and k_E, the effective coupling factor k_{eff} of a magnetostrictive transducer has many 'good' reasons to be much smaller than the material coupling factor k_{33} if the designer does not take care. As this factor is important, some of its aspects given by the lumped parameter model will be recalled. Introducing the dual effective coupling factor K_{eff} as follows:

$$K_{eff}^2 = k_{eff}^2 / (1-k_{eff}^2) \tag{11.12}$$

and, to simplify, assuming that Q_{aE} is higher than one (the treatment for the general case being detailed elsewhere (11.9), without any change in the conclusions), it can be shown that at the antiresonance frequency:

$$K_{eff}^2 = 1 / Q_e Q_m \tag{11.13}$$

where Q_m is the transducer mechanical quality factor and Q_e is the transducer over-voltage factor (the ratio of the imaginary part of the impedance to the real part).

These relations show that a low effective coupling factor will give either a high over-voltage factor or a high mechanical quality factor, neither case being desired. Because, classically in magnetostrictive transducers the mechanical quality factor is rather low and as the effective coupling factor is not large due to the previously mentioned reasons, the over-voltage factor will often be high. This implies that magnetostrictive transducers will need more reactive power than equivalent piezoelectric transducers.

Using the dual effective coupling factor definition, an important formula relating to the total elastic and magnetic stored energies, E_e and E_m, reveals the behavior of the coupling at the antiresonance:

$$E_m = (Q_m K_{eff})^{-2} E_e \tag{11.14}$$

First, it shows that due to the mechanical resonance phenomena, the 'output' elastic energy may be higher than the 'input' magnetic energy as soon as the resonance is not completely damped and the effective coupling is acceptable ($Q_m^2 >> 1$, $K_{eff} \approx 1$). Second, it permits to estimate the amount of reactive electric power (and so the power supply) to reach a desired acoustic power (this being defined by the stored mechanical energy, see above).

At last, when derived for the active material, it is equivalent to:

$$T_3 \approx Y^H S_3 \approx Q_{am} d_{33} Y^H H_3 \tag{11.15}$$

where T_3, S_3, H_3 are the amplitudes of the material internal longitudinal stress, strain and magnetic field, respectively, and d_{33}, Y^H are the material magnetostrictivity and Young's modulus.

This expression permits either to choose an optimized mechanical quality factor or, when this last term is fixed, to choose a magnetic bias field H_o well fitted to the prestress T_o. The interest of the lumped parameter model is shown through these formulae which provide a qualitative analysis of the role of the material and transducer parameters. Thanks to this model, it has been shown and experimentally checked that grain-oriented Terfenol-D is better than random material. A good compromise to get power and high coupling is to use the material biased at 80 kA/m and prestressed at 30 MPa. However, its drawbacks are too serious to permit its use as a design tool. It is difficult to get accurate formulae away from the resonance frequency. The model assumptions have a very strong effect on the magnetic field circuit and the mechanical behavior. The models depend on the transducer type. It is not a suitable model for transducers with a complex mechano-acoustic coupling, such as flextensional types.

11.2.2 The Plane Wave Model and the Transfer Matrix Method

Although the plane wave model does not permit to deal with more transducer types than does the lumped parameter model, because it can only be applied to longitudinal vibrators, it has the advantage, when associated with a transfer matrix method, to provide accurate results in a large passband. As a consequence, this model can be a useful tool to design tonpilz-type transducers. It has already been successfully applied to characterize Terfenol-D (11.14) and to model the quadripode transducer (11.8). As the details of this method have already been published (11.14, 11.9), only the principles will be recalled.

The transfer matrix method consists in a decomposition of the transducer in homogeneous individual items: the magnetostrictive rods, the prestress bolt, the headmass, the tailmass, etc. Each item is considered as a two-port cell (mechanical items) or a three-port cell (magnetostrictive items, see Fig. 11.2) which relates the forces F_i and the velocities v_i of its two faces (i=1,2). To each element e is associated a transfer matrix M and an excitation vector T which relate the face forces and the face velocities.

$$\begin{bmatrix} F_2 \\ V_2 \end{bmatrix} = M \cdot \begin{bmatrix} F_1 \\ V_1 \end{bmatrix} + T \tag{11.16}$$

The excitation vector is equal to zero for mechanical items, proportional to the excitation current for magnetostrictive rods The PWM is used to obtain the transfer matrix and the excitation vector expressions. It is based on the propagation equations written with longitudinal motion assumptions. It is possible to take losses into account using complex magneto-elastic coefficients and magnetic leakage flux by adding an appropriate inductive term.

To calculate a whole transducer, two operations are needed. Its equivalent matrix has to be obtained from chaining the matrix of each element using appropriate interface conditions. Two cases may arise: the elements are placed in parallel (the rods and the prestress bolt) or in series (the magnetostrictive motors, the headmass and the tailmass. As a good example of chaining, two elements e' (M', T') and e" (M", T") put in series (Fig. 11.3) are equivalent to a new element e, the matrix M and the excitation vector T which are simply defined as: $M = M" . M'$ and $T = T" + M" . T'$

The final system relating forces and velocities at the transducer faces by the equivalent matrix and excitation vector of the whole transducer is solved by applying boundary conditions such as F=0 on the free face of the tailmass and F = Z.v (face loaded by an impedance) on the emitting face of the headmass. The forces and velocities of the transducer faces being found, the forces and velocities at each interface are simply inferred using a backward substitution. Then, the electrical impedance, the powers, the source level may be completely determined.

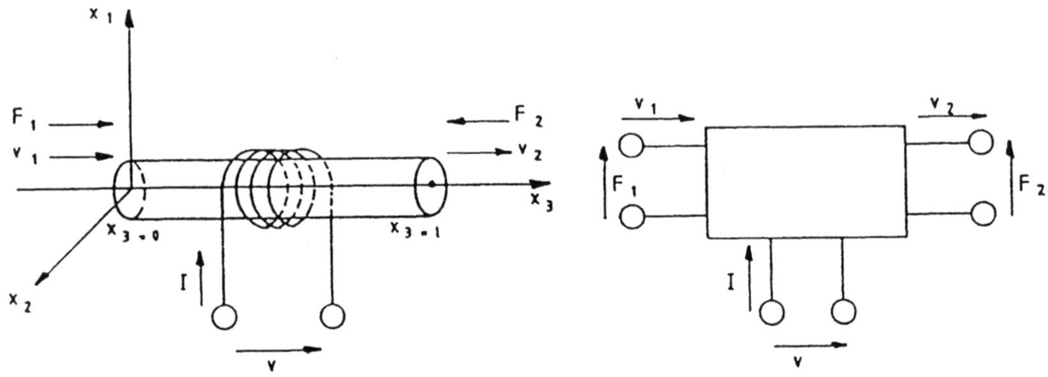

Fig. 11.2 Length-expander rod and equivalent three-port cell

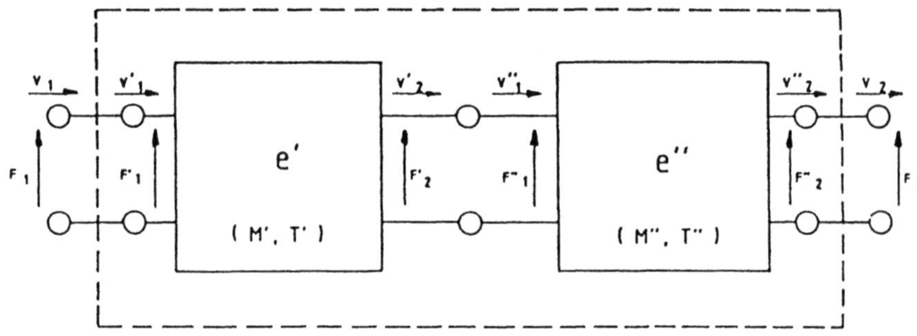

Fig. 11.3 Two elements in series

Even if the plane wave model associated with a transfer matrix method is practical and more accurate than the lumped parameter model, it suffers from the same limitation. It does not permit to optimize the magnetic circuit and the mechanical items, or to design transducers with a complex mechano-acoustic coupling.

11.3 FINITE ELEMENT MODELING

In order to be able to model any type of magnetostrictive transducer without any heavy restriction on the shape and the materials and without any strong assumption about the magnetic and displacement fields, a new model has been developed. It is based on an original variational principle which has been derived following a finite element approach. The theory has been established with the aim of implementing this model within the ATILA finite element code (11.22-11.24). The objective is to be able to compute the three dimensional dynamic behavior of any magnetostrictive heterogeneous structures in which the sources are coils excited either by given currents or voltages. The formulation of the model will be presented focusing on the electro-mechanical coupling (the mechano-acoustic coupling being an independent problem already solved by several methods). This will demonstrate the generality of this model. An application will be done on the characterization device in the following section.

11.3.1 The Basic Equations

Magnetostrictive transducers being used under bias conditions, the dynamic magnetostrictive effects are linear and are part of piezomagnetism. Mechanical items and magnetic elements will be considered as piezomagnetic elements with particular coefficients (zero coupling). The set of constitutive laws (11.16) written for all the materials, take into account three dimensional fields, and is therefore true for every point of the computation domain v, the transducer and its surrounding air:

$$\begin{aligned} T_{ij} &= c^{H}_{ijkl}\, S_{kl} - e_{nij}\, H_n \\ B_m &= e_{mkl}\, S_{kl} + \mu^{S}_{mn}\, H_n \end{aligned} \qquad (11.17)$$

where:

T and S are the stress and the strain tensors,

B and H are the induction and the magnetic field vectors,

c^H, e and μ^S are the constant-H stiffness, the piezomagnetic and the constant-S permeability. They constitute the magneto-elastic tensor and their terms are complex when taking losses into account.

The strain is expressed in terms of the spatial derivatives of the displacement u:

$$S_{kl} = \frac{1}{2}\left[\frac{\partial u_k}{\partial x_l} + \frac{\partial u_l}{\partial x_k}\right] \qquad (11.18)$$

The magnetic field is expressed using the reduced-scalar-potential formulation (11.25) which consists in a decomposition of the magnetic field into two parts:

$$H_n = H_n^{S} + H_n^{M} \qquad (11.19)$$

where:

$H_n^S = {}^{o}H_n^{S}.I$ is the magnetizing field, in which I is the coil current and ${}^{o}H_n^{S}$ is a term analytically calculated using the Biot and Savart law (11.26).

$H_n^M = -\partial ø/\partial x_n$ is the demagnetizing field, which can be expressed as the gradient of a reduced-scalar-potential ø.

The state variables of the problem are therefore u, I, ø.

Euler's equations, and the boundary conditions are then taken into account. The flux seen by the windings Q^V which will give the coil voltage and the impedance, is the partial derivative of the magnetic coenergy E_M (this term being naturally precise in the variational principle) with respect to the current I,

$$Q^V \cdot \delta I = \delta E_M. \tag{11.20}$$

The magnetic flux density has to be conservative.

$$\frac{\partial B_m}{\partial x_m} = 0 \tag{11.21}$$

In the solid parts, of density ρ, Newton's law is applied in harmonic analysis:

$$\frac{\partial T_{ij}}{\partial x_j} + \rho\omega^2 u_i = 0 \tag{11.22}$$

The two different types of boundary conditions are:

- the Neumann type conditions (prescribed normal stress T_N and prescribed normal induction B_N, on the surfaces S_{Tn} and S_{Bn}),
- the Dirichlet type conditions (prescribed displacement and prescribed potential, on complementary surfaces).

The possible electrical conditions are the prescribed current I or the prescribed winding flux Q^V which is equivalent to a prescribed voltage condition. These conditions fulfil all the possible requirements.

The main assumptions made in writing these equations are:

- the dynamic constitutive laws of the materials are linear,
- the shapes of the coils belong to a given library, (available shapes are sufficient to cover most practical cases),
- there are no strong eddy current effects (small eddy current effects can be taken into account by considering that the permeability is multiplied by a complex factor).

These assumptions, which may be eliminated if needed, are classical and do not restrict very much the application field of the model.

11.3.2 The Variational Principle

All the previous equations may be gathered in a variational principle. The stationarity of a proper quadratic functional L with respect to the state variables is mathematically equivalent to Euler's equations and takes into account the boundary conditions.

This principle is, in our particular case of magnetostrictive coupling, an extension of Hamilton's principle, where the stationary quantity is equal to a generalized Lagrangian *L* added to a virtual work *W*. This stationarity condition is obtained for small variations δ around the solution which can be written as:

$$\delta L + \delta W = 0 \qquad (11.23)$$

The generalized Lagrangian is defined by the difference between the kinetic energy E_c and the magnetic enthalpy *H* (11.17):

$$\delta L = \delta E_c - \delta H \qquad (11.24)$$

where:

$$\delta E_c = \int_V \rho\omega^2 \, u\delta u \; dv \quad \text{and} \quad \delta H = \int_V (T\delta S - B\delta H)\, dv + Q^V . \delta I$$

The virtual work *W* is the sum of the mechanical work done by the prescribed normal forces on the surface S_{Tn} and the work done by the prescribed induction on the surface S_{Bn}.

$$d\,W = \int_{S_{Tn}} T_n \delta u \;\; dS \; + \int_{S_{Bn}} B_n \, \delta\phi \;\; dS \qquad (11.25)$$

This thermodynamic interpretation of the proposed variational principle ensures the existence and the uniqueness of the solution.

11.3.3 The Finite Element Method

The main advantage of a variational principle is its global approach as opposed to Euler's equations which are local, which permits the problem to be treated by the finite element method (FEM). Thus the solving of the 3D magnetostrictive problem is equivalent, when discretized using the finite element method, to the numerical solution of a linear system, the unknowns of which are the discretized state variables.

Due to the proposed variational principle and to the reduced-scalar-potential formulation, strong algorithmic analogies have been developed with respect to the method implemented within the ATILA computation code for piezoelectric transducers (11.22-11.24).

The ATILA code has been therefore extended to the computation of 2D and 3-D magnetostrictive transducers under advantageous conditions (11.10-11.12). Different types of solutions have been developed: constant-current or constant-voltage harmonic analysis and modal analysis (antiresonances and resonances). The typical program outputs are the transducer impedance, the magnetic field maps and the displacements. Some other developments (B-field map) will be done.

The code has been tested successfully on more than 10 different structures (11.9,11.12), one of them being the characterization device.

11.4 CHARACTERIZATION

Mathematical modeling requires values for the magneto-elastic constants of the Terfenol-D under conditions normally used in transducers. These are, usually, with a magnetic bias, a prestress, a large dynamic strain at low frequency and a temperature near room temperature. It should be noted that finite element modeling creates new requirements. The whole magneto-elastic tensor is now needed. An attempt to get values for this tensor as well as its application on a characterization device are now presented.

Three types of measurement methods are available: the static methods, the resonance methods and the ultrasonic methods. The most appropriate method with respect to the previous specifications is the resonance method applied to a simple electro-mechanical resonator. The problem is that it is not possible to determine all the magneto-elastic constants with an equal accuracy in this way. The proposed process, until there is a better solution, is first to characterize carefully the longitudinal-mode constants (the constants related to the k_{33}) because these are the most important constants in transducers. Secondly, to find the missing magneto-elastic constants by other available methods. The process will be presented on the 'M5' materials biased at 60 kA/m and prestressed at about 20 MPa. At last it should be known that, thanks to a preliminary analysis (11.27), it has been theoretically proved that the non-zero coefficients, which have to be measured are:

$$\begin{array}{l} s_{11}^H \quad s_{12}^H \quad s_{13}^H \quad s_{23}^H = s_{13}^H \quad s_{33}^H \quad s_{44}^H \quad s_{55}^H = s_{44}^H \quad s_{66}^H = 2\left(s_{11}^H - s_{12}^H\right) \\ d_{31} \quad d_{32} = d_{31} \quad d_{33} \quad d_{15} \quad d_{24} = d_{15} \qquad \mu_{11}^T \quad \mu_{22}^T = \mu_{11}^T \quad \mu_{33}^T \end{array} \tag{11.26}$$

These coefficients are part of three families which can be called the k_{33}-mode constants, the k_{15}-mode constants, and the k_{31}-mode constants because these three coupling coefficients use nearly all the coefficients independently:

$$k_{33}^2 = \frac{d_{33}^2}{s_{33}^H \mu_{33}^T} \qquad k_{15}^2 = \frac{d_{15}^2}{s_{44}^H \mu_{11}^T} \qquad k_{31}^2 = \frac{d_{31}^2}{s_{11}^H \mu_{33}^T} \tag{11.27}$$

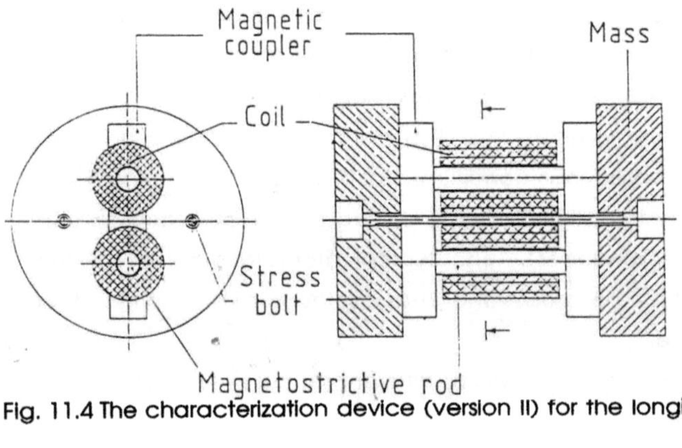

Fig. 11.4 The characterization device (version II) for the longitudinal constant measurement

11.4.1 The k_{33}-mode Constants

In order to characterize the longitudinal-mode constants, a mass-loaded dual-rod resonator (Fig. 11.4) has been devised (11.13,11.14). The proposed method involves the measurement of the electrical impedance and the fitting with a theoretical impedance model provided by a plane wave model associated with a transfer matrix method. The set of material parameters used for the fit contains the Young's modulus, the longitudinal permeability and the coupling coefficient.

This method has been applied on 5 materials from different origins (M1,...,M5) versus the magnetic bias. It has permitted a comparative study which has shown that Grain-Oriented materials (G.O.) were much better than random materials. The results for the M5 G.O. material under the conditions specified are:

$k_{33} = 0.69$, $\quad s_{33}^H = 4.2\ 10^{-11}\ m^2/N$, $\quad \mu_{33}^T = 4.4\ \mu_o$.

11.4.2 The k_{15}-mode Constants

An attempt has been made to get the k_{15}-mode (the twisting mode) constants with a symmetrical Wiedemann-effect twisting pendulum (Fig. 11.5). A Terfenol-D rod is prestressed and biased by a longitudinal H_o-field. A longitudinal dynamic current I is passed through the rod as shown, which creates a turning field H_θ generating, in principle, a twisting of the rod and an alternating angular rotation of a flywheel. According to the theory (11.9), the electrical impedance (given by the voltage V measured along the rod) should exhibit a low-frequency electrically-coupled k_{15} resonant mode. Unfortunately, the measured impedance has only shown a high-frequency resonance. This behavior has not been explained. However, this measurement has at least permitted to get the transversal free permeability under bias and prestress:

$$\mu_{11}^T = 6.9\ \mu_o \tag{11.28}$$

Thanks to the assumption of magnetization rotation which leads to the relation,

$$d_{15} = \frac{3\ .\ \mu_{11}^T}{2\ .\ \mu_{33}^T}\ .\ d_{33}, \tag{11.29}$$

the d_{15} constant may be estimated with a 20% correction, equal to:

$$d_{15} = 2.8\ 10^{-8}\ m/A. \tag{11.30}$$

The resonance not being found, it was not possible to measure the s_{44}^H elastic constant. For this constant, the value given by H.T. Savage (11.28) has been borrowed, waiting for better results.

$$s_{44}^H = 1.67\ 10^{-10}\ m^2/N \tag{11.31}$$

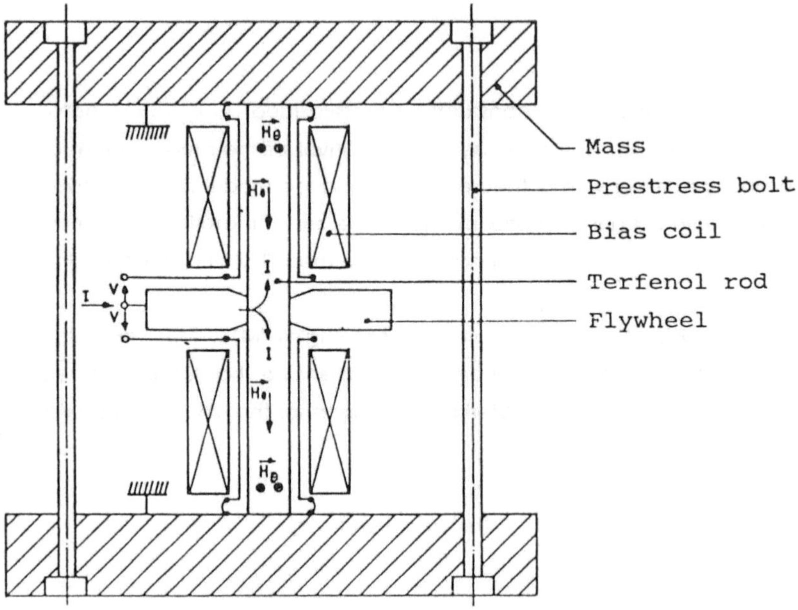

Fig. 11.5 The Wiedemann-effect twisting pendulum for the twisting constant measurement

This set of values leads to the twisting coupling k_{15} value, which is higher than the k_{33} value, as noticed by H.T. Savage (11.28):

$$k_{15} = 0.74 \tag{11.32}$$

11.4.3 <u>The k_{31}-mode Constants</u>

Direct strain gage measurements (11.27) seem to have proved that the assumption of the constant-volume magnetostriction giving:

$$d_{31} = - d_{33} / 2 \tag{11.33}$$

is well verified. This leads to:

$$d_{31} = - 0.52\ 10^{-8}\ \mathrm{m/A}. \tag{11.34}$$

These measurements have also given the constant-B Poisson's ratio equal to 0.42 which leads to:

$$s_{13}{}^{B} = -0.0924\ 10^{-10}\ \mathrm{m^2/N}. \tag{11.35}$$

Finally, to complete the elasto-magnetic tensor, the constant-B cubic-symmetry assumption has been used to write:

$s_{11}^{B} = s_{33}^{B}$ and $s_{12}^{B} = s_{13}^{B}$

Therefore the magneto-elastic matrix coefficients have been obtained. They have been summarized in different consistent sets, in Table 11.1.

There is a strong probability that some of the table values are good if considering their order of magnitude, but as it will be seen in the next section, this set constitutes a good working basis.

Table 11.1 Magneto-elastic coefficients in MKSA units of the M5 material biased at 60 kA/m and prestressed at 20 MPa

$k_{33} = 0.69$ $k_{31} = 0.43$ $k_{15} = 0.74$

$s_{11}^{H} = 2.70\ 10^{-11}$	$c_{11}^{H} = 10.70\ 10^{10}$
$s_{12}^{H} = -0.43\ 10^{-11}$	$c_{12}^{H} = -7.48\ 10^{10}$
$s_{13}^{H} = -1.90\ 10^{-11}$	$c_{13}^{H} = -8.21\ 10^{10}$
$s_{33}^{H} = 4.20\ 10^{-11}$	$c_{33}^{H} = 9.81\ 10^{10}$
$s_{44}^{H} = 16.70\ 10^{-11}$	$c_{44}^{H} = 0.60\ 10^{10}$
$s_{66}^{H} = 6.30\ 10^{-11}$	$c_{66}^{H} = 1.61\ 10^{10}$
$d_{31} = -0.52\ 10^{-8}$	$e_{31} = -0.90\ 10^{2}$
$d_{33} = 1.04\ 10^{-8}$	$e_{33} = 1.66\ 10^{2}$
$d_{15} = 2.80\ 10^{-8}$	$e_{15} = 1.68\ 10^{2}$
$\mu_{11}^{T} = 6.9\ \mu_0$	$\mu_{11}^{S} = 3.16\ \mu_0$
$\mu_{33}^{T} = 4.4\ \mu_0$	$\mu_{33}^{S} = 2.28\ \mu_0$

11.4.4 Analysis of the Characterization Device

The characterization device of Fig. 11.4 is a good test for the models and the constants. Measurements have been conducted on this device under conditions as specified above. As in the plane wave model (PWM) the impedance has been fitted to the measured impedance, this model is in perfect agreement with the measurements and may be considered as the reference for the comparison. The effective coupling factor is found equal to $k_{eff} = 0.530$ using the formula based on the antiresonance frequency ω_l and the resonance frequency ω_V:

$$k_{eff}^2 = (\omega_V^2 - \omega_l^2)/\omega_V^2$$

In order to provide an analysis with the lumped parameters model the values of the previously introduced parameters may be given:

Table 11.2 Values of the parameters

$m = 1.93$ kg	$e_t = 75.8\ 10^{-9}$ m/N
$e_a^H = 30.1\ 10^{-9}$ m/N	$e_a^B = 15.1\ 10^{-9}$ m/N
$L_o = 25.12\ 10^{-6}$ H	
$L_a^T = 60.8\ 10^{-6}$ H	$L_a^S = 30.4\ 10^{-6}$ H

This permits to calculate the antiresonance frequencies (Table 11.3) and the different coupling factors:

$$k_E = 0.913 \qquad\qquad k_M = 0.842$$

which gives by the simple product $k_M\ k_{33}\ k_E$ the resulting $k_{eff} = 0.544$, and by a slightly more complex formula (11.9) $k_{eff} = 0.530$, in better agreement with measurements. There is no difference between the LPM and PWM because these models are nearly identical in this situation, the rod mass being negligible compared to the end masses.

Finally, characterization device has been analyzed using the finite element method with the magnetostrictive version of ATILA and a set of magneto-elastic coefficients nearly identical to that of Table 11.1. The results (Table 11.3) are in good agreement with results from the other models. However, it can be noted that the computed coupling factor is slightly lower than the measured coupling factor. This means first, that the method used for the characterization gives a k_{33} value that is 3% smaller than the actual value and second that the formulae used by the PWM and the LPM to calculate k_M and k_E are not perfectly accurate. Even in this favorable situation, some three-dimensional effects exist (Fig. 11.6). The flux for the in-air return path may be observed. The magnetic field in the rod edges has components perpendicular to the rod axis which have never been mentioned before. These effects predicted by the FEM analysis still have to be measured. They are due to the fact that the coil length is smaller than the rod length

Table 11.3 Comparison on the PWM, the LPM, the FEM predictions for the normalized antiresonant and resonant frequencies and the effective coupling factor.

	ω_l/ω_o	ω_v/ω_o	k_{eff}
PWM	0.780	0.920	0.530
LPM	0.780	0.920	0.530
FEM	0.785	0.912	0.512

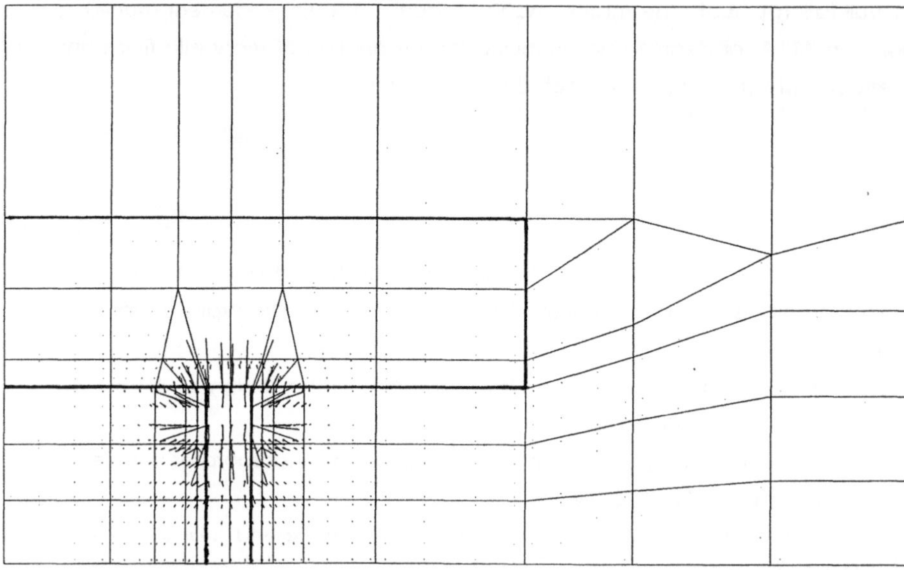

Fig. 11.6 Magnetic field map in the quarter of a symmetry plane containing the rods and the magnetic couplers, as obtained using ATILA

This comparative test for the three types of model is a self-consistent validation for both the theory and the implementation of the models and for the characterization method. Although some coefficients of the magneto-elastic matrix have not been determined with all the desired accuracy, the matrix presented here may be used successfully for the design of magnetostrictive transducers because the impedance and the longitudinal components of the fields and of the displacements are correctly computed. It is also supposed, pending experimental confirmation, that the transverse-field effects predicted by ATILA exist, at least qualitatively. This type of prediction is one of the assets of finite element modeling.

11.5 CONCLUSION

Three types of magnetostrictive transducers models have been presented. The lumped parameter model permits to focus attention on the main transducer parameters. It has shown the role of magnetic parameters, such as the magnetic coupling factor, which are not encountered in piezoelectric transducers. In spite of restrictive assumptions, the plane wave model associated with a transfer matrix method is useful in applications to the design of tonpilz transducers and to characterize the Terfenol-D longitudinal constants. The finite element modeling retains a much more general approach because it deals with the three-dimensional behavior of transducers. As a consequence, it also requires more data. To provide them, a characterization process has been proposed giving as a result the complete elastic-magnetic matrix. The comparative test of the methods, done on the characterization device has given a first but partial validation of both the matrix and the models. The finite element modeling, now possible thanks to ATILA, appears to be promising for the analysis of magnetic fields and 3-D mechanical effects but has still to be subjected to several tests.

REFERENCES

11.1 D. Boucher, 'Trends and problems in low frequency sonar projectors design,' in Power Sonic and Ultrasonic Transducers Design, Ed. B.Hamonic, J.N. Decarpigny, Springer-Verlag, 100-133 (1988).

11.2 L.H. Royster, 'The Flextensional Concept: a New Approach to the design of Underwater Acoustic Transducers,' Applied Acoustic, **3**, 117 (1970).

11.3 W.C. Greenlaw, and F.R. Hodges, 'Sonar transducer design incorporates rare earth alloy,' Defense System rev., 50-55, (Nov. 1984).

11.4 R.W.Timme, and S.W.Meeks, 'Magnetostrictive underwater sound transducers,' J. Phys.(Paris), **40**, C5-280 (1979).

11.5 A.E. Clark, 'Magnetostrictive rare earth Fe_2 compounds, Ferromagnetic materials,' Ed. E.P.Wohlfarth,.**1**, ch.7, North-Holland Pub. (1980).

11.6 S.W. Meeks, and R.W. Timme, 'Rare earth iron magnetostrictive underwater sound transducer,' J. Acous. Soc. Am., **62**, (1977).

11.7 F. Claeyssen, D. Boucher, and C. Pohlenz, 'Analysis of a magnetostrictive Tonpilz transducer,' J. Acous. Soc. Am., **81,** Sup.1, MM6,.S89 (1987).

11.8 F. Claeyssen, D. Boucher, and S. Faure, 'Improvement of a magnetostrictive length-expander transducer by use of a Grain-Oriented material,' J. Acoust. Soc. Am., **85**, Sup.1, LL5, S90, (1989).

11.9 F. Claeyssen, 'Conception et réalisation de transducteurs sonar basse fréquence à base d'alliages magnétostrictifs Terres rares-Fer,' Thesis n. 89 ISAL 0065, INSA Lyon Fr., in French (1989).

11.10 F. Claeyssen, and D. Boucher, 'Modelling of magnetostrictive transducers using the Finite Element Method,' submitted to the Int. J. Num. Meth. Eng. in 1990.

11.11 F. Claeyssen, D. Boucher, K. Anifrani, R. Bossut, and J.N. Decarpigny, 'Analysis of magnetostrictive transducers by the ATILA Finite Element code,' J. Acoust. Soc. Am., **85**, Sup.1, LL4, S90 (1989).

11.12 R. Bossut, J.N. Decarpigny, F. Claeyssen, D. Boucher, and K. Anifrani, 'Modélisation de transducteurs magnétostrictifs à l'aide du code éléments finis ATILA,' Proc. of the First French Conference on Acoustics, J. Phys. (Paris), FASC. 2, **1**, C2-617-621 (1990).

11.13 F. Claeyssen, D. Boucher, and S. Faure, 'Characterization of length-expander magnetostrictive rare-earth iron rods under normal use conditions in transducers,' J.Acous.So.Am., **83**, Sup.1, H7 S20 (1988).

11.14 F. Claeyssen, D. Boucher, and S. Faure, 'Comparative study of piezomagnetic constants,' Sec. Int. Conf. Giant Magnetostrictive Alloys, Spain, Oct. 1988.

11.15 F. Claeyssen, D. Boucher, A. Foggia ,and E. Sabonnadiere, "Analysis of the magnetic fields in magnetostrictive rare earth-iron transducers," Compumag Conf. 1989, to be published in IEEE Trans on Magnetics, 1990.

11.16 D.A. Berlincourt, D.R. Curran, and H. Jaffe, *Piezoelectric and piezomagnetic materials* Physical Acoustics, Ed. W.P.Mason, Academic Press, Ch.3, (1964).

11.17 O.B. Wilson, *An introduction to the theory and design of Sonar transducers*, Peninsula Pub. Co., Los Altos (Ca), USA (1989).

11.18 M.J. Earwicker, "Mathematical modelling of piezoelectric transducers," Admiralty Underwater Weapons Establishment report, Portland, USA, March 1973.

11.19 S.W. Meeks, and R.W. Timme, "Acoustic resonance of a Rare Earth Iron Magnetostrictive Rod in the presence of large Eddy-Currents," IEEE Trans. Sonic and Ultrasonic **SU-27** (1980).

11.20 J.L. Butler, "Application manual for the design of ETREMA Terfenol-D magnetostrictive transducers," Edge tech. Inc., Ames,1988.

11.21 R.S. Woollett, "Effective Coupling Factors of Single Degree-of-Freedom transducers," J. Acoust. Soc. Am., **40**, 1112-1123 (1966).

11.22 J.N. Decarpigny, "Application de la méthode des éléments finis à l'étude des transducteurs piézoélectriques," Doctoral thesis, Lille University Fr., in French (1984).

11.23 J.N. Decarpigny, J.C. Debus, B. Tocquet, and D. Boucher, "In-Air analysis of piezoelectric Tonpilz transducers in a wide Frequency Band Using a Mixed Finite Element Method," J. Acous. Soc. Am., **78**, 1499-1507 (1985).

11.24 B. Hamonic, "Application of the Finite Element Method to the Design of Power Piezoelectric Sonar Transducers," in Power Sonic and Ultrasonic Transducers Design, Ed. B. Hamonic, J.N. Decarpigny, Springer-Verlag 143-159 (1988).

11.25 O.C. Zienkiewicz, "Three dimensional magnetic field determination using a scalar potential - a Finite Element solution," IEEE Trans. on Mag., **13**,.1649-1656 (1977).

11.26 P. Labbie, and J.C. Sabonnadiere, "FLUXLAB library of inductors," LEG, ENSIEG, Grenoble Fr., 1988.

11.27 E. Du Tremolet Delacheisserie, and J. Rouchy, "La détermination de la matrice élastomagnétique et son domaine d'application dans le cas des alliages de Terfenol-D," GERDSM, DCAN Toulon, Fr., report n°015980, 1989.

11.28 H.T. Savage, and R. Abbundi, "Perpendicular susceptibility, magnetomechanical coupling and shear modulus in $Tb_{0.27}Dy_{0.73}Fe_2$," IEEE Trans. on Mag. **14**, 545-547 (1978).

12. A Magnetostrictive Transducer Design

G.A. Steel and J.R. Oswin

British Aerospace (Dynamics) Ltd,
P.O. Box 5, Filton, Bristol BS12 7QW, United Kingdom

A magnetostrictive sonar transducer operating at 1350 Hz is described. The device is a small double-ended piston containing three terfenol rods in independent magnetic circuits. The paper presents detailed aspects of the design and a summary of test results. An important observation is the effect of heat on terfenol, which causes changes in elastic modulus and hence resonant frequency. Experience with this device allows a course to be set for future studies, both into material properties and associated transducer designs.

12.1 INTRODUCTION

British Aerospace is currently involved in the study of terfenol for use in sonar transducer applications. The work includes measurement of material properties and development of suitable transducer designs. This paper describes the analysis and testing of one particular design during the period 1988 to 1989.

Unlaminated terfenol rods were obtained from two suppliers; one set from EDGE Technologies (previously AMES Laboratory), and one set from Rare Earth Products (REP), part of Johnson Matthey PLC. All rods were nominally 127 mm long and 6.35 mm diameter. Measurements of their magnetostrictive properties are reported in (12.1).

Three identical transducers were constructed, two using REP rods and one using EDGE rods. The specified resonant frequency was 1350 Hz. Only three rods per transducer were available. This was a relatively small volume of terfenol so the transducer design was only intended for low power operation.

12.2 TRANSDUCER DESIGN

The transducer design requirement was for a piston device operating at shallow depths at 1350 Hz and having a wide bandwidth. A double-ended configuration was chosen to give maximum radiating area for best bandwidth.

A diagram of the transducer is shown in Fig. 12.1. It was a symmetrical structure with a piston at each end. The pistons were bonded to three terfenol rods which were mechanically in parallel.

Each terfenol rod was surrounded by a 546 turn coil on a nylon bobbin. A cross-section is shown in Fig. 12.2. The ends of the bobbin and outside of the coil were covered with layers of mu-metal (12.2) a high permeability alloy, which formed the magnetic return path. In this way the three rods were contained in three independent magnetic circuits. There was a constriction at the centre of each bobbin which gripped the rod at its mid-point and acted as a nodal support.

During construction it was noted that the REP rods were straighter than the EDGE rods and could be gripped more securely. The REP transducers were therefore superior in terms of stress distribution, a fact which showed up during high power tests.

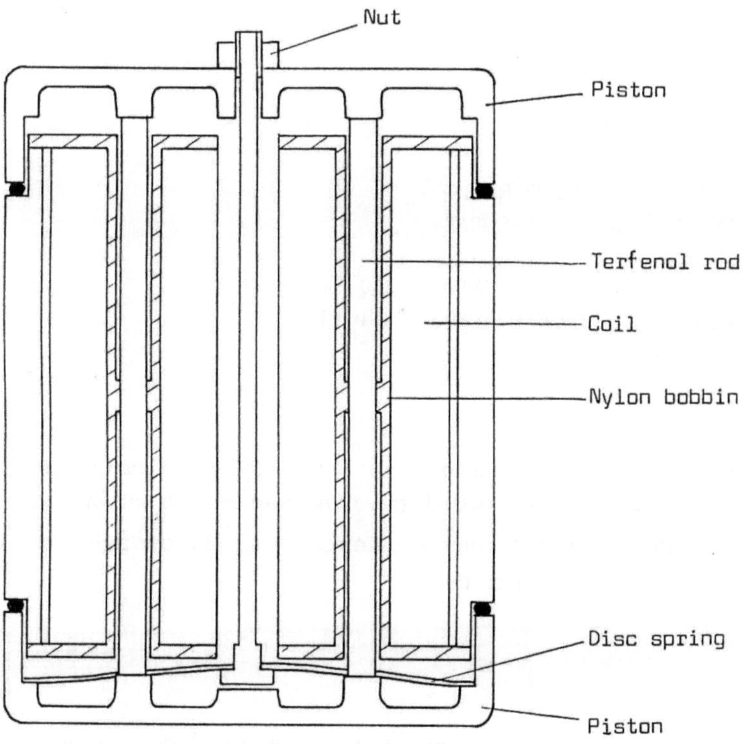

Fig. 12.1 Simplified diagram of complete transducer

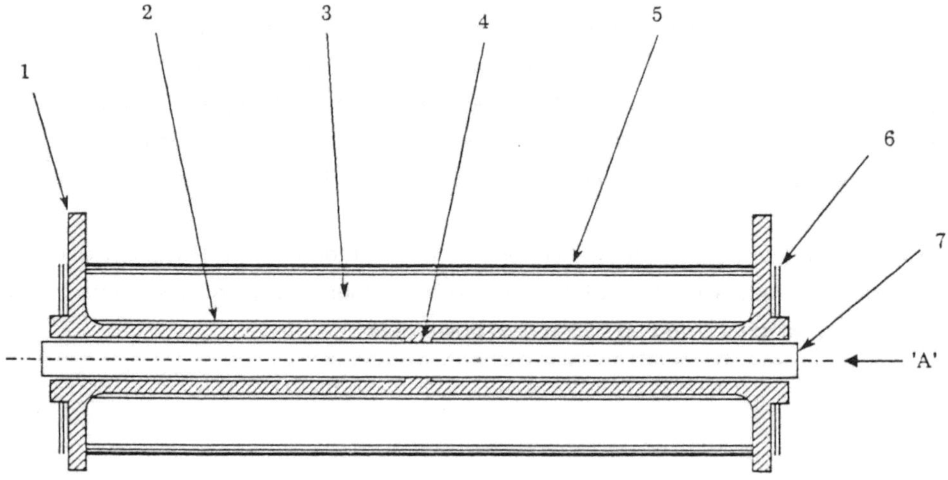

SECTIONAL VIEW

1. BOBBIN
2. SEARCH WINDING - 150 TURNS
3. MAIN WINDING - 546 TURNS (7/0.4)
4. ROD SUPPORT
5. Mu METAL FLUX CAGE - 117 WIDE x 600 x 0.1
6. Mu METAL POLE - 0.1 LAMINATIONS (14OFF)
7. TERFENOL ROD

Fig. 12.2 Coil details

The three terfenol rods had a combined compliance of 4.4 x 10^{-8} m/N (assuming Y_3^H = 30 GPa). For resonance at 1350 Hz a total mass of 0.63 kg was required at each end of the rods. This was the sum of piston mass plus radiation mass. The radiation mass per piston is approximately independent of frequency and is given by:

$$M_r = \frac{\pi \rho}{8\sqrt{2}} d^3 \tag{12.1}$$

where d is the piston diameter and ρ is the density of water. Increasing diameter allows greater output power (because of increased radiation resistance) but reduced the mass of the piston, which must not be allowed to flex. Hence output power was limited by the ability to design a piston of low mass and high stiffness.

Aluminium was chosen as a suitable material. An optimum diameter of 110 mm was selected to give a radiation mass of 0.37 kg and a piston mass of 0.26 kg. A diagram of one piston is shown in Fig. 12.3. The inner surface had radial and concentric ribs to achieve high stiffness. The outer surface was flat.

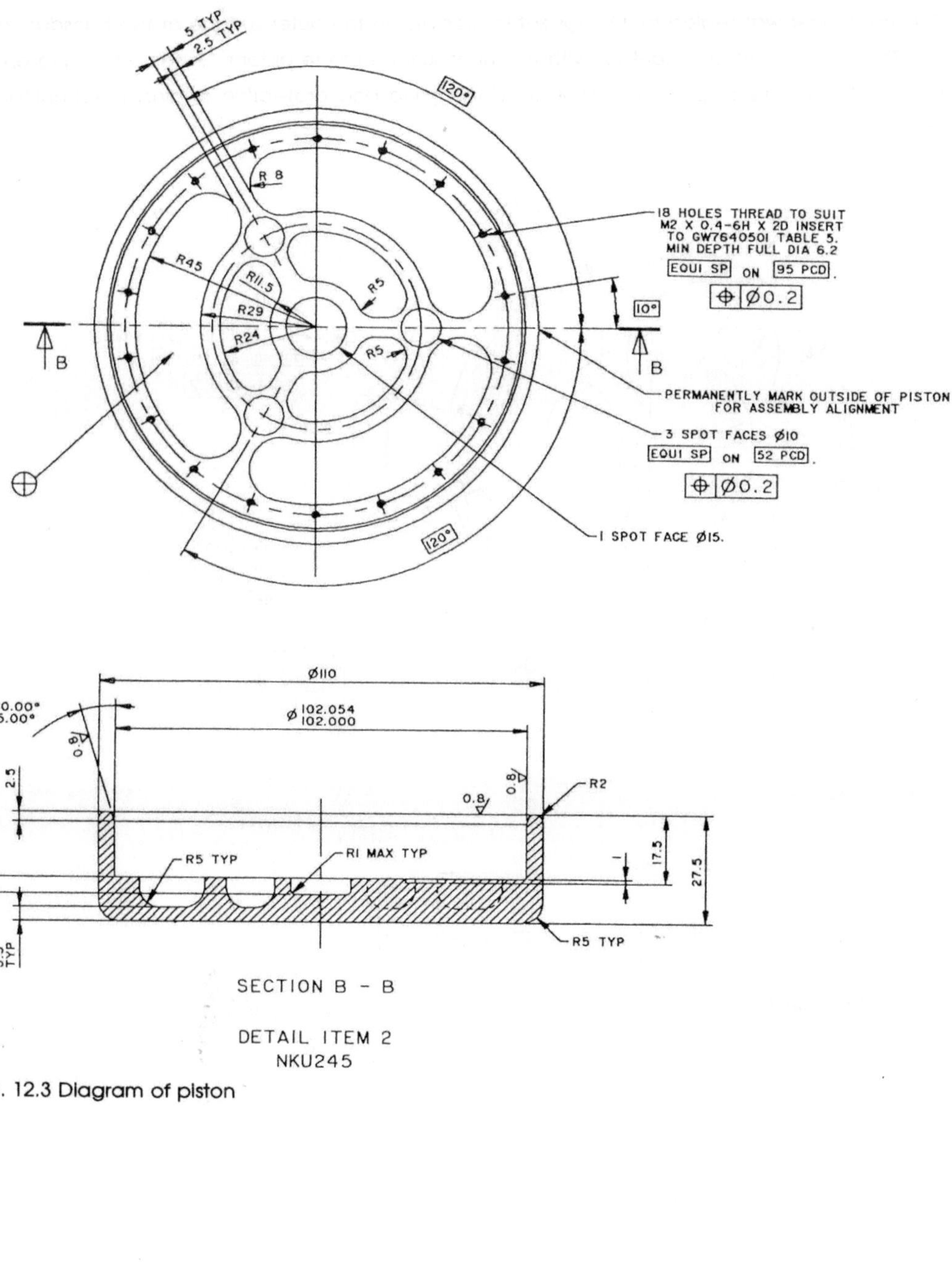

Fig. 12.3 Diagram of piston

Most transducer designs include a bolt to provide compressional stress bias on the terfenol. In this case a beryllium-copper disc spring was used instead because a very high compliance was required. The spring is shown in Fig. 12.4. The spring was bonded directly to one piston and attached to the other by a rigid bolt. Tightening the bolt caused increased pressure on the terfenol. The spring was set to provide a compressional stress of 15 MPa. This is equivalent to a compression of 500 parts per million (ppm) with the coils open-circuit.

The transducer was sealed by O rings set in grooves on the outer surface of the transducer body. These O rings were in contact with the inner surface of the pistons. A second O ring on the base of each piston provided acoustic decoupling and protection against mechanical shock.

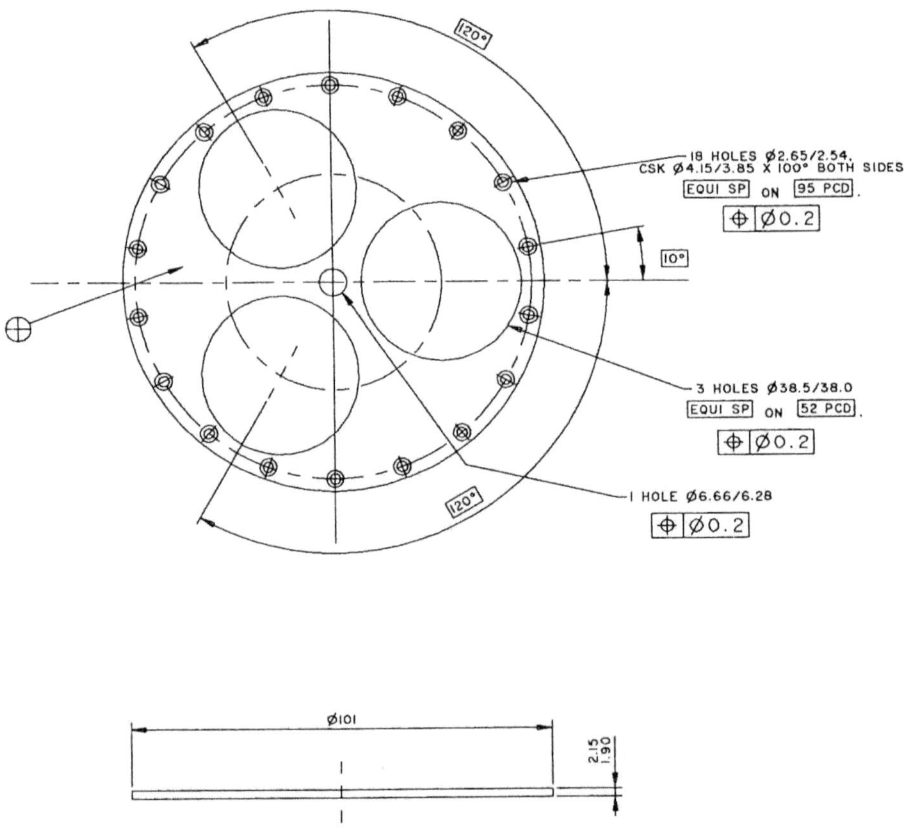

Fig. 12.4 Disc spring

12.3 TRANSDUCER TEST RESULTS

Input impedances at low power in air and water are shown in Figs. 12.5 and 12.6 for one of the REP transducers. Resonance occurred at 2100 Hz in air and 1600 Hz in water. The other two transducers showed almost identical performance. Polar plots indicated omnidirectional radiation at frequencies up to 1600 Hz.

The circuit of Fig. 12.7 was used to provide high power pulses for source level measurements. Bias current was switched off between pulses. A typical current waveform is shown in Fig. 12.8. Pressure levels were monitored on a Bruël and Kjaer 8105 ball hydrophone at a distance of 6.1 m. Results are shown in Fig. 12.9. At low power, resonance occurred at 1600 Hz as observed in impedance measurements. As power increased all three transducers showed a reduction in resonant frequency, tending towards 1300 Hz. The change in frequency was caused by heat. No temperature sensor was fitted, so the effect could not be accurately monitored, but a change in elastic modulus (Y_3^H) from 42 GPa to 28 GPa is indicated.

The Y axis of Fig. 12.9 is labelled both in Pascals and Watts. Output power of a small double-ended piston is related to strain by the equation:

$$W = \frac{\pi^3 f^4 \rho l^2}{2c} (SA)^2 \qquad (12.2)$$

where A is piston area, l is length of terfenol and S is peak to peak strain. At high power all three transducers showed strains greater than 1500 ppm, significantly higher than obtained in static measurements.

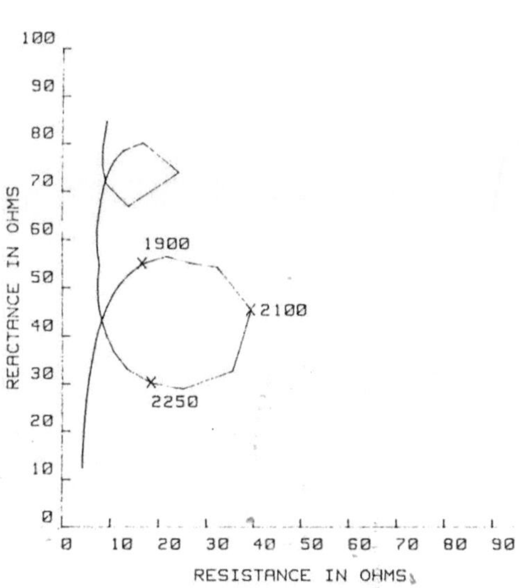

Fig. 12.5 Input impedance in air

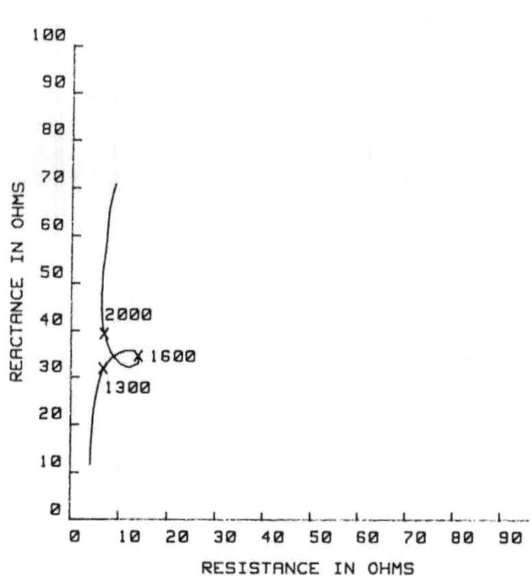

Fig. 12.6 Input impedance in water

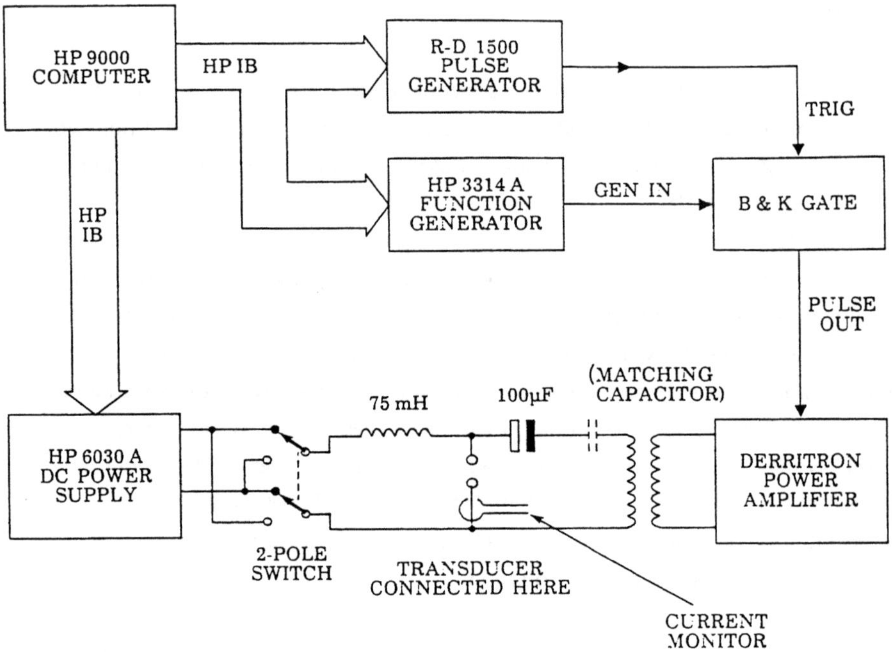

Fig. 12.7 Circuit diagram for high power tests

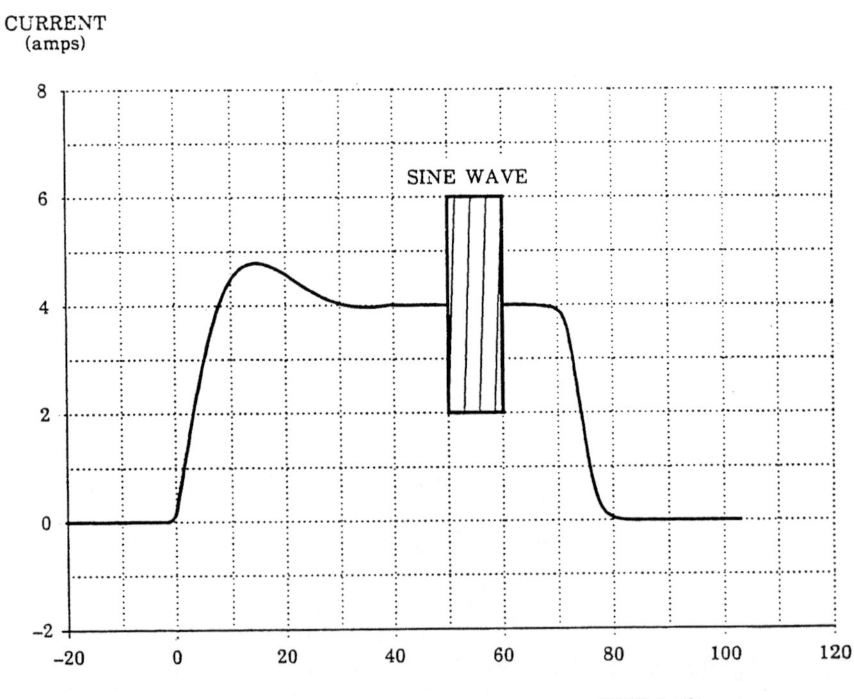

Fig. 12.8 Typical current waveform

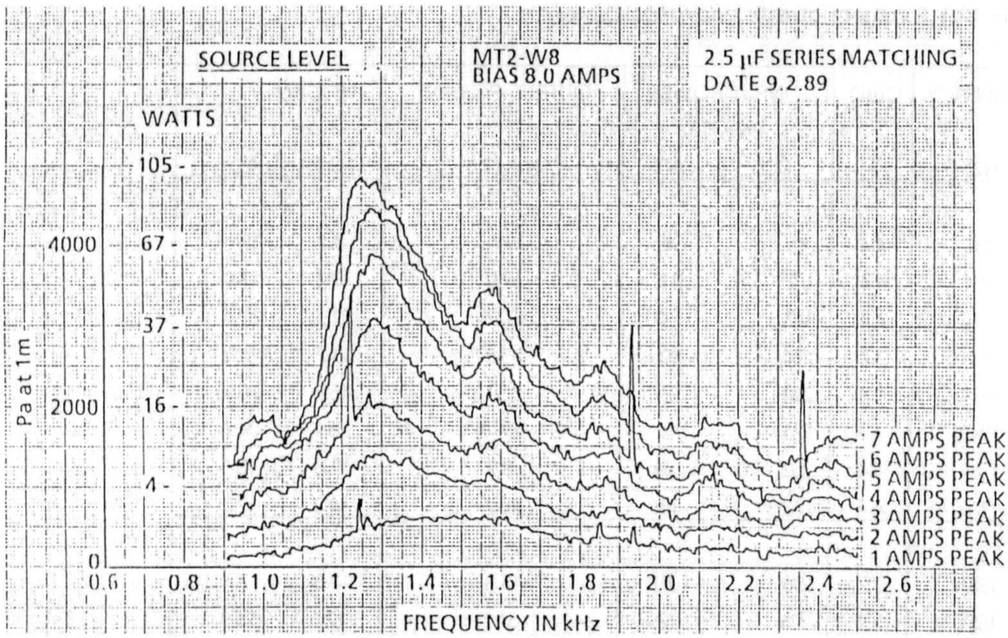

Fig. 12.9 Source level

Figure 12.9 shows a series of small maxima and minima superimposed on the main curves at intervals of 150 Hz. These were caused by surface reflections and were not related to transducer characteristics.

Efficiency at high power was measured by inserting a matching capacitor at the transformer output. Frequency was varied until voltage and current were in phase, allowing direct calculation of input power. Results are shown in Table 12.1. At resonance AC efficiency reached 30%. However, there was significant DC power loss in all cases, reducing the total efficiency to 20%.

The EDGE transducer failed during this test. Input power was 130 Watts, with a pulse length of 10 ms and repetition frequency 2 Hz. It is significant that the REP transducers, which had straighter rods than EDGE, survived the test with no damage. In a further high power test the pulse length was increased to 300 ms with a repetition frequency of 0.5 Hz. One of the REP transducers was damaged at a power level of only 33 Watts. Clearly this particular transducer design is most suitable for operation with short pulses.

Table 12.1 Summary of efficiency measurements

FREQUENCY HZ	C(μF)	INPUT POWER (WATTS) DC	AC	OUTPUT POWER WATTS	EFFICIENCY TOTAL	AC
MT1						
1417	4.4	45	48	9.4	8.3	19.6
1271	4.4	147	240	67.9	17.3	27.9
1274	4.4	147	237	102.6	21.1	30.4
1233	3.7	224	300	45.6	8.7	15.2
1284	4.4	224	287	81.1	15.8	28.2
1248	4.4	224	432	106.8	16.2	24.7
1180	4.4	262	44	8.2	2.6	18.6
1160	4.4	262	136	16.7	4.2	12.3
MT2						
1331	3.7	56	51	3.8	3.5	7.4
1652	3.7	126	125	31.7	12.6	25.3
1676	3.7	224	129	25.5	7.2	19.8
1679	3.7	224	175	28.3	7.1	16.2
MT3						
1813	2.5	58	16	3.0	4.0	18.9
1443	4.4	58	44	11.4	11.0	25.9
1195	4.4	230	46	7.6	2.7	16.6
1268	4.4	230	220	67.0	14.9	30.4
1266	4.4	230	390	124.0	20.0	31.9

12.4 CONCLUSION

The most important observation from this series of experiments was the effect of heat on terfenol. The main sources of heat were eddy currents and losses due to coil resistance.

Energy loss in the coil is partly due to DC bias current. This can be eliminated by use of permanent magnet biasing; however, the magnets involved will be subjected to large alternating flux which causes de-magnetisation, also, there will be energy loss due to hysteresis since permanent magnets always have "open" B-H loops. Hence there is much scope for investigation of the choice between DC and permanent magnet biasing.

Reduction of eddy currents can be achieved by laminating the terfenol, but this process is expensive and causes reduced strength. An alternative approach may be to include additives in terfenol to increase its electrical resistivity - a problem for material scientists rather than transducer designers. A further problem which needs attention is the very low permeability of terfenol. The resulting leakage flux causes reduced coupling, so the high coupling of pure terfenol is difficult to exploit. There is a trade-off here. Additives in terfenol to increase its permeability may cause lower coupling in the pure material but lead to transducers of superior performance because of reduced leakage flux.

12.5 FUTURE DEVELOPMENTS

The double-ended piston design would not be optimum for all conditions. Indeed, if the bandwidth requirement were reduced, balanced single-ended piston designs would be more suitable, as the Rare Earth material would not be subject to hydrostatic pressure.

Transducer design is also being eased by the availability of the rare earth material in forms other than slender rods. This now allows much greater versatility in transducer design.

Terfenol is being produced in significantly increased quantities. This is helping to overcome its image as an exotic high-priced material of questionable availability. Continuing improvements in material properties and production techniques point the way to an alloy of real commercial interest rather than just a scientific curiosity. As in many markets there is a degree of positive feedback here. The more terfenol we use the more must be produced so the least expensive it should become. Development of cheaper low frequency transducers, with higher power handling capability, naturally follows.

At British Aerospace the current aim is to achieve close monitoring of thermal effects, with resulting accurate control of resonant frequency and bandwidth. Piston transducers remain the focus of attention. They are simple to analyse and construction is relatively easy. Use of terfenol in flextensional transducers is obviously the next step and is likely to become increasingly important over the next few years as more people want more power at lower frequencies for less money.

12.6 ACKNOWLEDGEMENT

This work was undertaken with the support of the U.K. Ministry of Defence (Procurement Executive).

REFERENCES

12.1 K.C.Pitman, A.Branwood, and A.R.Pierce, "Device Oriented Measurements of Terfenol Materials," Proceedings of Second International Conference on Giant Magnetrostrictive and Amorphous Alloys for Sensors and Actuators. October 1988.

12.2 *Mu-metal Alloys* Publication TP11-576, Telecon Metals Ltd, Crawley, Sussex.

13. Magnetohydrodynamic and Thermoacoustic Mechanisms for Generation of Sound in Seawater

S.L. Garrett
Physics Department - Code 61 Gx, Naval Postgraduate School,
Monterey, CA 93943, USA

T.B. Gabrielson
Naval Air Development Center - Code 5044, Warminster, PA 18974, USA

Conventional techniques for producing high-amplitude sound rely on the oscillatory motion of solid materials to generate the requisite volumetric velocity. Material properties limit the displacement of these solids; hence, further increases in radiated power require transducers of larger surface area. This paper examines magnetohydrodynamic and thermoacoustic methods for moving fluid directly as a means of producing more compact projectors with source strengths that are comparable to conventional projector designs. In addition, test results from small prototypes of each transducer are summarized.

13.1 INTRODUCTION

The analysis of sound generation is simplified when the sound source is much smaller than the wavelengths of sound, λ, that are being generated. For this reason, the geometrical details of the projector are unimportant and only the volumetric velocity of the source, Q, as defined below, is significant in determining the amount of acoustic energy that will be radiated into the medium (13.1):

$$Q = \int_{\text{Surface}} \mathbf{v} \cdot \mathbf{n} \; dA \qquad (13.1)$$

The integral (13.1) is evaluated over a surface where **n** is the unit vector normal to the element of surface of area dA, which is moving with velocity **v**.

The pressure amplitude, P(r), of a spherical wave radiated by such a source as measured at a distance r from the acoustic center of the source will depend upon the acoustic impedance of the surrounding medium and its boundaries. For the purposes of this discussion, we will assume that the source is within an infinite medium of density ρ and sound speed c and is, therefore, radiating spherical waves. By the reciprocity principle (13.2), such a source will have a constant pressure-distance product under free-field conditions given by the expression below:

$$P(r) \cdot r = \frac{\rho c}{2\lambda} Q \qquad (13.2)$$

If we consider conventional sound sources that rely on the motion of a solid surface to generate the volumetric velocity, then the maximum displacement of those solid surfaces will limit the amplitude of the radiated sound. For piezoelectric or magnetostrictive materials, this displacement is the maximum allowable strain of the material. Similarly, flextensional sources are limited by the strain in the shell or by the excursion of the seals. The maximum excursion of flexible seals also limits the displacement of electrodynamic sources. Under the assumption of a displacement limitation, the surface velocity $v = \omega d$, where $\omega = 2\pi f$ is the angular frequency of the oscillating surface and d is the displacement. At fixed distance $r >> \lambda$, the net effect of equations (13.1) and (13.2) is that the pressure amplitude due to a constrained surface displacement amplitude will be proportional to ω^2 and the radiated acoustic power will be proportional to ω^4. This analysis suggests that, if one wishes to generate large-amplitude sound using an oscillating solid surface, the surface area must increase quadratically with decreasing frequency to maintain a constant pressure amplitude in the farfield.

Alternative strategies could be developed if one were to move the fluid by applying forces directly rather than through an intermediary oscillating surface. Since only the fluid would move, the limitations to the fluid velocity would be hydrodynamic (*e.g.*, turbulent) rather than elastic. Several such mechanisms exist and it is the purpose of this paper to examine two mechanisms of contemporary interest in some detail.

Before doing so, it is interesting to mention a powerful sound source (thunder) that has existed since the time of creation and has undergone extensive scientific investigation over the past 110 years. If a volume of fluid is heated and its temperature is changed with time, it will generate a volumetric velocity if its isobaric coefficient of thermal expansion, $\beta = (1/V)(\partial V/\partial T)_p$, is non-zero. When the atmosphere is heated by the passage of a large electrical current (lightning), the magnitude of the sound wave is impressive. The first quantitative analysis of thermally generated sound was that of Rayleigh (13.3) commenting on the invention of the photophone by Alexander Graham Bell in 1881 (13.4). In Bell's case, light heated a solid surface causing the adjacent gas to expand. In 1907, a Russian engineer named Gwozdz, in Lodz, Poland, generated sound due to heating of electrical wires (13.5). A careful analysis of this effect lead to the construction of the thermophone (13.6, 13.7) which was used as a primary acoustical calibration standard for many years. The mechanism of the direct effect of heating to generate volumetric velocity by thermal expansion is far less efficient in water because the thermal expansion coefficient is so small in comparison with that of air and, in fact, vanishes at 4°C under atmospheric pressure. In spite of this limitation, there have been extensive studies of this mechanism (13.8) since lasers can remotely generate spatially extended source regions in the water (13.9).

This paper will examine a different thermoacoustic sound generation process that can realize much greater conversion efficiencies by the construction of a heat engine that exploits acoustic resonance and the interaction of the working fluid with a second thermodynamic medium. Before doing so, an entirely different mechanism, which relies on the electrical conductivity of seawater, will be presented.

13.2 MAGNETOHYDRODYNAMIC (MHD) SOUND GENERATION

Another mechanism that is capable of applying forces directly to electrically conducting fluids is the hydrodynamic Hall effect. Although most of the research in this area is done by plasma physicists (13.10), there has been a fair amount of recent interest in this mechanism for acoustical generation and detection in both seawater and liquid metals. At the most simplistic level, the force **F**, on a conductor of length **l**, in a magnetic field of induction **B**, carrying a current i, is given by $\mathbf{F} = i\,\mathbf{l} \times \mathbf{B}$. In a fluid, this force generates a pressure gradient ∇P, due to a current density $\mathbf{J} = q\mathbf{u}$, where **u** is the velocity of the current-carrying fluid with charge density q in Coulombs per cubic meter.

$$\nabla P = \mathbf{J} \times \mathbf{B} \tag{13.3}$$

Similarly, by Faraday's Law, a fluid moving with velocity **u** within a magnetic field generates an electric field, **E**, given by

$$\mathbf{E} = \mathbf{u} \times \mathbf{B} \tag{13.4}$$

Although the above pair of equations are by no means complete since they ignore both electrical and viscous dissipation, boundary conditions, and the time dependent effects of the electromagnetic fields described by Maxwell's equations, the general picture provided by (13.3) and (13.4) of a reciprocal set of transduction equations (13.11) is basically correct. Schemes that exploit this effect for measurement of quasi-static flow velocity (13.12) and as a fluid pumping mechanism (13.13) have been in existence since at least as early as 1919 (13.14) and include a 1931 British patent issued to Albert Einstein and Leo Szilard (13.15).

The earliest applications of this effect to acoustical transduction were by Campanella (13.16) in 1955 and by Adjisaka and Hixson (13.17,13.18) in 1975. A brief flurry of interest was generated in the use of this effect to detect sound waves in the ocean by measuring the electric field generated by acoustic fluid oscillations in the Earth's magnetic field when two experiments (13.19,13.20) in the early 1980's observed signals which were an order-of-magnitude larger than that predicted by (13.4). These results were later shown to be an experimental artifact involving the motion of the detecting electrodes within the magnetic field (13.11,13.21). The development of thermoacoustic heat engines (13.22), which will be described in following sections, also contributed to the interest in MHD acoustic transduction (13.11,13.23) when thermoacoustic engines using liquid metals as a working fluid were developed for direct conversion of thermal to electrical energy.

The work of Adjisaka and Hixson (13.18) was extended to produce an under-water sound source by Shreppler and Busch-Vishniac (13.25). Their projector used two gold-plated copper electrodes that were 10.0 cm long and 3.81 cm high, separated by 3.18 cm with a magnetic field within the electrodes of 4.3 kgauss (0.43 tesla). The space between the electrodes was filled with a 6% NaCl-water solution that was confined within the transducer by polyurethane acoustic windows. The transmitting current response of the transducer increased approximately linearly with frequency, as would be expected for a dipolar source. In comparison, the standard USRD Type J-11 electrodynamic projector (13.26) has about a 120 dB advantage over the Schreppler/Busch-Vishniac transducer after correcting for the relative area difference between the two systems. The above comparison does not take into account the higher magnetic field strength in the gap of the J-11, which is a result of better magnetic structure and a much lower gap reluctance due to its much smaller size. It also ignores the difference in the maximum electrical power handling capacity of the two transducers. Although this MHD projector produced experimental results in good agreement with theory, the transduction efficiency makes such a design practically useless as a high power, lower frequency projector.

In order to estimate the ultimate potential for such an MHD-powered transduction scheme, it is necessary to recognize that, unlike the conventional displacement-limited (i.e., non-compliant) transduction mechanisms that produce their maximum output driving an acoustical load of high impedance, an MHD transducer is ultracompliant (13.11), having fixed force and unlimited displacement. Such an ultracompliant transducer would therefore have its optimum output when driving an acoustical load of minimum impedance. Having recognized the orthogonality of these two mechanisms with respect to the optimal load, an experiment was performed at the Naval Postgraduate School that place the MHD generator at the neck of a novel Helmholtz resonator (13.27) the inertia of which was controlled by the mass of the fluid in the neck but the elasticity of which was controlled by the dilatational modulus of a polyvinylchloride (PVC) plastic body. In addition to reducing the impedance of the load seen by the MHD generator, this geometry had the additional advantage of producing a monopolar source when the PVC body was enclosed by the projector case. It also allowed the transducer to resonate at a much lower frequency than would have been possible if the Helmholtz resonator's restoring force were provided by the bulk modulus of water contained in a rigid container of the same volume.

The magnetohydrodynamically driven Helmholtz projector constructed for this test consisted of a cylindrical pipe of PVC approximately 20 cm long and 5 cm in diameter. A piezoelectric hydrophone (Celesco LC-10) was placed within this volume to monitor the magnitude of the internal pressure oscillations. The neck of the resonator was placed within the pole pieces of a permanent magnet and was constructed from 3 mm thick plexiglass. It had a rectangular cross-section of 1 cm in the magnet gap by 3.8 cm and contained a water column 10 cm long. The magnetic field in the gap was measured to be 2.5 kgauss (0.25 tesla).

Figs. 13.1 and 13.2 provide a schematic diagram of the transducer. Figure 13.1 shows the neck and pole-piece geometry as viewed in cross-section through the neck and Fig. 13.2 shows the arrangement of the neck, cylindrical PVC volume, and horseshoe magnet as viewed from above the neck looking down toward the PVC volume.

Measurements of volumetric velocity as a function of current were made both by direct observation of the motion of the free surface of the water in the neck when operated in air using a fiber-optic displacement probe (MTI Fotonic™ Sensor) and by measurement of the pressure within the volume using the calculated acoustic impedance to relate volumetric velocity through the neck to pressure in the volume. These two methods were found to agree to within about 20% and produced a measured source strength of $5 \pm 1 \times 10^{-6}\ m^3$/A-s.

This value can be used to calculate the freefield transmitting current response from (13.2) and the fact that $c = f\lambda$. If one compensates for the smaller radiating area (3.8 cm^2 vs. 24 cm^2 → 16 dB) and lower magnetic field strength (2.5 kgauss vs. 4.3 kgauss → 5 dB) of the MHD Helmholtz projector relative to the Schreppler/Busch-Vishniac design, the Helmholtz design has a normalized transmitting current sensitivity advantage of over 100 dB ! This is still nearly 20 dB lower than the Type J-11 (again neglecting magnetic field strength, which is about 20 dB higher in the J-11), and only gets this close by operating at resonance. This result does demonstrate that a projector design that is more carefully tailored to the peculiarities (*e.g.*, ultracompliance) of the MHD transduction mechanism could approach practical efficiencies although considerations of weight (due to the magnets) and maximum current capacity (from Joule heating of the salt water) have still to be considered.

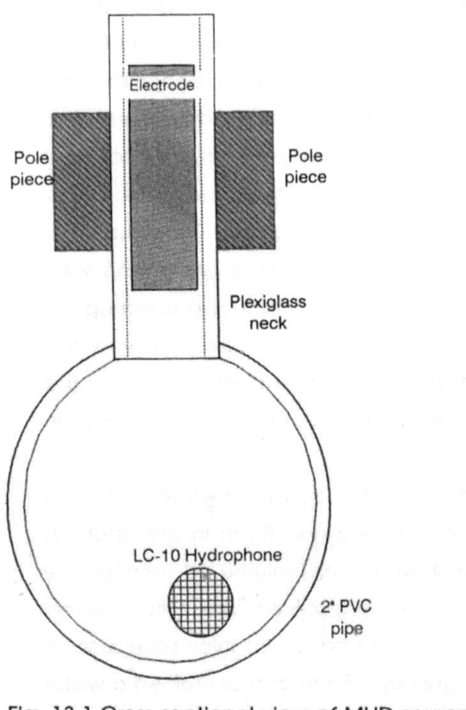

Fig. 13.1 Cross sectional view of MHD source

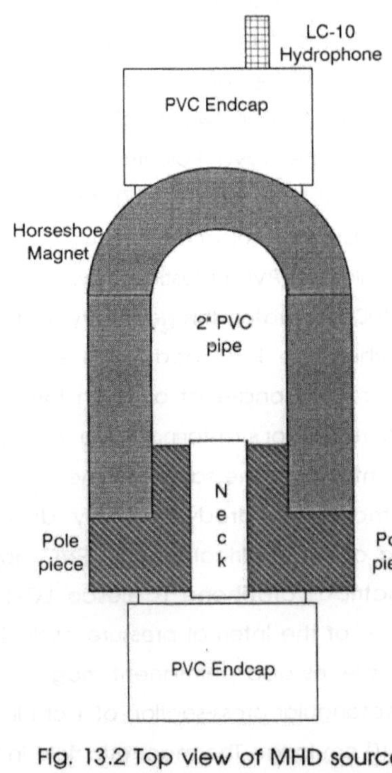

Fig. 13.2 Top view of MHD source

Even given improved magnets such as Nd-Fe-B or possibly high-temperature superconductors, one would still have to provide large amounts of electrical power, just as is required of conventional piezoelectric or magnetostrictive transducers. High energy densities are easily available in chemical fuels; however, for an electroacoustic source, the overall transduction efficiency would have to be reduced by the efficiency of the conversion of thermal energy to electrical energy. The remainder of this paper will explore the direct conversion of heat energy into sound in a standing-wave resonant thermoacoustic engine.

13.3 THERMOACOUSTIC SOUND GENERATION

Thermoacoustic generation of sound is made possible by establishing a thermodynamic cycle in which the acoustic vibration itself provides the mass transport and a second stationary medium controls the heat transport. This process exploits the particular phase relationship between the local pressure oscillations and the local oscillations in elemental volume (or temperature) in an acoustic wave. By appropriately positioning the second medium in the acoustic flow and imposing a large temperature gradient, heat can be transferred to and from the cycle with the phase required to generate high-amplitude oscillations.

The thermal interaction between an acoustic wave in a primary fluid and a stationary second medium has been observed and modeled by Rayleigh (13.28), Weston (13.29), Merkli and Thomann (13.30) and Rott (13.31,13.32) both for heat pumping along the second medium (a tube wall, for example) and for acoustic generation. More recently, the design of the second medium has been optimized for sound generation by Wheatley, Hofler, Swift, and Migliori (13.22,13.23,13.33-13.36) and some acoustic sources capable of substantial power outputs have been constructed (e.g., 13.36). The history of thermoacoustics is reviewed in many of these references.

A thermoacoustic source designed and built at the Naval Air Development Center for radiating sound underwater is shown in Fig. 13.3 and, in exploded view, in Fig. 13.4. The device consists of a stack of parallel plates (the stationary second medium) inside an air-filled acoustic resonator. The stack is thermally attached at the upper end to either ambient air or water or a hot-water reservoir and at the lower end to a liquid-nitrogen reservoir. The resonator is a brass tube of rectangular cross-section closed at the top and joined to an acrylic tube that is open at the bottom. The device is designed to be completely submerged once the liquid-nitrogen reservoir is filled; the internal air (or gas) pressure is compensated to keep the air-water interface near the bottom of the acrylic tube. The acoustic energy radiates from the open end of the acrylic tube. Since the dimensions of the open end are much smaller than an acoustic wavelength in water, the radiation is essentially omnidirectional and adequately described by equations (13.1) and (13.2) once the motion of the air-water interface is known.

Liquid nitrogen was used in this model because it is safe, easy to handle and permits the use of simple construction techniques. In practice, a high-power source would probably use an exothermic chemical fuel or waste heat from some other process on the hot side and sink the cold side to the ocean. (If size, weight, and expense of the power source are not critical, electrical resistance heating could also be used on the hot side.) However, it is not the absolute temperature of the hot side that is important; instead, the critical factors are the temperature gradient maintained on the stack plates and the ratio of hot to cold (absolute) temperature.

The essential elements of a thermoacoustic source are: some mechanism for providing the required phasing for heat transfer, a suitable working fluid, an appropriate second medium to support the temperature gradient, some means for radiating sound, and "hot" and "cold" temperature reservoirs. The remainder of this paper will review these aspects after a brief introduction to the theory of thermoacoustic generation of sound.

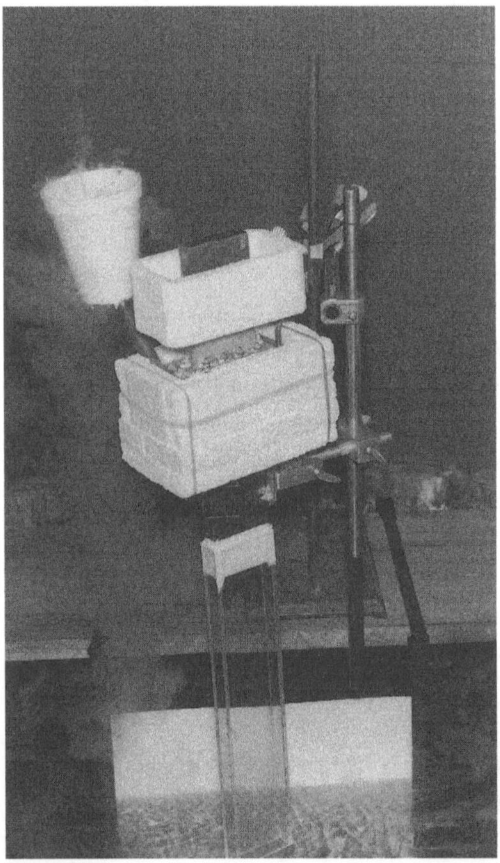

Fig. 13.3 A thermoacoustic source designed for radiation underwater

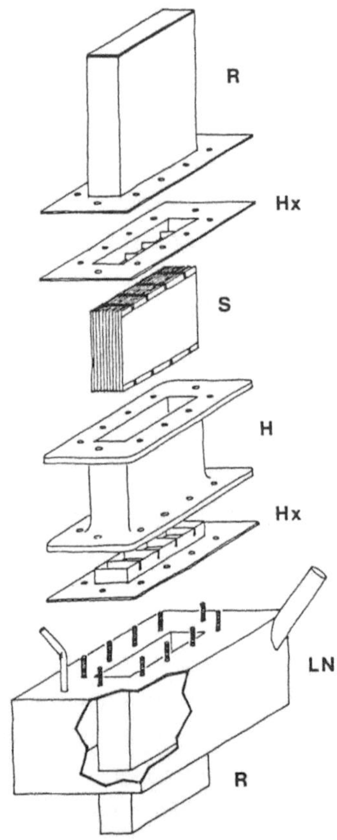

Fig. 13.4 Exploded view of upper section of source. Shown are: resonator tube (R), heat exchangers (Hx), stack (S), stack housing (H), and liquid-nitrogen reservoir (LN)

The prototype shown in Fig. 13.3 is a standing-wave thermoacoustic oscillator. The resonator encourages the formation of a standing-wave. In this standing-wave, the working fluid oscillates along the tube axis moving alternately toward and away from the closed end (Fig. 13.5a). As a particle of the fluid moves toward the closed end it increases in pressure, decreases in volume, and increases in temperature; as this particle moves toward the open end the opposite changes occur (Fig. 13.5b). Except at extremely high frequencies or near boundaries, acoustic waves are very nearly isentropic; consequently, the cycle curve on the P-V diagram encloses zero area: there are no net energy transfers to or from the fluid particle.

If the fluid movement occurs immediately adjacent to a surface having a temperature gradient, then the fluid temperature oscillation is affected as shown in Fig. 13.5c. Because the heat transfer takes place virtually instantaneously, the P-V curve still encloses no area and there is no net energy transferred into the acoustic oscillation. Farther from the surface, the thermal conduction takes time so there is a phase lag in the energy transfer. As long as the gradient on the stationary surface is large enough so that the heat transfer takes place from the hot side to the compressed fluid and from the expanded fluid to the cold side, the phase lag is in the proper direction to add energy per cycle to the acoustic flow (Fig. 13.5d). The length scale for this effect is the thermal penetration depth, δ_k given by

$$\delta_k^2 = 2k / \omega \rho c_p \tag{13.5}$$

where k is the thermal conductivity, ω is the radian frequency, ρ is the density, and c_p is the specific heat at constant pressure of the fluid.

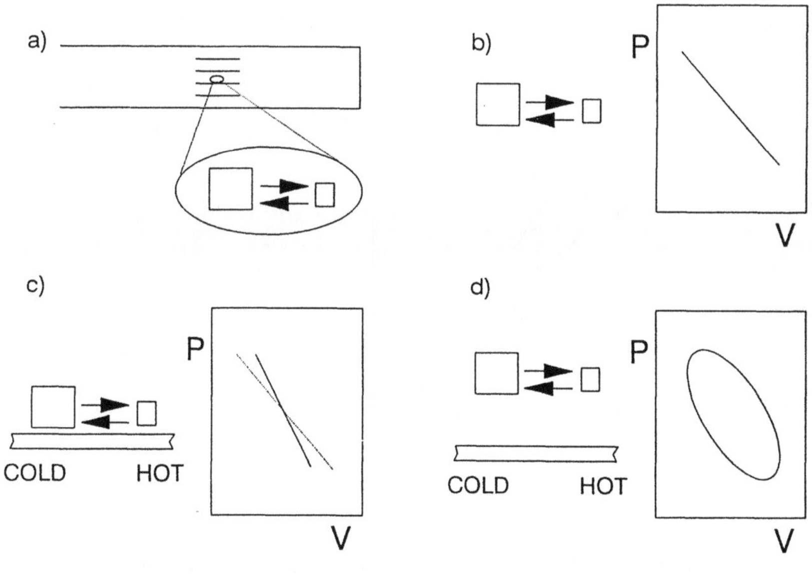

Fig. 13.5 Schematic illustration of thermoacoustic sound generation: a) fluid motion in a standing wave, b) isentropic cycle of acoustic oscillation, c) instantaneous heat transfer, d) delayed heat transfer through a distance of about a thermal penetration depth

The surface must be oriented so that the hot end is toward the closed end of the resonator (in the direction of temperature increase for a fluid particle). Also, the temperature gradient on the plate must be greater than the equivalent gradient calculated from the total adiabatic temperature change in a fluid particle divided by the peak-to-peak displacement of that particle. In the macroscopic view, the "individual" fluid particles act as many overlapping cycles and produce a net heat transport down the stack (in addition to normal longitudinal conduction) and net acoustic work (see Fig. 13.6).

As shown below, instantaneous transfer of heat to the acoustic flow would neither enhance nor degrade the cycle but the natural time required for thermal conduction from the plate some distance into the flow properly delays the temperature changes with respect to the pressure changes resulting in energy input to the acoustic cycle. The useful distance away from the plate over which the influence of the plate is significant is about two times the thermal penetration depth in the fluid. Unfortunately, for most common fluids, thermal conduction and viscosity are intimately related and their effects are of roughly the same order. While thermal conduction produces a lag in fluid temperature, viscosity causes a lag in the fluid motion (as well as a reduction in amplitude) that reduces the effective phase shift in the cycle. This also reduces the energy transfer into the acoustic oscillation.

The length scale for viscous effects is the viscous penetration depth δ_v given by

$$\delta_v^2 = 2\mu / \omega\rho \tag{13.6}$$

where μ is the fluid viscosity. The relevant fluid parameter for thermoacoustics is the Prandtl number, which expresses the ratio of viscous effects to thermal conduction effects:

$$\sigma = \delta_v^2 / \delta_k^2 \tag{13.7}$$

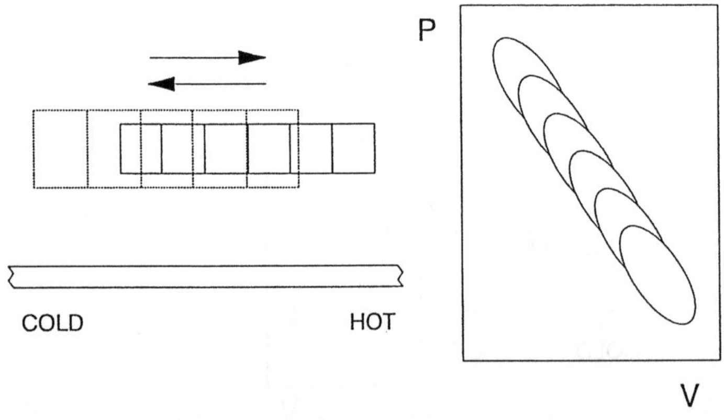

Fig. 13.6 Macroscopic thermoacoustic cycle composed of many overlapping fluid particle cycles

For ideal gases, the Prandtl number is about 0.7. Room temperature air has a Prandtl number close to this value and several thermoacoustic oscillators have been made that use air as the working fluid so the Prandtl number does not have to be much smaller than one for the fluid to be useable.

From the above explanation, it should be evident that it is easy to provide the proper conditions for thermoacoustic oscillation. However, since a standing wave is required to provide the proper base phasing of the cycle, the device has a naturally high Q with small amounts of power radiated from the open end relative to the energy per cycle in the standing wave. This means that control or modulation of the acoustic output without mechanically changing the configuration of the resonator would be difficult.

In order for the standing wave oscillator to radiate any energy at all, there must be a small travelling-wave component in the direction of the open end. The hot end of the second medium is toward the closed end, which attenuates the travelling-wave component somewhat. The proper configuration for amplifying travelling waves is to arrange the hot end of the second medium to be in the direction of travel (13.37). This conflicting requirement can be resolved by using a gas-filled resonator underwater. In this case, the open end of the resonator is almost a velocity node (see below for a discussion) so the fundamental resonance is a half-wavelength resonance and the stack can be placed near the open end with the hot end both in the direction of the travelling-wave component and toward the pressure maximum of the standing wave. In fact, two stacks with opposing temperature gradients can be used in this resonator, one toward the closed end and the other toward the open end.

An acoustic standing wave can be set up in any fluid but, in order to support thermoacoustically driven oscillations, the working fluid must change in specific volume with changing temperature and the thermal penetration depth must be larger than the viscous penetration depth so that the phase lag in temperature is not cancelled by the phase lag of the fluid motion. In particular, the normalized coefficient of volume expansion $T\beta$ should be large and the Prandtl number should be less than one. Most gases at temperatures well above their liquification temperature have $T\beta$ of order one and Prandtl number slightly less than one, which are both acceptable. Mixtures of gases with widely differing molecular weights and fluids with free-electron conduction (liquid metals) have very low Prandtl numbers; however, liquid metals, at least, would probably not be practical for an underwater sound source.

In theory, either a liquid or a gas could be made to function as the working fluid. Several gases have been used successfully (air, several of the inert gases singly or as mixtures) and a successful thermoacoustic source has been constructed using liquid sodium as the working fluid (13.36). Other than the liquid sodium source, liquids have not generally been exploited. Liquid water at room temperature has a prohibitively large Prandtl number (7 at 20°C) and a small $T\beta$ (zero at 4°C !); however, its properties improve substantially with elevated temperature (13.38). At atmospheric pressure, the temperature range from melting to boiling is only 100°C for water; consequently, a useful water-based source would probably have to be pressurized (that is, operated at some reasonable depth). In addition, the thermal penetration depth in water is very small at typical frequencies, which means that the optimum plate spacing in the stack would be much smaller than for a gas-filled source. Operation near the critical point of a fluid may be advantageous since the expansivity is large there; however, no attempt has been made to exploit this yet. ("Near the critical point" is necessarily vague since the range of temperature in the stack is typically several hundred Kelvin.)

Many gases would be suitable as working fluid for an underwater source. Water vapor could be produced by boiling the seawater out of the resonator but this scheme has several drawbacks: If the upper temperature of operation is limited (by materials, for example), then the temperature range of the water vapor engine is smaller than for other gases because at least 100°K is used to maintain the vapor. This increase in cold-side temperature becomes greater with depth and would reduce both the maximum power and the efficiency. In addition, the vapor-liquid interface at the exit of the resonator would be subject to large pressure variations around the saturation pressure (perhaps 10% of hydrostatic pressure) leading to cyclic condensation/evaporation and probable relaxation losses. Corrosion could also be a problem for water vapor as a working fluid.

Alternately, once the source has been deployed to its operating depth, a gas could be generated by chemical reaction to drive the seawater out of the resonator. If a mixture of gases with widely different molecular weights could be generated, then a reduction in the Prandtl number may be possible. Of course, a "neutral" gas could be supplied at hydrostatic pressure, however, this would not be practical at any but very shallow depths.

The requirements for the second medium (making up the stack) are straightforward. Since the stack plates must act as constant temperature reservoirs (locally anyway), the plates must be selected so that the following relationships hold (13.22):

$$\varepsilon = \sqrt{\frac{(k\,\rho\,C_p)_{\text{fluid}}}{(k\,\rho\,C_p)_{\text{plate}}}} \ll 1 \qquad (13.8)$$

and (plate thickness) >> $\delta_{k\text{-plate}}$.

If the working fluid is a gas, the stack material can be selected to have a much higher ρc_p than the gas. This allows use of materials with small thermal conductivities thereby reducing heat losses from normal conduction along the plates. If the working fluid is water, this is not generally possible, so a material with a relatively high thermal conductivity must be selected and the conduction losses must be accepted as part of the design.

The thermoacoustic source radiates through the open end of a resonator. The boundary condition here affects both the form of the standing wave inside the resonator and the efficiency with which the resonator radiates. If the resonator is filled with the same fluid as that into which it is immersed, then the fundamental resonance is that of a closed-open tube: approximately one-quarter wavelength. On the other hand, if the tube is gas filled, then the fundamental resonance is about one-half wavelength where the wavelength is calculated in the gas. The gas-filled resonator is not intuitively attractive since the "open" end of such a resonator would be almost a node for fluid motion and there is a large acoustic impedance difference between water and low-pressure gases. This suggests poor radiation performance but this intuition is misguided.

As long as the exit is small with respect to a wavelength (a wavelength outside the tube, that is), the radiated power is (13.39),

$$E_{RAD} = \frac{\pi R^2}{8} (k_1 R)^2 \rho_1 c_1 U_{EXIT}^2 \tag{13.9}$$

The curve in Fig. 13.7 shows the ratio of radiated power for a source with the given impedance ratio (outside/inside) to radiated power for a source having the same impedance inside and out. While the ratio of exit speed to internal speed amplitude (in the standing wave) is low for a tube with a large impedance ratio, the internal speed amplitude itself increases with the impedance ratio; the net effect is a substantially greater radiated power as long as the pressure amplitude inside the tube is kept constant. The case shown in Fig. 13.7 is for a cylindrical resonator with an exit radius given by kR = 0.2 where k is ω/c in the external medium.

For two resonators, then, one gas-filled with a large acoustic impedance ratio and one water-filled, operating at the same internal pressure amplitude:

- the internal particle speed amplitude is much larger in the gas-filled tube;
- the stored energy in the resonator is much larger in the gas-filled tube;
- the ratio of exit speed to speed amplitude is smaller in the gas-filled tube; however,
- for low hydrostatic pressure, a gas-filled resonator radiates more power than a liquid-filled by a factor of about $1 / (0.36 (kR)^2)$.

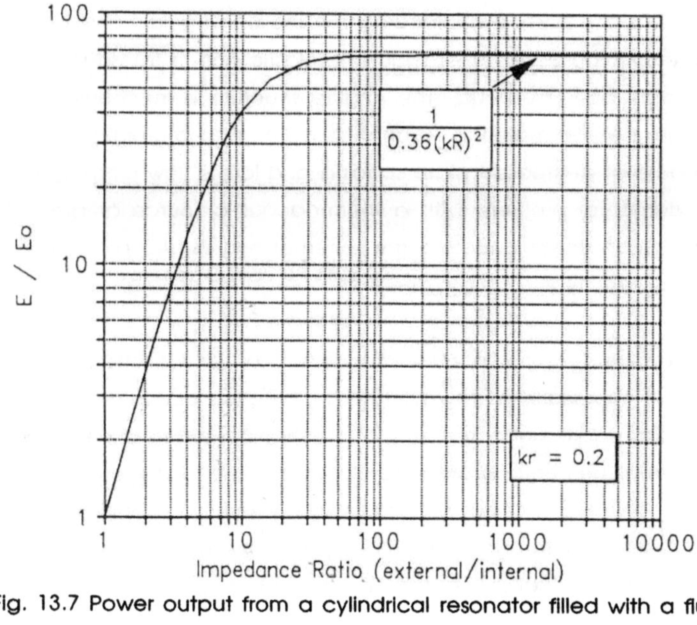

Fig. 13.7 Power output from a cylindrical resonator filled with a fluid having a smaller specific acoustic impedance than the fluid into which it is radiating

From the standpoint of total radiated power, the gas-filled source is preferable. These characteristics are independent of the mechanism used to drive the resonator.

The ultimate efficiency of a thermoacoustic source depends on the ratio of the absolute temperatures of the hot and cold reservoirs, the rate of energy extraction from the cycle, and the effects of irreversibilities in the cycle. Because of materials limitations, the hot-to-cold temperature ratio would be unlikely to exceed 3.5 (275°K cold side and 1000°K hot side). This gives a zero-power (Carnot) efficiency of 0.7. By allowing for the effects of the heat exchangers and the desire for maximum power, the efficiency reduces to 0.45 (13.40). The gradient on the stack plates must exceed the critical gradient (that gradient corresponding to the adiabatic temperature change of the fluid) by some factor in a real source; for a gradient-to-critical-gradient ratio of two, the efficiency would be reduced to about 0.2 (13.22). When other losses (such as viscous losses in the stack, thermal relaxation and viscous losses in the resonator, and normal heat conduction losses) are considered, this efficiency may be reduced to 0.1.

The efficiency will determine the size (and weight) of fuel required for a given energy output. One of the advantages of thermoacoustics is that high-energy exothermic chemical fuels can be used. Some of these fuels have energy densities of 20×10^9 J/m^3 (13.41) whereas high-current lithium/sulfur-dioxide batteries have energy densities of about 0.2×10^9 J/m^3.

While the design of the heat exchangers is probably important, it is not clear how critical this is for at least two reasons: first, the theory of lateral heat transfer in the stack resulting from nonuniform heating of the edges is not well understood; and, second, the performance calculations for these devices (reference 13.22, for example) are very sensitive to the heat exchanger performance and real heat exchangers are difficult to model.

Since the power output is limited by the particle displacement in the working fluid and that displacement is limited by linearity or turbulence rather than any elastic limit in transduction material, the potential exists for large power output. The engine structure is mechanically simple and probably would not require exotic materials. Except for an increasing problem with stack impedance matching (because of decreased plate spacing and larger flow blockage) in the resonator, there is no fundamental problem with a thermoacoustic source at great depths. (The pressure inside the resonator is equal to the external hydrostatic pressure.) Furthermore, chemical fuels have very high energy densities compared to high-current batteries.

On the negative side, the requirement for a resonator has two serious effects on performance: the high-Q resonator is necessarily narrow bandwidth therefore acoustic control would either be limited or rely on the additional complexity of mechanical controls; and the radiating surface must be a rather small percentage of the total surface area of the resonator, which limits the maximum power achievable. Also, a fluctuating-pressure mode has been observed in a gas-filled engine radiating into water (13.42). As the depth of operation is increased, larger stack blockage and internal reflections may be a problem.

Beyond the problems of efficiency and heat exchangers, if the source is intended to be used intermittantly, then some means for control of the heat flow must be devised unless the supply of heat is unlimited.

Initial tests of the prototype thermoacoustic source give a radiated acoustic power of 40 mW at a depth of 7 meters. While this is not, in itself, impressive, it is roughly the level predicted for that source by the radiation theory (13.39). Projecting these results to a source with a larger cross-sectional area and operated at, for example, 5 MPa hydrostatic pressure, yields an acoustic power in the tens of megawatts.

Although these initial results are promising, there is clearly much work to be done -- both theoretical and experimental -- before the thermoacoustic source can be considered as a practical means for generating large power output at low frequency underwater. While the theory for thermoacoustic behavior near an isolated surface is well developed, the more practical case in which two surfaces are within a few thermal penetration depths has not been adequately treated.

The minimum Q required for efficient operation should be established as this will determine the modulation bandwidth (without mechanical valves or slides) and the amount of power required to phase lock or modulate the output. Also, the influence of the stack impedance on the standing wave should be ascertained and, if possible, some means of matching the impedance to the resonator should be explored. This will be particularly important for sources designed for greater depths. (The increase in gas density with depth decreases the thermal penetration depth so the optimum plate spacing is reduced meaning more stack interference in the flow.)

These questions primarily reflect the fact that the field of thermo-acoustics as applied to underwater generation of sound is in its infancy. The potential advantages are, however, fundamental: the fluid displacements that can be achieved are only limited by turbulence or nonlinearity; and, chemical fuels with very high energy densities can be used without intermediate conversion to electrical power.

REFERENCES

13.1 J. Lighthill, *Waves in Fluids*, (Cambridge, 1978), S1.4.

13.2 I. Rudnick, "Unconventional reciprocity calibration of transducers," J. Acoust. Soc. Am., **63**, 1923 (1978).

13.3 J.W. Strutt (Lord Rayleigh), "The photophone," Nature **23**, 274 (1881).

13.4 A.G. Bell, "Upon the production of sound by radiant energy," Philos. Mag. (Ser. 5) **11**, 510 (1881).

13.5 L.L. Beranek, *Acoustic Measurement*, (Wiley & Sons, New York, 1949), p. 161.

13.6 H.D. Arnold, and I.B. Crandall, "The thermophone as a precision source of sound," Phys. Rev. **10**, 22 (1917).

13.7 W.M. Wright, 'Generation of sound within a closed cell by an alternating current in a straight wire,' J. Acoust. Soc. Am., **82**, 654 (1987).

13.8 L.M. Lyamshev, and L.V. Sedov, 'Optical generation of sound in a liquid: Thermal mechanism (review),' Sov. Phys. Acoust. **27** (1), 4 (1981).

13.9 Y.H. Berthelot, and I.J. Busch-Vishniac, 'Laser-induced thermoacoustic radiation,' J. Acoust. Soc. Am., **78**, 2074 (1985).

13.10 D.D. Holm, 'Hall magnetohydrodynamics: Conservation laws and Lyapunov stability,' Phys. Fluids **30**(5), 1310 (1987).

13.11 G.W. Swift, and S.L. Garrett, 'Resonance reciprocity calibration of an ultracompliant transducer,' J. Acoust. Soc. Am., **81**, 1619 (1987).

13.12 A. Kolin, 'An alternating field induction flow meter of high sensitivity,' Rev. Sci. Inst. **16**, 109 (1945).

13.13 R. Panholzer, 'Electromagnetic pumps,' Electrical Eng., **82** (2), 128 (1963).

13.14 J.F.G.P. Hartmann, 'Improvements in or relating to apparatus for producing a continuous electrically conducting liquid jet,' British Patent Specification No. 126,947 (December. 1919).

13.15 A. Einstein, and L. Szilard, 'Pump, especially for refrigerating machine,' British Patent Specification No. 344, 881 (1931).

13.16 A.J. Campanella, 'Generation and detection of acoustic waves by the Hall effect in electrolytes,' Ph. D. Dissertation, The Pennsylvania State University, June, 1955.

13.17 H. Ajisaka, and E.L. Hixson, 'Liquid state electromagnetic transducer,' Acoustic Research Laboratory, Technical Report No. 172, Dept. of Elect. Eng., U. Texas, Austin, June, 1975.

13.18 H. Ajisaka, and E.L. Hixson, 'Transduction principles of a liquid state electromagnetic transducer,' J. Acoust. Soc. Am., **77**, 1933 (1985).

13.19 V.V. Bogorodskiy, A.V. Gusev, A.M. Zufrin, A.P. Polyakov, A.N. Ponomarev, and B.M. Stepanov, 'Synchronous measurements of infra-low-frequency acoustic and electromagnetic waves in a marine medium,' Arktic. Antarktic. Nauchno-Issledovatel Inst. **374**, 80 (1980).

13.20 P.H. Moose, and R.F. Klaus, 'Experimental observations of magneto-acoustic fields,' J. Acoust. Soc. Am., **74**, 1066 (1983).

13.21 A.D. Pierce, 'Discrepancy between experiment and theory for magneto-acoustic fields in salt water,' J. Acoust. Soc. Am., **75**, S35 (1984).

13.22 G.W. Swift, 'Thermoacoustic engines,' J. Acoust. Soc. Am., **84**, 1145 (1988).

13.23 G.W. Swift, A. Migliori, T. Hofler, and J. Wheatley, 'Theory and calculations for an intrinsically irreversible acoustic prime mover using liquid sodium as primary working fluid,' J. Acoust. Soc. Am., **78**, 767 (1985).

13.24 G.W. Swift, 'A liquid-metal magnetohydrodynamic acoustic transducer,' J. Acoust. Soc. Am., **83**, 350 (1988).

13.25 S.C. Schreppler, and I.J. Busch-Vishniac, 'Magnetohydrodynamic underwater acoustic transducers,' J. Acoust. Soc. Am., accepted for publication; see also S.C. Schreppler,

Technical Report No. ARL-TR-86-30, Applied Research Laboratories, U. Texas, Austin, December, 1986.

13.26 *Underwater Electroacoustic Standard Transducer Catalog*, (Transducer Branch, Naval Research Laboratory, Underwater Sound Reference Detachment, Orlando, FL, May, 1982), p. 129-132.

13.27 C.E.D. Haney, J.T. Newmaster, and S.L. Garrett, "A magnetohydrodynamically driven Helmholtz resonant projector," J. Acoust. Soc. Am., **83**, S19 (1988).

13.28 J.W. Strutt (Lord Rayleigh), *Theory of Sound*, (Dover Publications, New York, 1945.), Vol. II, S322g-j.

13.29 D.E. Weston, "The theory of the propagation of plane sound waves in tubes," Proc. Physical Society, London, **B66**, 695 (1953).

13.30 P. Merkli, and H. Thomann, "Thermoacoustic effects in a resonance tube," J. Fluid Mech. **70**, 161 (1975).

13.31 N. Rott, "Thermoacoustics," in Advances in Applied Mechanics, ed. Chia-Shun Yih, (Academic Press, New York, 1980) Vol. 20, p. 135.

13.32 N. Rott, "Thermoacoustic heating at the closed end of an oscillating gas column," J. Fluid Mech. **145**, 1 (1984).

13.33 J.C. Wheatley, and A. Cox, "Natural engines," Physics Today, **38**(8), 50 (1985).

13.34 J C.Wheatley, T. Hofler, G.W. Swift, and A. Migliori, "Understanding some simple phenomena in thermoacoustics with applications to acoustical heat engines," Am., J. Phys. **53**, 147 (1985).

13.35 J C. Wheatley, G.W. Swift, and A. Migliori, "The natural heat engine," Los Alamos Science **14** (Fall 1986).

13.36 A. Migliori, and G.W. Swift, "Performance of a liquid-sodium thermoacoustic engine," J. Acoust. Soc. Am., **84**, S37 (1988).

13.37 P.H. Ceperley, "A pistonless Stirling engine -- the traveling wave heat engine," J. Acoust. Soc. Am., **66**, 1508 (1979).

13.38 G.W. Swift, and A.M. Fusco, "Seawater as a working fluid for thermoacoustic engines," J. Acoust. Soc. Am., **84**, S37 (1988).

13.39 T.B. Gabrielson, "Radiation performance of a submerged thermoacoustic source," J. Acoust. Soc. Am., **85**, S48 (1989).

13.40 F.L. Curzon, and B. Ahlborn, "Efficiency of a Carnot engine at maximum power output," Am., J. Phys. **43**, 22 (1975).

13.41 Herbert Ellern, *Military and Civilian Pyrotechnics*, (Chemical Publishing Co., New York, 1968), tables 22, 25, 26.

13.42 T.B. Gabrielson, "Transition from acoustic oscillation to irregular pressure fluctuation in a thermoacoustic engine," J. Acoust. Soc. Am., **87**, S85 (1990).

14. A Low-Frequency, Tow-Powered Sound Source

J.E. Blue, A.L. Van Buren
Naval Research Laboratory, Underwater Sound Reference Detachment, P.O. Box 568337, Orlando, Florida 32856-8337, USA

LT P.A. Semper
Navy Nuclear Power School, Orlando, Florida, USA

This paper discusses resonant cavities and their potential sound pressure output when they are modulated hydrodynamically. Rough calculations show that in the frequency range of 10 to 100 Hz source levels in excess of 200 dB re 1 μPa should be achievable. Experimental data from three suboptimum devices showed source levels of 182 dB at 16.6 and 50 Hz and 168 dB at 34.5 Hz. It is expected that with appropriate designs one can achieve greater than 200 dB from 10 to 100 Hz with fairly small, lightweight devices. Further, with innovative designs one should be able to obtain frequency diversity and perhaps broad bandwidth with a single device.

14.1 INTRODUCTION

Sound generation is known to be inefficient when the radiating device is small compared to a wavelength of the frequency one wishes to generate. Production of high sound pressure levels at low frequency from devices small compared to a wavelength requires large amounts of acoustic power. For example, to produce 200 dB re 1 μPa at 10 Hz requires 1 kW of acoustic power and a volume displacement of 10^4 cm^3. With efficiencies of the order of 1% for many low-frequency devices, the power amplifier requirement would be about 100 kW. If one could develop a suitable tow-powered source to derive power primarily from a ship's propulsion system, the power might be available without having to provide extra generating equipment on-board the vessel. A vessel such as the U.S. Naval Ship Stalwart (TAGOS-1) has a power plant that delivers 3,200 brake horsepower or about 2 MW of power. This vessel would be capable of towing such a device with about 5% of its power going to sound generation. A highly resonant device might do even better. Here we will consider resonant air cavity devices and give some results from theory and experiment.

14.2 SOME CONSIDERATIONS FOR PULSATING SPHERES AND RESONANT BUBBLES

From the first paper in this meeting by Timme, Young, and Blue, let us examine Fig. 14.1. At 10 Hz a spherical source driven uniformly over its area would require 1 kW of acoustic power in the water to achieve a sound pressure level of 200 dB re 1 μPa. The question now is whether or not that source level is achievable by causing the flow past a 10 Hz resonant bubble to sinusoidally modulate the pressure field in the fluid surrounding the bubble. A candidate source design that we examined previously (14.1) is shown schematically in Fig. 14.2. In this case we attached a rubber membrane to a rigid hemisphere to produce the resonant bubble or air cavity. The use of a rigid hemisphere was intended to reduce multipole excitation of the resonant cavity. Our objective was to produce a robust, inexpensive sound source capable of high (>200 dB) sound pressure levels at low (<100 Hz) frequency and whose primary source of power is derived from being towed through the water. The approach was to investigate theoretically and experimentally a new sound source concept involving a resonant air cavity excited by a modulated flow field.

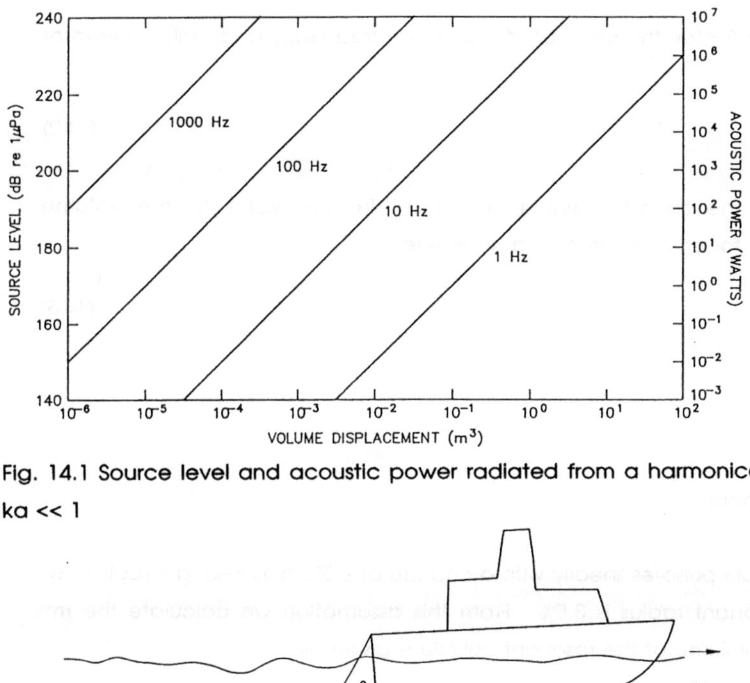

Fig. 14.1 Source level and acoustic power radiated from a harmonically pulsating sphere with ka << 1

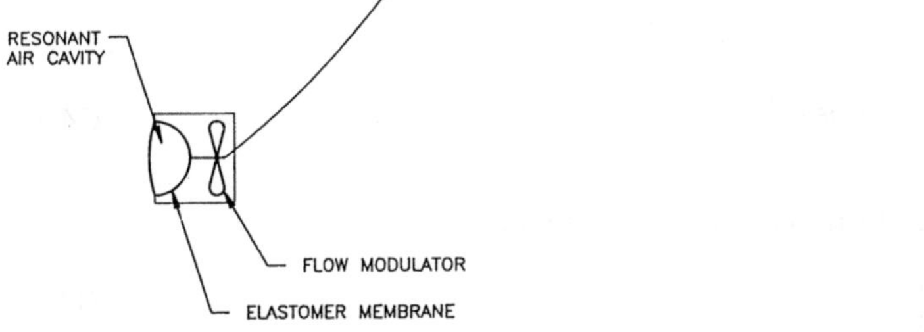

Fig. 14.2 Tow-powered, low-frequency sound source

Let us now examine the size and volume displacement necessary to produce this 200 dB source level from a resonant bubble. They are determined from Minnaert's equation (14.2) for a resonant gas bubble in water:

$$f_0 r_0 = \sqrt{\frac{3 \gamma P_h}{4 \pi^2 \rho}} \qquad (14.1)$$

where:

f_o, resonant frequency,

r_o, radius at resonance,

γ, adiabatic constant,

P_h, hydrostatic pressure,

ρ, density.

Using appropriate conversion factor, the product of resonance frequency and radius in term of water depth h is:

$$f_0 r_0 = 3.26 \sqrt{\frac{h + 9.8}{9.8}} \qquad (14.2)$$

The relationship between the sound pressure produced in the water to the volume displacement of the bubble, for an omnidirectional source is:

$$|P|_{1m} = \frac{\omega^2 \rho}{4\pi} \delta V \qquad (14.3)$$

with:

ω, angular frequency,

ρ, density,

δV, volume displacement.

If it is assumed that the bubble pulsates linearly with a change of ± 5% in its resonant radius; i.e., the rms change in the resonant radius is 3.5%. From this assumption we calculate the rms volume displacement. The volume of the resonant bubble is given by:

$$V_0 = \frac{4}{3} \pi r_0^3 \qquad (14.4)$$

so that

$$(\delta V)_{3.5\%} = \frac{4}{3} \pi \left[(1.035)^3 - 1\right] r_0^3 = 0.457\ r_0^3 \qquad (14.5)$$

Combining these equations, we obtain the acoustic pressure:

$$|P|_{1m} = \frac{4.95 \times 10^4}{f_0} \left[\frac{h + 9.8}{9.8}\right]^{3/2} \qquad (14.6)$$

the sound pressure level is:

$$SPL = 213.9 - 20 \log f_0 + 30 \log\left[\frac{h+9.8}{9.8}\right] \text{ dB re } 1\mu\text{Pa at } 1\text{m} \quad (14.7)$$

Here we see that a 10 Hz resonant bubble at a depth of 9.8 m pulsating at 3.5% rms of the resonant radius (r_o = 0.46 m) will give a sound pressure level of 203 dB re 1 μPa at 1 m. To obtain 200 dB at 100 Hz maintaining the 3.5% rms displacement criterion one would have to place the resonant bubble at a depth of 62.3 m. Its radius at that depth would be 0.088 m.

14.3 SOME RESULTS FROM SUB-OPTIMUM DEVICES

We have tested three different prototype tow-powered sources at the Naval Research Laboratory. None of the prototypes is close to being optimum, however, the results do show that producing fairly high sound pressure levels from such sources is not that difficult. Figure 14.3 shows the concept for a 16.6 Hz resonant source.

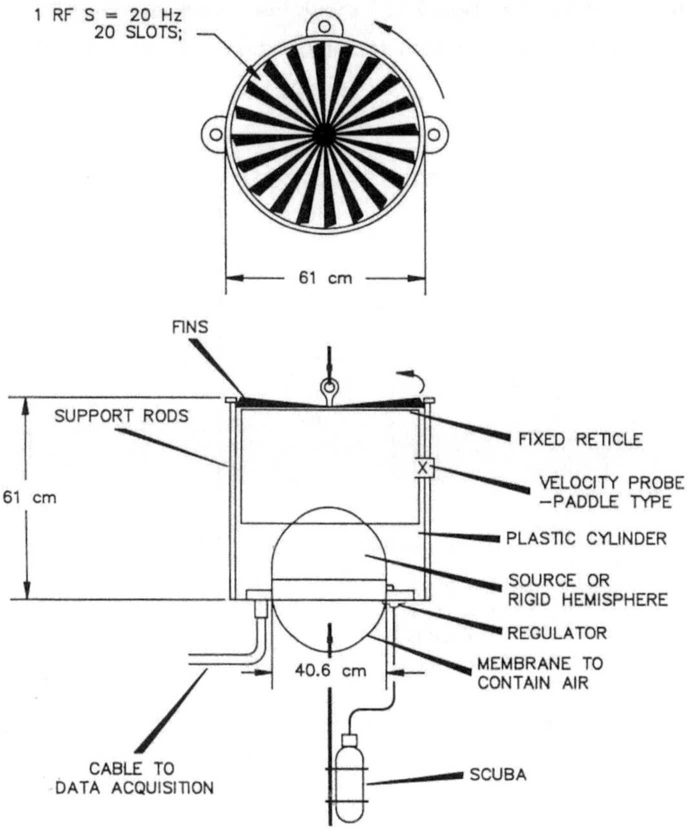

Fig. 14.3 Flow modulator

The salient features here are the flow modulator, plastic shroud, and rigid hemisphere. For this device and for a 50 Hz resonant device that we built, the stiffness and loss in the membrane as well as the design of the flow modulators were factors in making the device suboptimum. Figures. 14.4 and 14.5 show results from these two devices. Source levels of about 182 dB were obtained at a tow speed of 1.85 km/h (1 knot) for both of these devices. Lack of linearity with tow speed causes us some concern. Figure 14.6 shows the devices were omnidirectional. No measurements of harmonic distortion were made. Figure 14.7 shows another concept for a tow-powered source that was tested. This device was modeled by LT Semper for his Master's thesis (14.3) in mechanical engineering at the University of Central Florida. This device corrects a deficiency of the first devices by providing a variable speed motor to control the modulation frequency instead of depending on tuning the flow modulator to rotate at the proper resonant rate. We will not go into detail about the modeling but it involves the elastic moduli of the elastomeric membrane, adds the radiation mass in an ad hoc manner, and calculates the peak modulation pressure as a function of tow speed. Figure 14.8 shows the calculated source levels as a function of the peak modulation pressure. Since the model used is a linear one, we expect the linear response shown here. For a peak modulation pressure of 0.1 psi, theory predicts a sound pressure level of 168 dB at 34.5 Hz. Figure 14.9 shows the experimental results which agree well at resonance with the theory after selection of appropriate parameters to put into the theory.

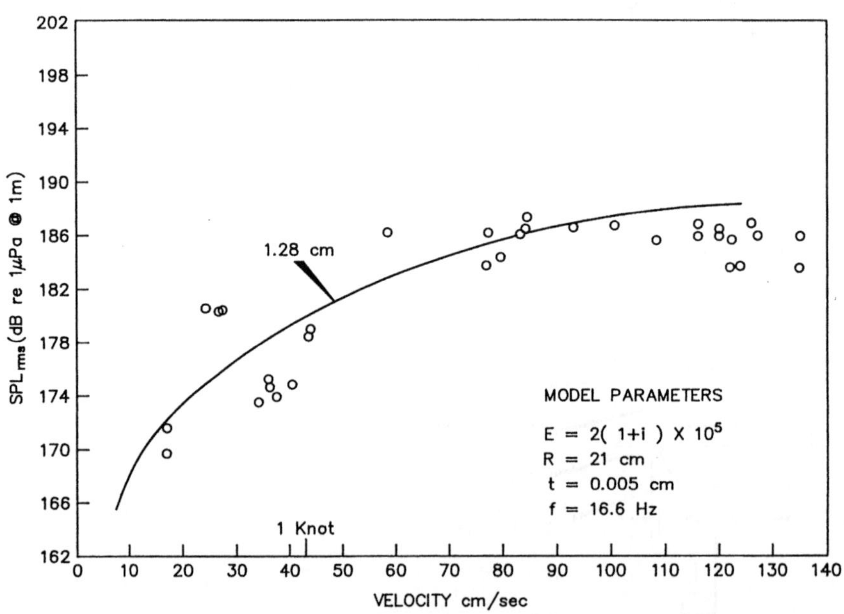

Fig. 14.4 Sound pressure level vs. tow speed

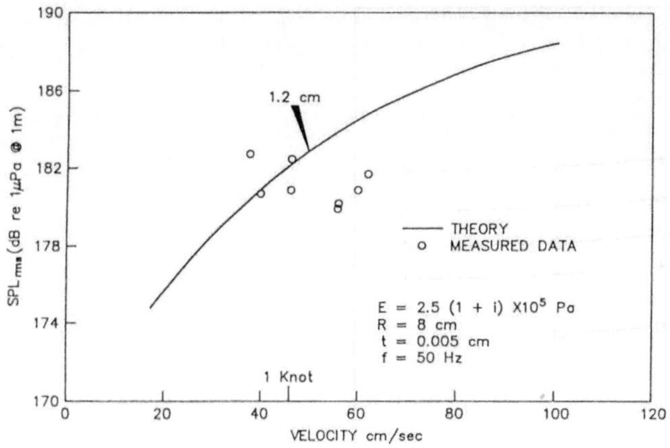

Fig. 14.5 Sound pressure level vs. tow speed

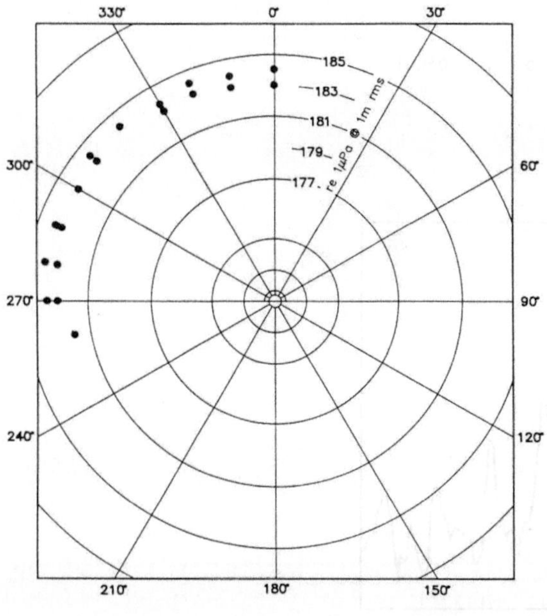

Fig. 14.6 Directivity of 16.6 and 50 Hz sources

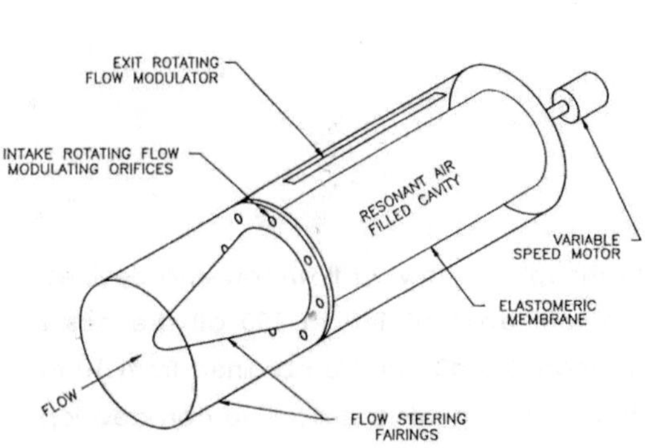

Fig. 14.7 Tow-powered sound source

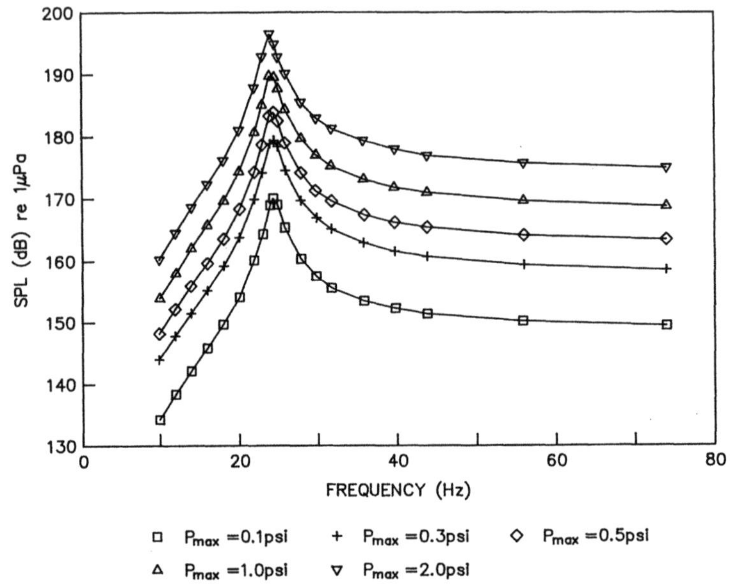

Fig. 14.8 Effects of the magnitude of forcing

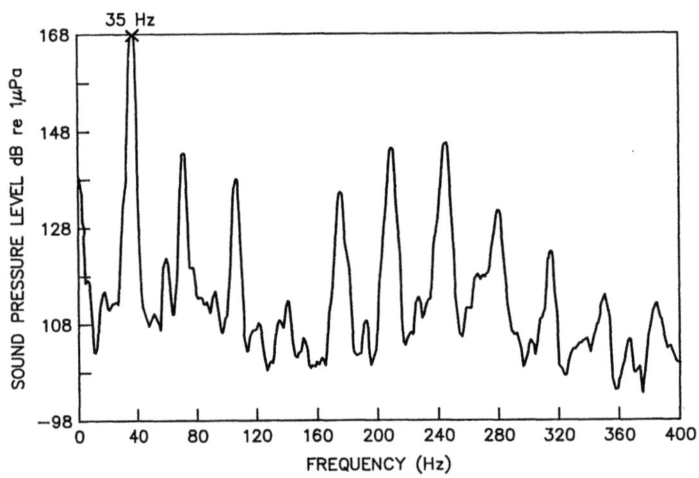

Fig. 14.9 Sound pressure level vs. frequency

14.4 CONCLUSION

The results presented here were meant to stimulate thought on tow- or flow-powered devices. We have shown with very crude devices that source levels of 170 to 190 dB are easily obtainable. Our belief is that source levels greater than 200 dB can be obtained from 10 to 100 Hz with rather small, lightweight devices and that with innovative design we can develop devices that have frequency diversity and maybe even fairly broad bandwidth.

REFERENCES

14.1 H.C. Schau, A.L. Van Buren, and F.J. Rodosta, "Low-frequency, tow-powered sound source," J. Acoust. Soc. Am., **74**, 583(A) (1983).

14.2 M. Minnaert, Phil. Mag. **16**, 235 (1933).

14.3 P.A. Semper, "Mathematical Modeling of a Low-Frequency Cylindrical Sound Transducer," Master's thesis, University of Central Florida, Orlando, FL, 1990.

15. Electronic Needs for High-Power Transducers

G.G. Dixon and S.R. Leyland

Marconi Command and Control Systems Ltd,
New Parks, Leicester, LE3 1UF, United Kingdom

Recent trends in active sonar design have called for large increases in the acoustic power output of sonar transducers. This paper discusses the impact of these trends on the power requirements of the electronics to drive the transducers. The effects of taking large power pulses from the supply are discussed and solutions to the consequential problems are outlined, for a variety of different systems. The application of various amplifier configurations and control techniques are examined in relation to the problems of matching the output of the power amplifier to the input of the sonar transducer. This is further expanded to consider the drive requirements of both magnetostriction and ceramic (PZT) transducers. Finally, the paper gives an example of the relationship between the required acoustic power and the rating of the power amplifier showing why apparently overrated amplifier power outputs are necessary.

15.1 INTRODUCTION

Over the past few years the power requirements for active sonars have increased dramatically, mainly due to the interest being shown in high-power, low-frequency systems. The total system power requirement can range from tens of kilowatts to hundreds of kilowatts with the rating of individual transmitter amplifiers up to 60 kW. The design of such a transmitter system must also take into account the effect on the power source and types of load i.e. the complete system as shown in Fig. 15.1. This paper considers the needs of the total system which includes the power source, switching amplifiers, balancing inductors (or matching) and the type of loads to be driven.

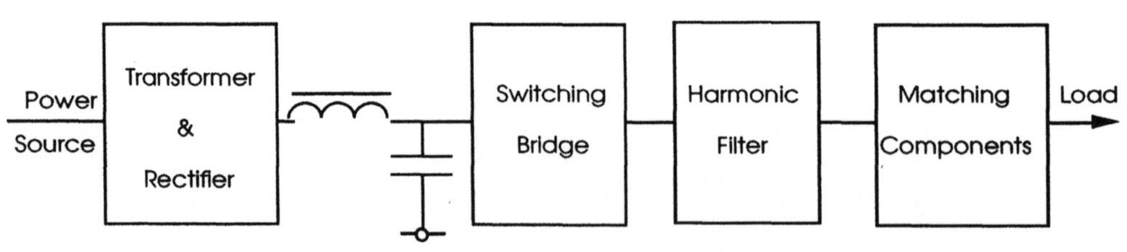

Fig. 15.1 Typical Sonar System

15.2 THE SONAR TRANSMITTER SYSTEM

Sonar systems have become increasingly complex with the expansion of data processing technology, requirement for low-frequency and much higher power levels. Typical performance requirements for a sonar system may include:

- variable power output from 1% to 100% of full load,
- minimal harmonic distortion at all output levels,
- minimal phase shift between input and output,
- frequency hopping,
- FM (frequency modulation) and CW (continuous wave modulation) pulses,
- accurate power control over the full range of output,
- constant power over a range of load impedances,
- small physical size,
- high-power at low-frequency.

The power for the system can be taken from the ship's general power source or from a dedicated source. The voltage from the power source is then rectified to provide a DC supply for the switching amplifiers. The output from the switching amplifiers is then fed through a filter to reduce the harmonic current flow. Finally the output is fed through the transformer to the transducer. The transducer self-capacitance term presents a low impedance to any harmonic current which the series filter allows to flow, hence resulting in a sinusoidal waveform at the load terminals.

Signals defining the transmitter output requirements are fed into a "Control and Interface" module, where they are processed and passed to the amplifier to generate the required output power, phase, shape and frequency. These defining signals can also be used to give advanced information to the power source if it is advantageous to prime it.

15.3 POWER SOURCES FOR SONAR APPLICATIONS

For very high-power systems, the power source may be the ship's general power system or a dedicated generator. The fundamental problem when considering power sources for sonar applications is that large power pulses are demanded. This gives rise to a number of related problems:

- the introduction of voltage transients each time the system transmits,
- distortion of the supply due to harmonic currents.,
- distortion of the waveform due to the process of rectification,
- radio frequency interference (RFI).

The relative influence and importance of each of these features is dependent upon whether the power is taken from the ship's main electrical system or whether it comes from a dedicated source.

15.3.1 Ship's Main Electrical System

15.3.1.1 Voltage Transients

A typical ship's generator is rated at 2 MW and will suffer from transients of typically 25% when throwing 100% loads on or off. Under normal operation, transients are generally less than 12.5% because practical switched loads are less than 50% of the machines full load rating. This is shown in Fig. 15.2. A sonar system of some 200 kW represents 10% of the full load rating of the typical 2 MW generator and as such would generate transients in the order of +3% and -3% at the pulse frequency of the sonar. This level of fluctuation may be acceptable for some ship systems, where there are no sensitive pieces of equipment on the same supply, but often it is not. Obviously smaller vessels have smaller generators and would suffer worse transient interference.

When this level of disturbance becomes unacceptable then it is necessary to have some form of energy storage (or buffer) to smooth out the pulses. Capacitors, batteries or a motor-generator set with a flywheel may be used to provide the energy required during the transients and may be re-charged during the long interpulse period (Fig. 15.3).

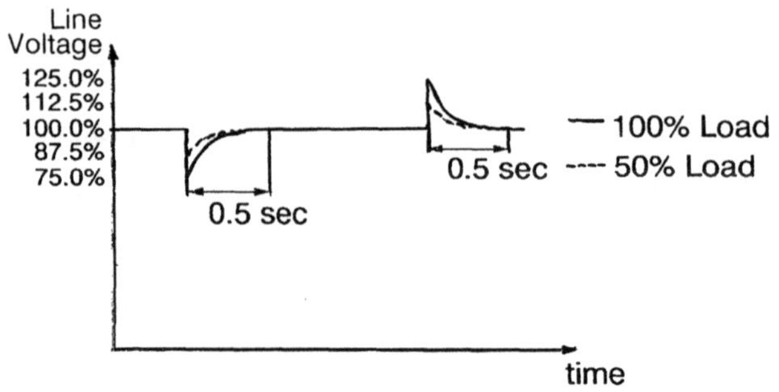

Fig. 15.2 Typical line voltage transients on a ship's supply

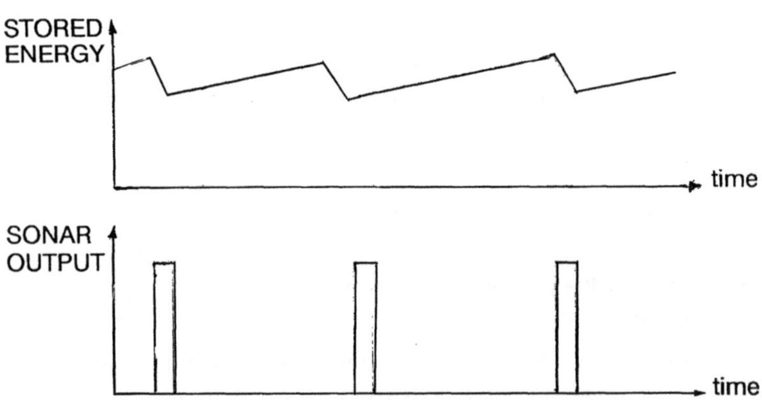

Fig. 15.3 Energy recovery during interpulse period

None of the above energy storage methods is particularly good for long pulses which cause supply droops and therefore have to be supported by energy taken from the better-regulated main ship's supply. They also suffer from being physically large for the desired level of energy storage, which is a major drawback in the present generation of space conscious ships. Other drawbacks for the battery option are the twin safety hazards of gaseous emissions and acid spillage with the attendant requirement for high levels of maintenance.

An alternative approach is to reduce the rate of change of power demanded from the ship's system, as shown in Fig. 15.4. This allows the ship's regulators more time to react to the situation.

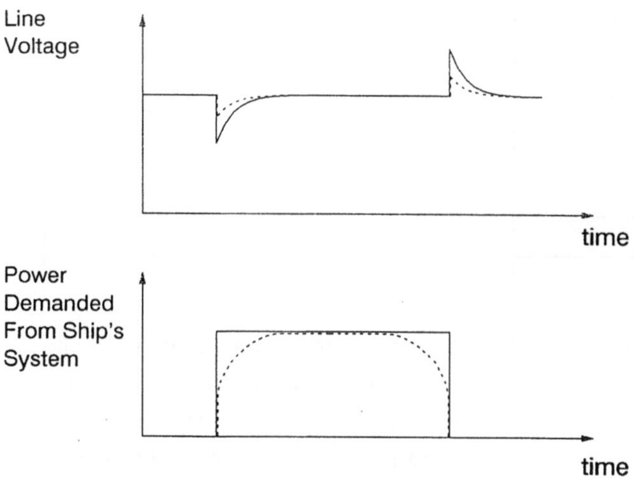

Fig. 15.4 Transient reduction by demand shaping

15.3.1.2 Harmonic Distortion

Figure 15.5 shows a typical AC line current when loaded by a six pulse, 3 phase rectifier.

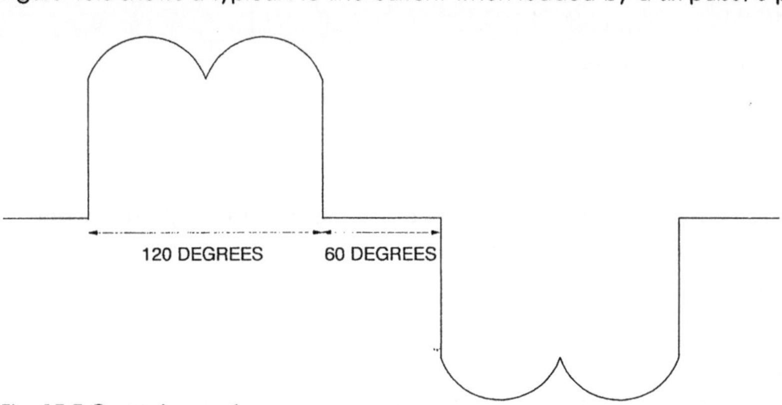

Fig. 15.5 Current waveform

Table 15.1 shows the theoretical harmonic current consumption of a typical 6 pulse, 3-phase rectified load. This has a total harmonic distortion of approximately 28% which is well in excess of the 5% total distortion and the 3% individual harmonic distortion required by some specifications.

Table 15.1 Harmonic currents (as a percent of fundamental) in a rectified supply

FREQUENCY	HARMONIC CONTENT (%)
F_0	100.0
3 F_0	0.0
5 F_0	19.6
7 F_0	14.3
9 F_0	0.0
11 F_0	9.1
13 F_0	7.7
15 F_0	0.0
17 F_0	5.9
19 F_0	5.3

The most common solution to this problem is to fit a multiwound transformer to enable 12 pulse rectification (rather than the more common 6 pulse rectifier) and then to filter the remaining harmonics. This is not only expensive but it also requires physically large components to provide the degree of filtering needed to meet the imposed limits. However, the addition of a mains transformer has several distinct advantages namely:

- it enables a 12 pulse rectifier system to be used,
- fault current levels from the ship's mains are reduced,
- harmonic currents from the supply are reduced,
- the supply voltage may be reduced to provide a safety margin for power switching devices,
- the suppression of RFI is made easier.

15.3.1.3 Voltage Distortion

Figure 15.6a shows the distortion that will occur on the line voltage of a three phase system when loaded by the rectifier set shown in Fig. 15.6b. This assumes that the inductance L is predominant in the DC smoothing circuit.

As the current transfers from one diode to another diode in the 3 phase rectifier there is a period of time when both diodes are conducting. This results in the line voltage being presented with a short circuit for the duration of the transfer period. The rate of transfer of current between the diodes and the AC source impedance determines the amount of distortion seen in the line voltage.

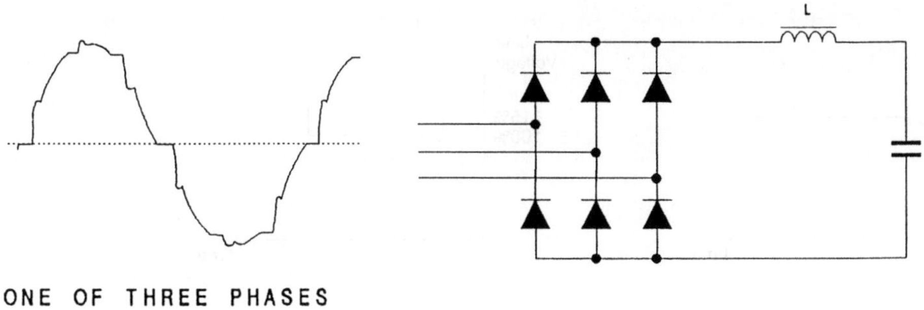

Fig. 15.6 Distortion of line volts due to rectification

15.3.1.4 Radio Frequency Interference (RFI)

RFI is a particular problem for sonar systems using switched-mode amplifiers: every sharp edge in the switching waveform produces a string of high frequency harmonics which cause interference on other electrical systems. Again, an inductor or transformer provides a buffer between the sonar and the ship's supply, but very careful filtering and screening is still required to bring the interference down to acceptable levels.

15.3.2 Dedicated Supply

The biggest advantage of having a dedicated supply is that all aspects of the system design and performance can be directly under the control of the transmitter designers and the parameters can be defined to give optimum performance. Its disadvantage is that it requires an additional generator system to be fitted.

15.3.2.1 Voltage Transients

A dedicated supply will suffer from the same problem of voltage transients as the ship's supply but it is made worse by the fact that the dedicated supply is much smaller. For a 200 kW sonar, we may consider using a 300 kW supply which would give 20% transients every time the sonar transmits. Since the sonar is the only equipment subjected to these transients their effects may be taken into account at the design stage. It is possible to reduce the overall effect of the transient by changing the generator control method (Fig. 15.7).

Consider a generator which has a 20% transient depression when full load is applied and is controlled to give 120% output in the no load state. When the load is applied, the output volts will transiently fall to 100%. The generator control loop will be changed to hold this new level. The above procedure is reversed when the load is removed. The nett effect is that the transmitter has to work from a supply which has a 20% variation as opposed to the 40% variation with normal control.

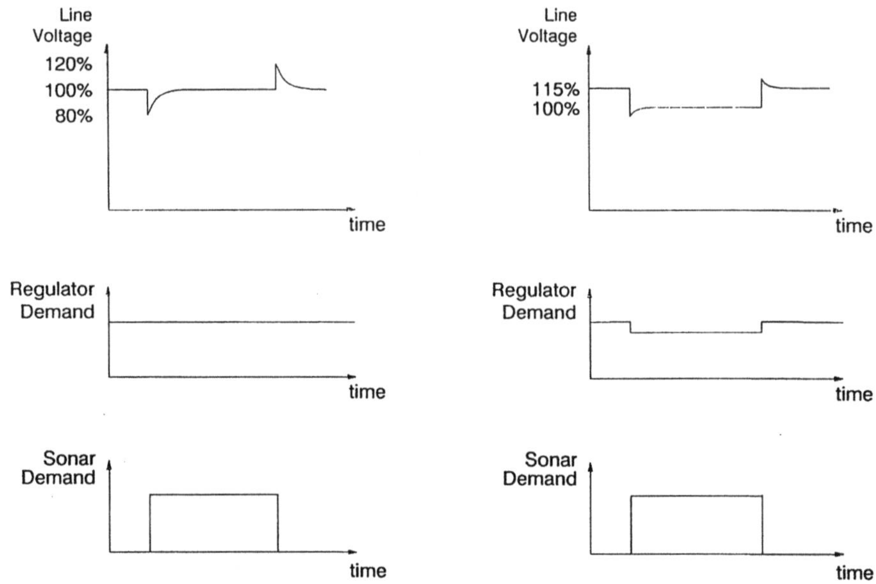

Fig. 15.7 Transient reduction by regulator demand shaping

Further enhancements may be achieved by feeding the generator prime mover advance information of the impending load change.

Whatever the type of prime-mover, the motor-generator set may be designed with a built-in flywheel effect which gives two further degrees of freedom to the designer. The machine may be chosen such that it is over-rated in order to achieve a given flywheel effect with its associated transient reduction or it may be electrically under-rated so that the smoothing effect of the flywheel is maximized.

15.3.2.2 Harmonic Distortion

With no other users on the dedicated supply, the level of harmonic distortion is no longer as critical.

15.3.2.3 Radio Frequency Interference

The radiated RFI must still be kept within the limits for shipborne equipment. Similarly the equipment must not be susceptible to the prescribed levels of RFI radiated from other equipments.

From the preceding discussions it can be seen that the precise choice of power source is dependent on many factors and will usually be a trade-off between system performance, efficiency, cost, size and interaction with other shipborne equipment. Each type of sonar system and each type of vessel has its own optimum which must be sought with all due care.

15.4 AMPLIFIER CONTROL

The power amplifiers may require many parameters to be controlled, the choice of which are dependent on a given system requirement. Feedback loops may be selected to control the following parameters:

- output voltage,
- output current,
- phase of input signal to output,
- output power,
- energy (volt-seconds).

15.4.1 Voltage

Voltage feedback control can take two forms: average or real time. Simple phase control amplifiers will regulate the output voltage to give an average level by sampling the magnitude over a number of cycles. A more complicated control system would utilise a high frequency carrier (PWM pulse width modulation) to give real time control of output voltage magnitude, waveform and phase.

15.4.2 Current

Current feedback may be used in exactly the same way as voltage feedback to control the output waveshape or amplitude in a current-fed amplifier. The current feedback may also be used to give protection to the power amplifier or load in the event of one of them failing.

15.4.3 Phase

The phase error between input signal and output may also form part of the feedback system. The phase error can be used to reduce any phase shift due to filter components in the amplifier output. The control of phase would usually be an average over a few cycles of the output waveform.

15.4.4 Power

By using both output current and voltage feedback, power control may be achieved. The product of current and voltage can be filtered to give a DC voltage that is proportional to the output power. The DC voltage can be fed into a narrow bandwidth (a few cycles) feedback loop to enable the output power to follow the input power demand signal.

15.4.5 Energy

The energy input or output of an amplifier may also be measured to provide a form of feedback, or feed-forward control. By measuring the voltage-time integral of the switching amplifier, it is possible to determine the shape of the output waveform being fed to any filter networks. The main advantage of this type of control is that it avoids the double pole that most filter networks introduce into the feedback loop and hence produces a more stable system. When using this form of control, consideration must be given to any regulation due to the impedance of the output filter.

15.5 SWITCHING AMPLIFIERS

Semiconductor switching amplifiers for sonar transmitter application have been in use since the late 1960's. Many types of circuits and control methods have been devised, utilising semiconductors such as

- thyristors,
- bipolar transistors,
- power FET transistors (field effect transistor),
- GTOs (gate turn-off thyristor).

The present day semiconductors most suitable for high-power switching amplifiers range up to:

a) Bipolar transistors
(Example V_{CEX} = 1000 V, I_{CONT} = 200 A),

b) power FET transistors
(Example BV_{D-S} = 500 V, I_{CONT} = 20 A).

with :

V_{CEX} :maximum permissible collector-emitter voltage for a reverse biased bipolar transistor.

I_{CONT} maximum continuous current carrying capability,

BV_{D-S} maximum permissible drain-source voltage for a FET,

The selection of which device to use is dependent on the required output power and frequency. For high-power, low-frequency amplifiers the most suitable devices are the bipolar transistors. Switching amplifiers achieve the required output by switching a DC input voltage in such a manner as to give an output voltage at the desired frequency. Two such methods of switching are considered below.

15.5.1 Phase Control (Class D)

A simple bridge circuit is shown on Fig. 15.8a. The drive signals to the switches are such as to produce an output waveform as Fig. 15.8b.

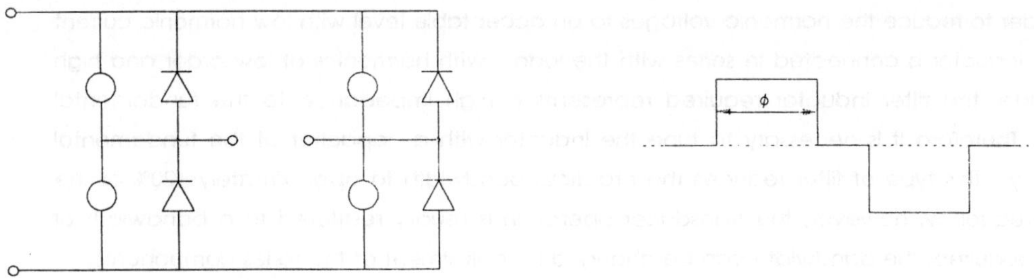

Fig. 15.8 a) Simple bridge circuit b) Phase controlled output waveform

The output waveform is made up of the required fundamental output voltage and unwanted harmonics. Figure 15.9 shows the proportions of fundamental and harmonic magnitudes for different output levels. The harmonics are of high magnitude and high frequencies. These are not difficult to attenuate but do require large filter components.

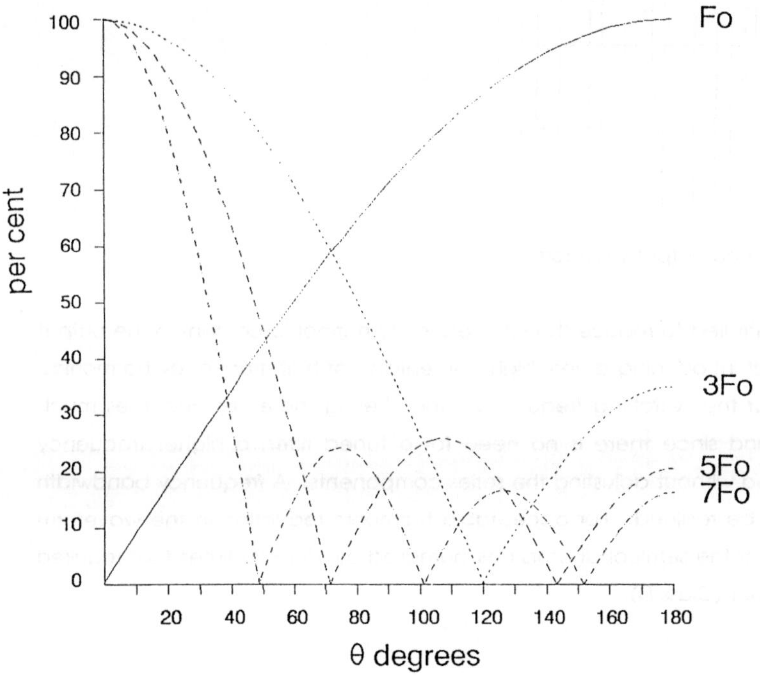

Fig. 15.9 Harmonics in a phase controlled output expressed as a percentage of the fundamental component

In order to reduce the harmonic voltages to an acceptable level with low harmonic current flow, an inductor is connected in series with the load. With harmonics of low order and high magnitude, the filter inductor required represents a high impedance to the fundamental voltage. Therefore it is necessary to tune the inductor with a capacitor at the fundamental frequency. This type of filter reduces the practical bandwidth to approximately ±20% of the centre frequency; however, the transducer operation is usually restricted to a bandwidth of ±15%. If required, the bandwidth can be changed by adjustment of the series components.

15.5.2 Pulse Width Modulation (Class S)

The circuit used is essentially the same as that shown in Fig. 15.8a with drive signals to the switches to produce a waveform as shown in Fig. 15.10.

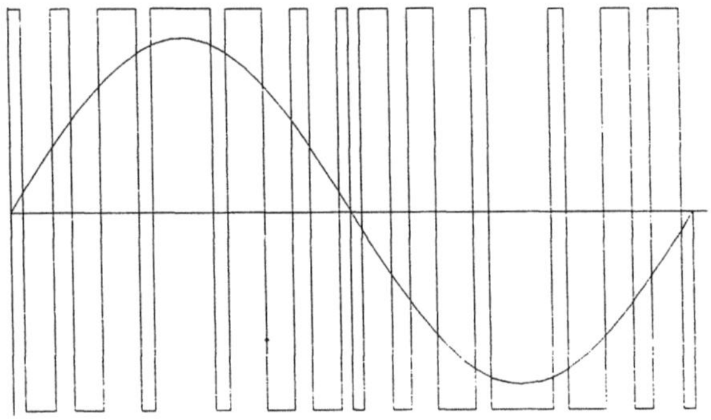

Fig. 15.10 Pulse width modulated output waveform

The switching pattern is controlled to reduce the low-frequency harmonic content in the output waveform at the expense of introducing a very high percentage of high-frequency harmonics, with the major harmonic at the switching frequency. The filtering therefore becomes much simpler than for Class D and since there is no need for a tuned filter, a higher frequency bandwidth can be achieved without adjusting the series components. A frequency bandwidth from DC to many kHz can be realised. For acceptable harmonic reduction in the waveform the frequency of switching of the semiconductors has increased by 15 to 20 times that required for the phase control amplifier (Class D).

15.6 DRIVE FOR MAGNETOSTRICTIVE TRANSDUCERS (MS)

The MS transducer requires a DC bias to enable it to operate in its linear range. This bias can be obtained by the following methods:

- fitting permanent magnets,
- passing DC current in a dedicated coil,
- passing DC current in the main coil.

To explain the drive requirements of the amplifier the simplified parallel equivalent circuit as Fig. 15.11a will be considered. Figure 15.11b is the series equivalent circuit and a similar argument would follow. The element R represents the load resistance and the inductor impedance X_L represents the magnetisation component of the transducer. The typical value of X_L is one third of R leading to an input current of $(3I)^2 + I^2 = (3.16I)^2$. For the required output power of VI, the input supply must be capable of delivering 3.16 times VI. The input supply requirement can be reduced by fitting a capacitive impedance of X_C in parallel with the transducer terminals (i.e. $X_C = X_L$ at the centre resonant frequency).

Fitting of the capacitive impedance has reduced the input current from 3.16 I to I. Hence the supply has to deliver I amps at V volts. The rating of the capacitor is 3VI (VARs volt-amps reactive,).

Alternatively, a capacitor of 3VI VARs could be fitted in series with the transducer resulting in the required input supply voltage being reduced.

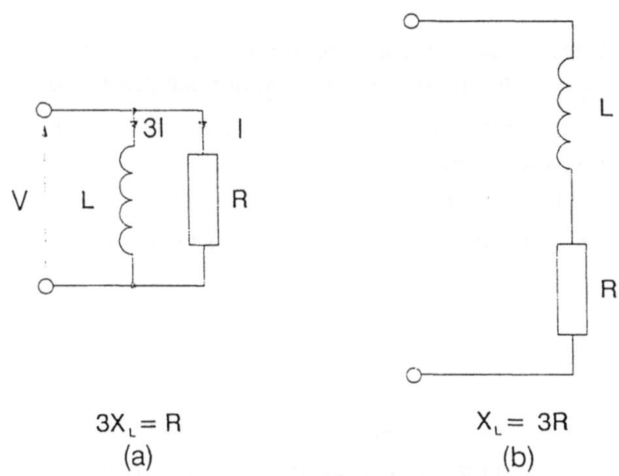

Fig. 15.11 Equivalent circuits for MS transducers

Three conditions for driving the transducer now exist as follows:

1. Drive the transducer with no matching capacitor:
   - Voltage Input = V.
   - Current Input = 3.16 I.
   - Volt Amp Input = 3.16 VI.

2. Drive the transducer with a parallel matching capacitor:
   - Voltage Input = V.
   - Current Input = I.
   - Volt Amp Input = VI.

3. Drive the transducer with a series matching capacitor:
   - Voltage input = 0.316V.
   - Current Input = 3.16 I.
   - Volt Amp Input = VI.

The above methods of matching are satisfactory for a sinusoidal power supply, but are unsatisfactory on their own when energised from a switching amplifier because the impedances involved give rise to high harmonic currents.

Now we will consider various methods of driving the transducer from a switched-mode amplifier.

15.6.1 Phase Control Amplifier (Class D, AC Output)

This type of switching amplifier will give an output that contains high magnitudes of low order harmonics (i.e. 3rd, 5th, 7th). In order to restrict the flow of these currents it is necessary to fit a series inductor of value approximately $X_L = 0.7R$. However, this inductor will present a significant impedance to current at the fundamental frequency, hence a series capacitor is also necessary to tune out this effect. The values of the inductance and capacitance are such that $X_L = X_C$ (i.e. tuned to the fundamental frequency).

The above, in conjunction with the parallel matching capacitor, will reduce the harmonics to an acceptable level. The matching circuit is shown in Fig. 15.12.

15.6.2 Pulse Width Modulation (Class S, AC Output)

This type of switching will give an output which contains mainly harmonics of high order, peaking at the carrier frequency (approx 20 times fundamental frequency). In order to restrict the flow of these harmonics, a series inductor will again be necessary but its impedance to the fundamental frequency will be of such a value that a series of tuning capacitance is not required. The matching circuit is shown in Fig. 15.13.

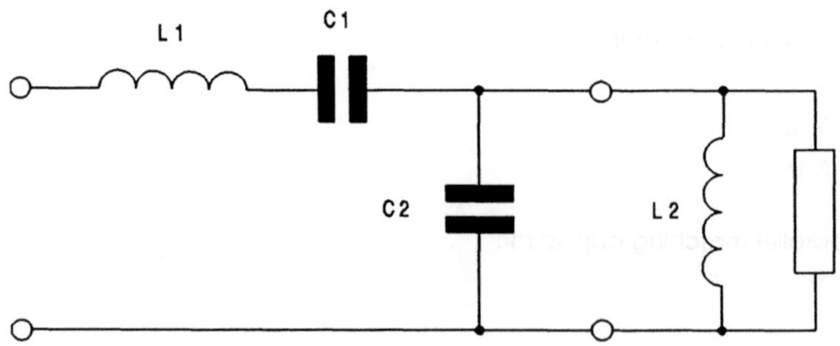

Fig. 15.12 Matching circuit for a phase controlled amplifier ($X_{l1} = X_{c1} = 0.7$ R, $3X_{L2} = 3X_{C2} =$ R)

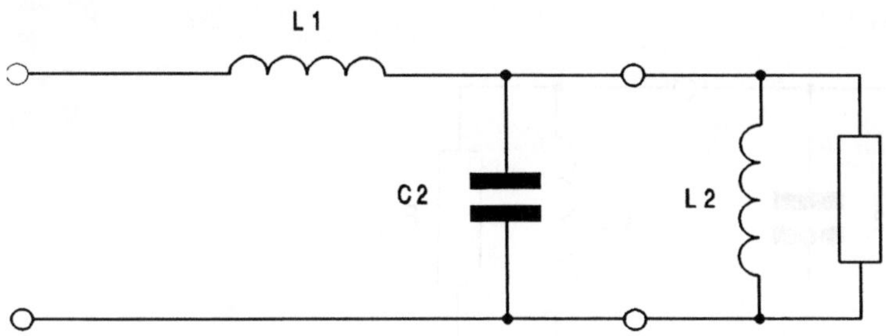

Fig. 15.13 Matching circuit for a pulse width modulated amplifier (X_{l1} = 0.15 R, $3X_{L2} = 3X_{C2}$ = R)

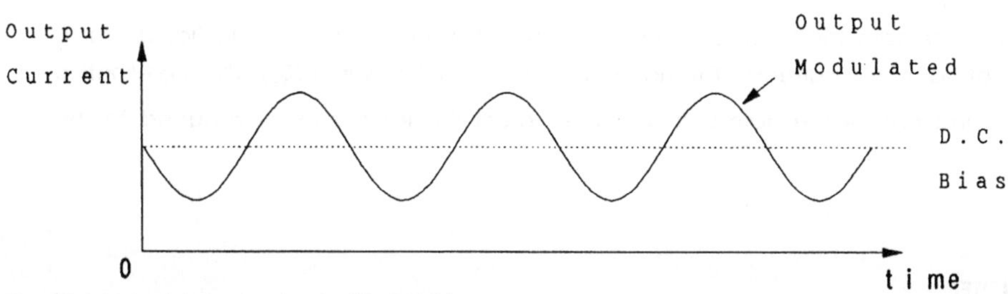

Fig. 15.14 Modulated output with DC bias

15.6.3 Pulse Width Modulation (Class S, DC Output)

This type of amplifier can be controlled to give a DC output current. The current can be set to give any required mean bias and also be modulated with the required output frequency, as shown in Fig. 15.14.

If required, the output current stage can be controlled to take into account any lack of symmetry (if it exists) in the MS transducer. The high frequency harmonic currents in the waveform will be removed by fitting the same series inductor and parallel matching capacitor as for the AC output case. The DC bias prohibits the use of a series matching capacitor or an isolation transformer since neither permit DC current to flow.

15.7 DRIVE FOR PZT TRANSDUCERS (CERAMIC)

The balancing and harmonic filter for PZT transducers are very similar to that used for MS transducers. An approximate equivalent circuit of the transducers is as shown in Fig. 15.11a. A balancing inductor-transformer is connected to resonate the load at the centre frequency of operation. This resonant circuit can have currents of 3 to 4.5 times the current in the load R.

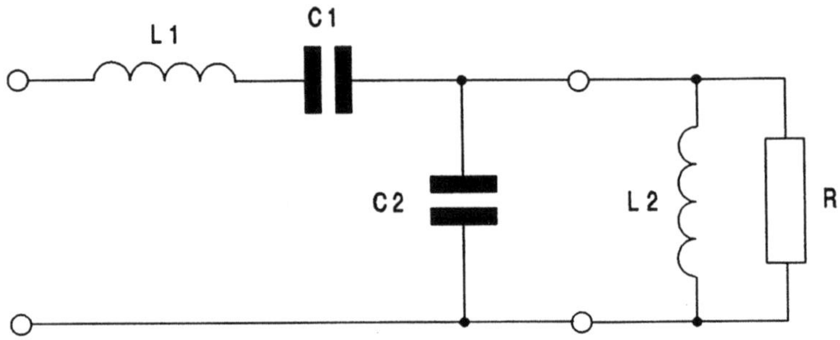

Fig. 15.15 Matching Circuit for PZT transducer (X_{l1} = X_{c1} = 0.7 R, $3X_{L2} = 3X_{C2} = R$)

To filter the harmonics from the load, a series filter has to be fitted (as shown in Fig. 15.15) to restrict the flow of harmonic currents. The harmonic currents will flow into X_{C2} which presents a low impedance and hence the harmonic voltage across the load can be reduced to the required level.

15.8 GENERAL NOTES

15.8.1 MS/PZT Transducer Comparison

The MS type of transducer has a number of differences to the PZT as follows:

- the MS transducer is more flexible because its supply voltage can be determined by the number of turns used in the existing coil. With the availability of high voltage and high current transistors an amplifier can be designed to remove the need for a matching transformer. An amplifier output voltage of 440 Volts rms can be achieved without the need for series connected transistors,
- the efficiency of the MS transducer is significantly less than the PZT transducer (e.g. 50% for MS and 80% for PZT) which results in the need for higher rated amplifiers,
- the matching components for MS are smaller, cheaper, lighter and more efficient than those required for PZT transducers.

15.8.2 Amplifier Rating

This is an example to show the relationship between the required output power from a transducer and the kVA output requirement from the amplifier. Consider the system requirement in Table 15.2.

The kVARs circulating in the matching circuit are three times the output power, i.e. 3 x 20 = 60 kVARs. At the frequency F_0 the net flow into the Inductor/Capacitor is zero. However, at $F_0 \pm \delta f$ the flow is +17 kVARs.

Table 15.2 Example of sonar system requirements

Output power acoustic	20 kW
Frequency bandwidth	+/- 15% (+/- δF)
Output control	constant power
Max. output phase angle	+/- 60 degrees
Matching Q	3
Efficiency	see Table 15.3
Centre frequency	F_0

Neglecting the efficiency of the transducer, the kVA in the load impedance Z at a phase angle of 60° is 40 kVA made up of 20 kW and 34.6 kVARs.

The output requirement from the amplifier at $F_0 \pm \delta f$ and a phase angle of -60° is:

20 kW and (34.6 + 17) kVARs

which leads to a kVA output of

$$(20^2 + 51.6^2)^{0.5} = 55.4 \text{ kVA}$$

Table 15.3 shows the amplifier rating required for various conditions. Note that Table 15.3 shows the amplifier rating required for the conditions shown in Table 15.2. The phase angle is defined as the phase angle into the load Z of Fig. 15.16. However, if we interpret the change in efficiency from 80% to 50% as only resulting in a change in kW and not kVAR, then the overall effect is small in the amplifier design.

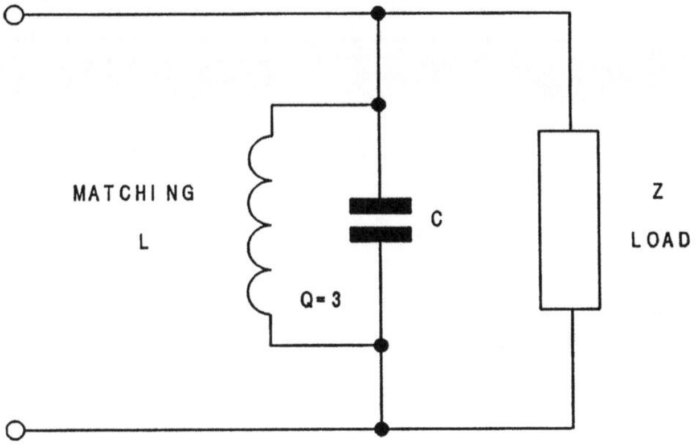

Fig. 15.16 Equivalent circuit of the sample load to be driven

Table 15.3 Amplifier ratings

EFFICIENCY (PERCENT)	FREQUENCY	PHASE ANGLE (DEGREES)	kW	kVars	kVA
100	F_0	0	20	0	20
100	F_0	60	20	34.6	40
100	F_0 +/- δF	60	20	51.6	55.4
80	F_0 +/- δF	60	25	64.5	69.3
50	F_0 +/- δF	60	40	103.2	110.7
50	F_0 +/- δF		40	64.5	75.9

The values in the table are only an indication. Factors such as tolerance of transducer parameters have not been allowed for. From the above figures it may be seen that transducer parameters must be clearly defined in order to obtain an amplifier of the correct rating for the application. It is imperative that array manufacturers, sonar system designers and transmitter amplifier designers discuss the required performance parameters in detail so that confusion does not arise resulting in either an underpowered or an overpowered amplifier.

16. Electronic Systems Modeling

S. FAURE
GERDSM -DCAN Toulon, DCN, Le Brusc, F-83140 Six Fours Les Plages, France

J.M. CAPRON
ISEN, 41 Bd Vauban, F-59046 Lille CEDEX, France.

This article first presents the needs to simulate an entire transmission chain including the electronic parts. Then it describes previously developed works for modeling power amplifiers driving the transducers, as well as transformers and their associated magnetic cores, the analog simulator beeing a SPICE derivative. Several sections show how the studies can be carried out at different levels in order to permit observation of different aspects, such as global behavior or component behavior, etc. Comparisons are made between simulations and measurements. Finally, a section on transmission line optimisation and efficiency shows how simulation could benefit their design.

16.1 INTRODUCTION

In the active sonar field, the current trend is to use lower frequencies. This obliges designers to build up arrays with much smaller relative masses and volumes. In this case, the effect of acoustical interactions is extremely important. This point is not restricting when working far from resonance. But in the neighbourhood of resonance, abrupt near-field variations can lead to changes in the electrical impedances of various elements in an array, as shown in Fig. 16.1.

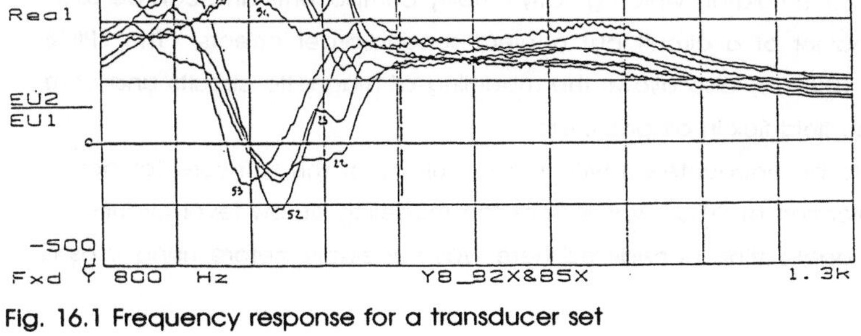

Fig. 16.1 Frequency response for a transducer set

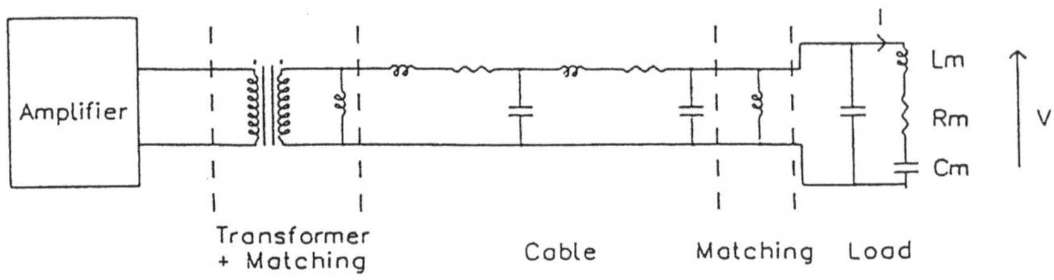

Fig. 16.2 Acoustical transmission chain

In order to prevent those changes and be adaptable, simulation becomes necessary, for the acoustics as well as the electronics. The design of a transmission chain intended to deliver the desired acoustical signals with high reliability and efficiency will be better carried out using simulation tools. Using the equivalent electrical schematic of a transducer, the full transmission chain can be represented as above (Fig. 16.2).

On the acoustical level, a main objective is to control the velocity of the radiating face, proportional to the current value in the motional branch. This variable is a function of the strength applied (voltage across the equivalent electrical schematic). The difficulty in analysing the whole system comes from changes in values of R_m and L_m elements, according to frequency and array configuration.

Since the modeling of the behavior of the acoustical part already exists, why not simulate as well the part before the load, namely the power amplifier? This notion induced our investigation of the possibility of simulating the whole electronic chain required to drive the transducer elements.

16.2 SIMULATION TOOL

The simulator used for electrical modeling is a SPICE (Simulation Program with Integrated Circuit Emphasis) derivative. This software, first developed during the seventies at the University of California at Berkeley, gained an international reputation and allows DC or transient analysis of non linear circuits, as well as AC analysis of linearised circuits. It is a good compromise, on the one hand, between device simulators, which are complex and computation time consuming, and which are more suitable for study at the component level instead of the whole electronic circuit, and, on the other hand, simulators which greatly simplify components and can be used to describe the global behavior of a circuit but neglect second-order effects. The SPICE derivative used in our case (DSPICE) also allows the modeling of magnetic circuits and core losses due to hysteresis, fringe field flux in air gaps, etc.

However, problems could be encountered with the accuracy of this software for power electronics. The initial application of SPICE was in fact the modeling of low level circuits. A software validation for the application considered here was necessary before using it as a design-aid tool.

The different points that will be interesting to study while designing or validating an acoustical transmission chain can be deduced from its synoptic schematic:

- supply: studying of supply operation e.g.: energy transmitted back into the source,
- emitter: electronics circuit optimisation,
- transformer: + matching size and weight evaluation of magnetic circuits, studying of losses,
- line: line studying and matching to improve efficiency,
- load: studying using the equivalent electrical circuit given by acoustical computer aided design (CAD) tools.

We first took interest in emitters, transformers and lines. The aim of this prior investigation was to evaluate the simulator in order to determine whether it satisfied our needs and to highlight its advantages and limitations. To verify the validity of simulation results, these have been compared with experimental measurements.

16.3 ELECTRONIC CIRCUITS SIMULATION

In the design of acoustical signal power amplifiers intended to supply high powers (greater than one kilowatt), one should not think of using class A or B amplifiers that provide low efficiency, mainly with out of phase loads, because they use transistors in linear mode. Therefore, it is normal to use inverters similar to that shown on Fig. 16.3 (H bridge) and ensuring very high efficiency because transistors are now used in a blocked-saturated state and losses occur mainly during the switching times.

The switches can be MOS or bipolar transistors. The advantages of the first type are that they are easily driven by the gate voltage and they have a negative temperature coefficient, which irons out the difficulties when they are placed in parallel. However, until recently, it was more difficult to find MOS transistors capable of withstanding high current densities, but with technological improvements, this point begins to be debatable. The bipolar transistors have an equivalent resistance when saturated that is less than that of MOS transistors. Nevertheless, they are more critical to drive and have a positive temperature coefficient, which can lead to component damage through second breakdown. In our case, we will use MOS transistors because they can be driven easily.

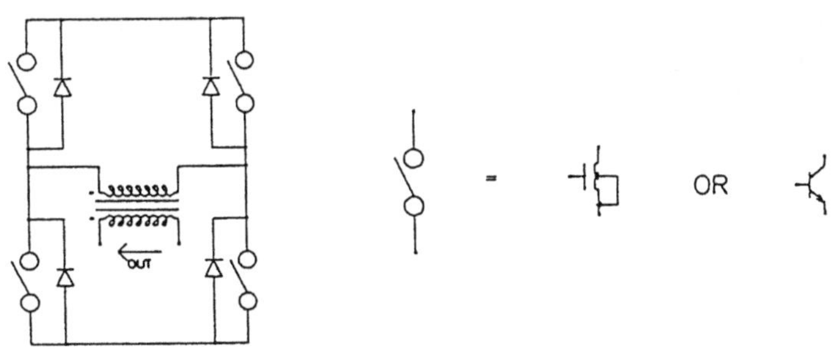

Fig. 16.3 H bridge structure

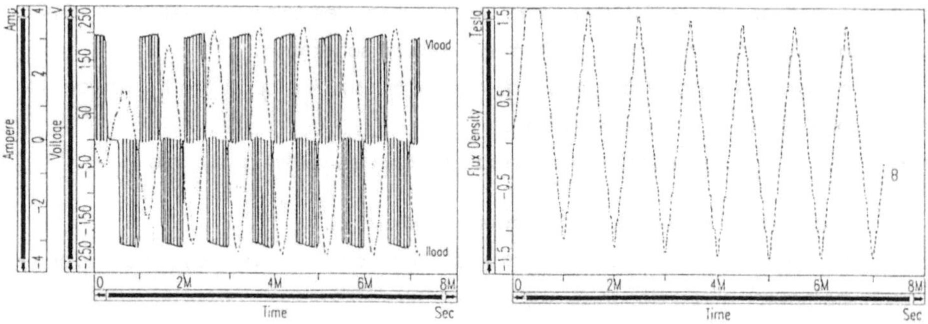

Fig. 16.4 Simulation: voltage, load current and magnetic induction in transformer

First of all, the circuit we studied applies a square wave signal across a tuned oscillatory circuit, which causes the current through the load to be sinusoidal. To reduce the distortion rate of the signal thus generated, the applied voltage has a Pulse Width Modulation (PWM) shape, through action on driving the transistors. This circuit has been simulated and computation results are reported above (Fig. 16.4). Several features can be seen from these first results. In particular, it is possible to evaluate the setting time of steady-state operation, depending on the quality factor of the oscillatory circuit. On the other hand, a residual oscillation can be seen on the voltage across the transformer secondary, shifted in phase by $\pi/2$ with respect to load current. Measurements on a real similar circuit the current and voltage waveforms of which are given in Fig. 16.5 show the same phenomenon.

This oscillation is due to leakage inductances in the transformer. Indeed, magnetic field lines do not follow as a whole the magnetic path of the core, but some of them cross the surrounding air. This is equivalent to an inductance in series with transformer windings and creates this parasitic effect, found both in reality and in simulation.

Moreover, simulation has been run in this case, assuming an initially demagnetized core ($B = 0$ at $t = 0$). A strong core saturation can then be noticed after starting; the magnetic induction B reaches its saturation value B_{sat} which is equal to 1.5 Tesla in the present case. The secondary voltage which is proportional to the time rate of flux density change becomes zero as shown on Fig. 16.5. This point, just mentioned for the moment, will be developed later in more detail.

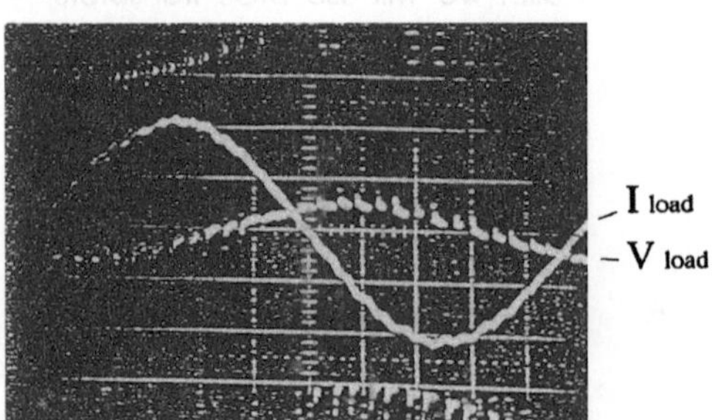

Fig. 16.5 Measurements: load voltage and current

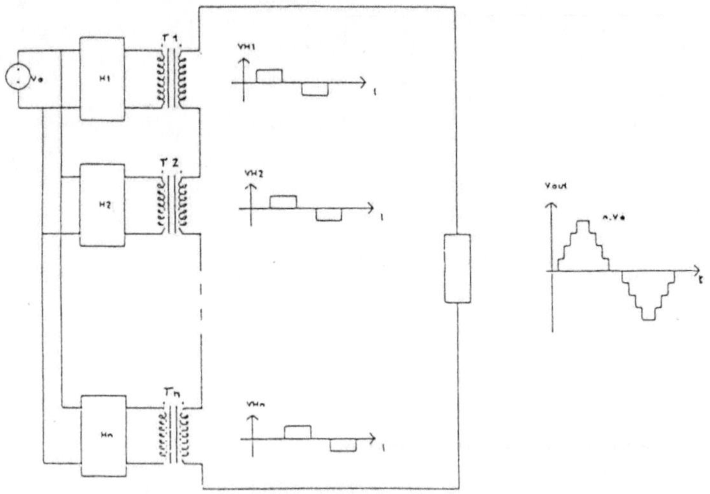

Fig. 16.6 Step generator

We then simulated the operation of a step generator. This kind of circuit uses several inverters such as the one previously described, and adds their output voltages by means of transformers, which permits to obtain an output voltage higher than the supply voltage. This principle is illustrated in Fig. 16.6.

A simulation has been run for a resistive load and Fig. 16.7 shows the result for the output current. For comparison, an oscillogram of the experimental measurement of the supply current for one of these inverters, equal to the load current during the operation time of the stage, is shown on the right of Fig. 16.7.

Figure 16.8 shows B values in transformers T1 and T2 and verifies the fact that B actually remains lower than B_{sat} (1.5 Tesla).

If the simulation is run again specifying operating conditions that cause core saturation, the result (Fig. 16.9) shows an output signal collapse.

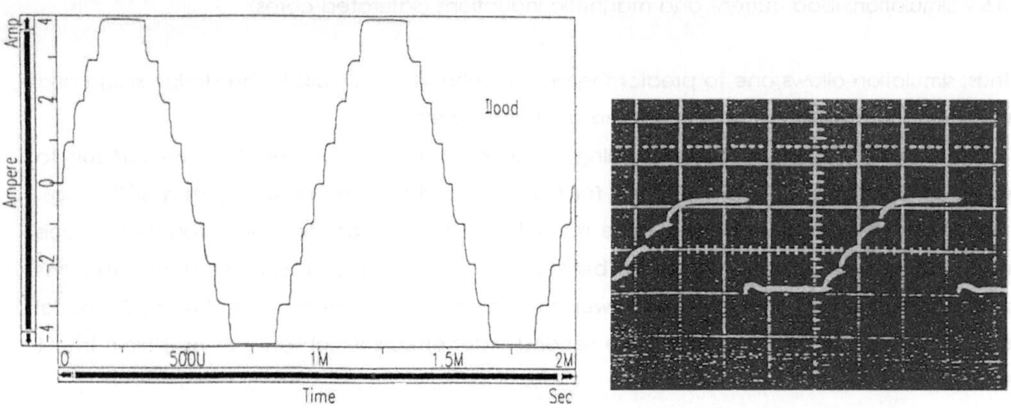

Fig. 16.7 Simulation of load current and measurement of supply current of one inverter

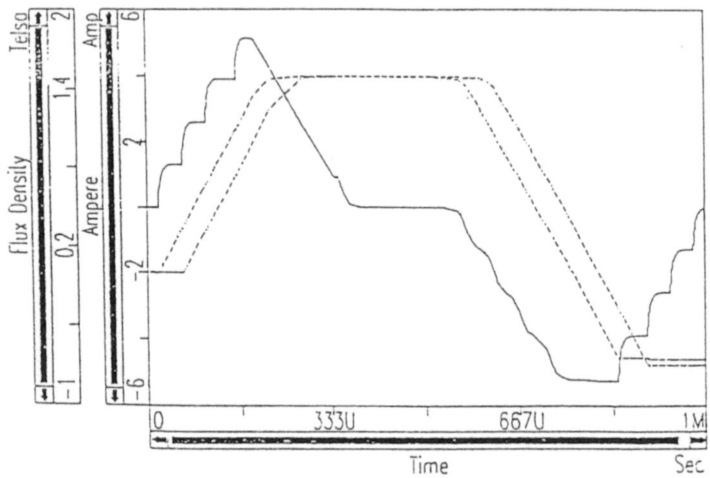

Fig. 16.8 Simulation: magnetic inductions (non saturated cores)

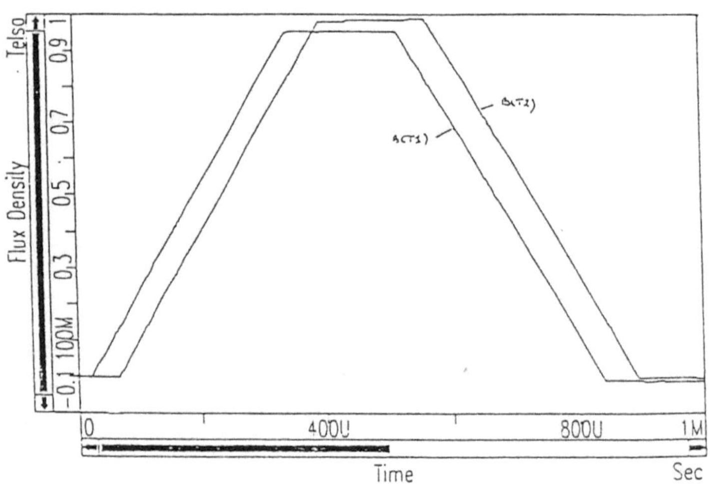

Fig. 16.9 Simulation: load current and magnetic inductions (saturated cores)

Thus, simulation allows one to predict the global behavior of circuits in the design stage and permits the reduction of the prototype and adjustment steps.

Another benefit provided by modeling is the estimation of values that are difficult to measure in practice. This is the case for losses in active components during switchings. Simulation of these losses requires precise modeling of the components and standard models found in SPICE or its derivatives need to be modified because they use component structures designed for low power and the high-power components are different. Thus, the diode model was modified to take into account some second-order effects which are not negligible in our case.

16.4 LOSSES DURING COMPONENT SWITCHING: CASE OF THE DIODE

16.4.1 Diode Modeling

Power diodes must be able to withstand high reverse voltages without junction breakdown. To attain this goal, they are made with 'wide' areas (some 10 μm) that are weakly doped (about 10^{14} at/cm^3), and thus are highly resistive. First this involves a conductivity modulation because, as soon as the junction becomes forward-biased, charges carrying current become non-negligible with respect to doping impurities and contribute to diode resistivity decrease. On the other hand, these carriers constitute a stored charge in material that will have to be evacuated before the component recovers blocking ability, causing a reverse current through the device. These effects that can have important consequences here are negligible for low power devices and thus are poorly modeled by the SPICE standard models.

Figure 16.10 shows current-voltage waveforms during turn-on and -off for an ideal diode (close to what would be modeled by SPICE) and for an actual diode. When the diode is turned on, current begins to flow and consequently the resistivity starts to decrease, which can lead to an overvoltage across the device if switching is fast. In the same way, during turn-off, the diode has to evacuate the stored charge before the junction becomes reverse-biased and this creates a reverse current that can be important (several amperes). This phenomenon, called 'reverse recovery', can lead to important losses. Moreover, with 'fast' diodes, this effect can be difficult to measure experimentally because it is very brief (a few 10 ns). Modeling is therefore a useful tool.

We had to develop a diode model able to simulate these effects and the results are presented in Fig. 16.11. In order to validate the model, we compared simulation and experimental results in the case of a step-down regulator.

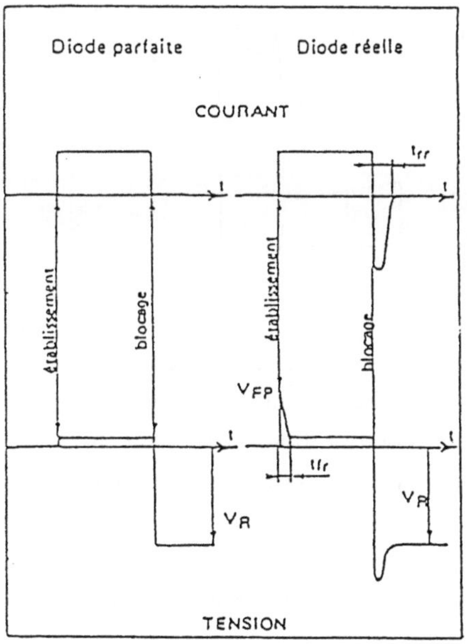

Fig. 16.10 Diode switching

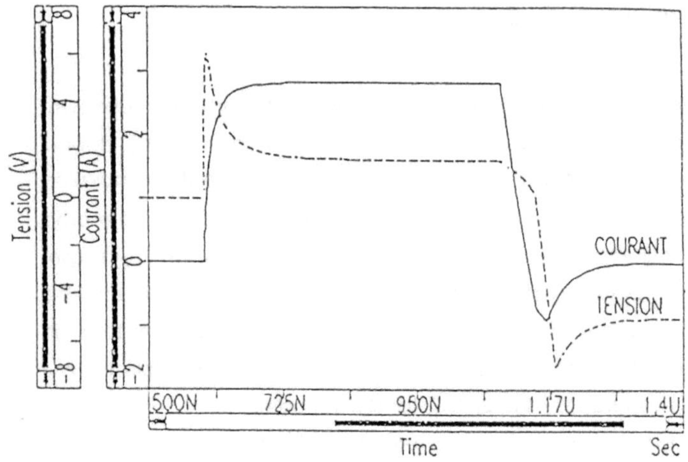

Fig. 16.11 Diode switching simulation

16.4.2 Step-down Regulator

This DC-DC converter uses a switching transistor to control the energy transfer from source to load. Its operating principle is illustrated on Fig. 16.12.

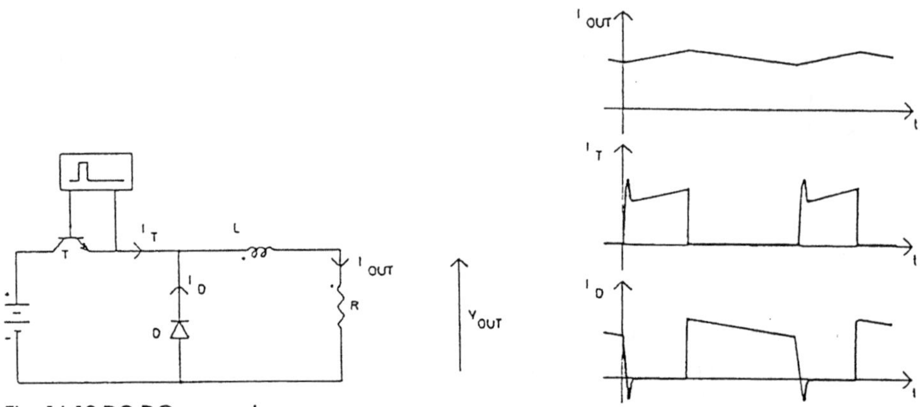

Fig. 16.12 DC-DC converter

For each turn-off of the free-wheeling diode, the reverse recovery current must be supplied by the source and consequently must cross the transistor. These overcurrents can involve additional losses in active components. We recorded experimentally these effects in case of a slow diode (i.e. the reverse current going up slowly) for measurement convenience. Figure 16.13 shows the experimental and simulated results for the current through the transistor.

NB: *With a faster diode, the maximum reverse current would be less important but its abrupt variations could involve high overvoltages in load or wire inductances.*

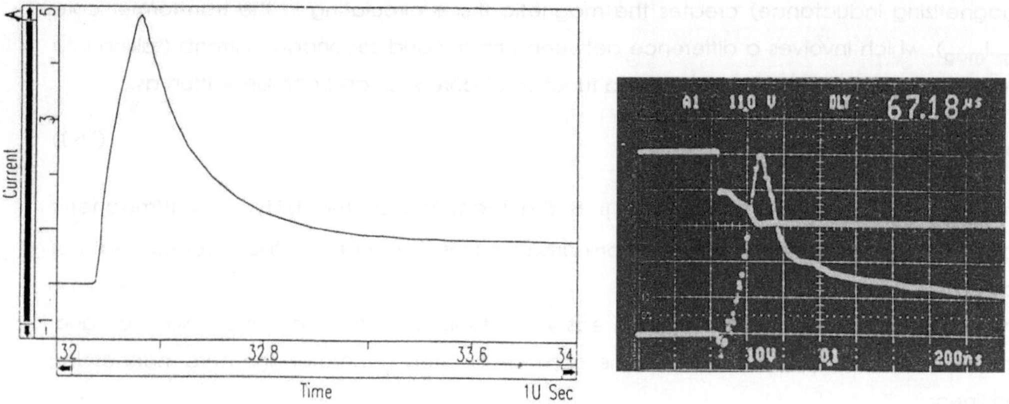

Fig. 16.13 Measured and simulated current through the transistor

16.5 TRANSFORMER MODELING

16.5.1 Core Size Influence

Although the use of electrical simulation as a tool is widespread, this is not the case for magnetic circuit modeling. These elements are, however, the connecting part between generator and load. The state-of-the-art for their design consists of choosing magnetic materials with reduced weight and space requirements, without undersizing the cores, which could become saturated, thus causing important losses.

A good way to illustrate the transformer operation is to represent it in the following manner (Fig. 16.14):

N1 = PRIMARY WINDING TURNS

N2 = SECONDARY WINDING TURNS

Fig. 16.14 Transformer loss representation

L_p (magnetizing inductance) creates the magnetic flux ϕ circulating in the transformer core ($\phi = L_p.I_{mag}$), which involves a difference between primary and secondary currents (taking into consideration the transformer ratio). L_p is a function of core sizes and can be written as:

$$L_p = N_1^{\ 2} . \mu A / l \tag{16.1}$$

where N_1 is the primary winding turns, μ is the magnetic permeability (μ = B(magnetic induction) / H(magnetic field)), A is the core area and l is the length of the magnetic path of the core.

The required magnetizing current can easily be displayed when applying a square signal across the transformer primary, which is the case for the step generator: the core magnetizing is then linear.

$$\begin{aligned} V_1 &= N_1\, d\phi / dt \\ \phi &= B.A \\ V_1 &= N_1.A.dB/dt \end{aligned} \tag{16.2}$$

and:

$$B = (V_1.t)/(N_1.A)$$

while the constant voltage V_1 is applied. On the other hand, $B = \mu.H$ where μ can be considered constant for low values of induction, and

$$H = N_1\, I_{mag} / l \quad \text{(Ampère's theorem)} \tag{16.3}$$

Thus, B is proportional to I_{mag} which varies linearly with time. I_{mag} is given by the difference between the primary and secondary currents (for unity transformer ratio), which is then easy to display when comparing them. This can be shown as well by simulation (Fig. 16.15) and by measurement (Fig. 16.7) where step slopes give the ratio I_{mag} / I_{oad} .

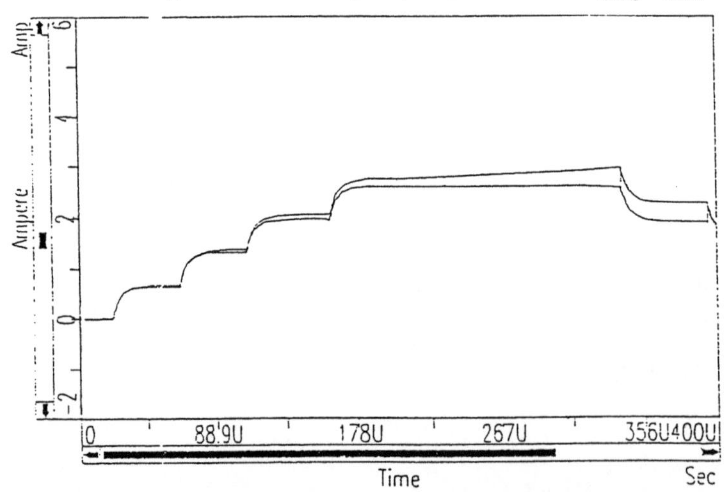

Fig. 16.15 Step generator simulation: primary and secondary currents of a transformer

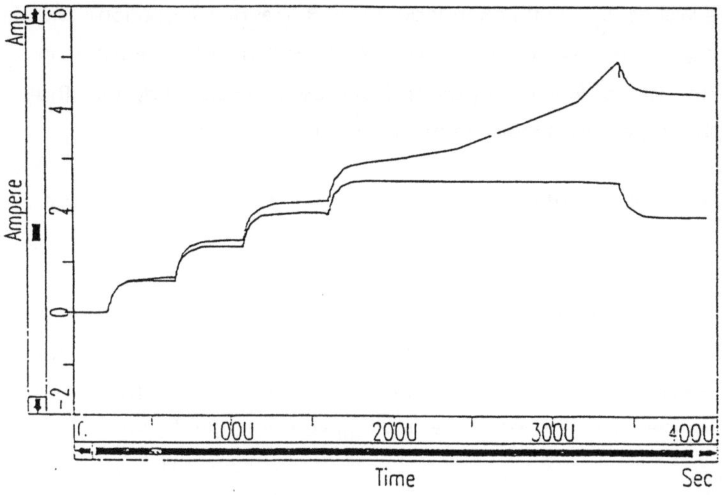

Fig. 16.16 Primary and secondary currents of a transformer in case of core saturation

To show the influence of the core area, the problem has been simulated again with the area reduced to one half. The new result is given on Fig. 16.16. In this case, I_{mag} is large, equal to several amperes, which leads to important losses in the core. Its variation is no longer linear. This last point is characteristic of magnetic material saturation. Indeed, when B becomes large, the B(H) curve is no longer linear (see Fig. 16.17). When B approaches B_{sat}, the permeability becomes less and results in an apparent decrease of L_p. Consequently, $I_{mag} = V_1.t/L_p$ increases. Moreover, hysteresis losses, which are equal to the area within the B-H cycle, become larger.

16.5.2 Parameter Extraction

As in all simulation work, the problem relative to the extraction of parameters needed to build component models arises. Some parameters are easily obtained from measurements or data sheets (wiring resistances, core sizes, etc). A more critical point concerns magnetic hysteresis B = f(H) as shown typically on Fig. 16.17.

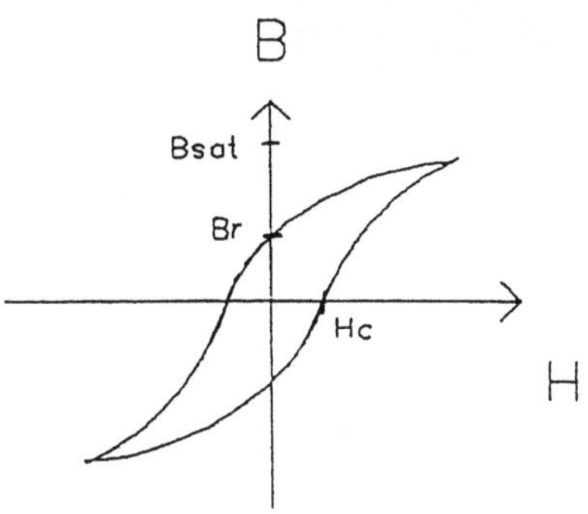

Fig. 16.17 B(H) hysteresis

Hysteresis is defined by three values: B_{sat} (saturation induction), B_r (residue induction) and H_c (coercive field). Determining these values is possible with an experiment illustrated in Fig. 16.18. A transformer the secondary of which is open, is driven by a sinusoidal signal. The measured current flowing through the primary is then the magnetizing current.

$$V_1 = V \,.\, \sin \omega t = N_1 \,.\, d\phi/dt = N_1 \,.\, A \,.\, dB/dt \tag{16.4}$$

Thus :

$B = -V/(N_1.A.\omega) \,.\, \cos \omega t = V/(N_1.A.\omega) \,.\, \sin(\omega t - \pi/2)$

B and V_1 are proportional and out of phase of $\pi/2$, while H and I_1 are proportional and in phase (see eq. 16.3). In this way, it is possible to obtain curves similar to those shown in Fig. 16.19. The values of B_r and H_c may be determined from the left figure and B_{sat} from the right figure.

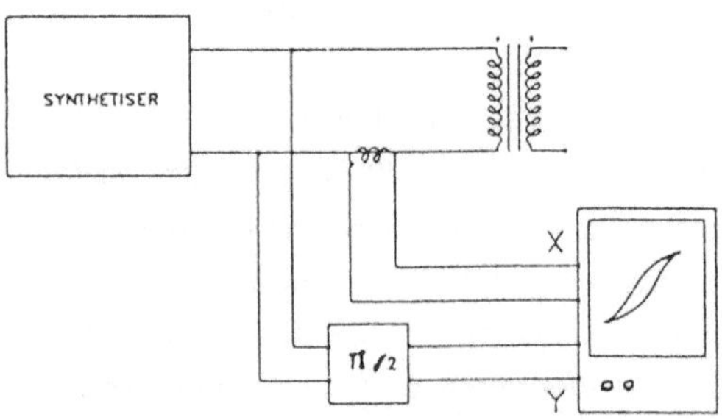

Fig. 16.18 Hysteresis recording setup

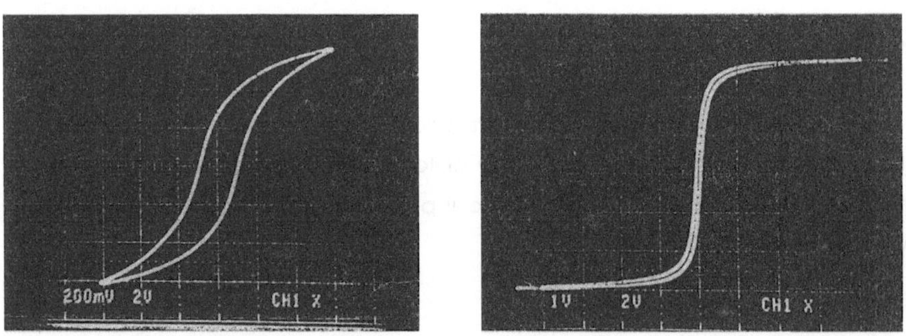

Fig. 16.19 Hysteresis records

16.5.3 Initial Transformer Magnetization

It is shown in Fig. 16.4 that the PWM. transformer is highly saturated after starting. The initial induction in the core was then assumed equal to zero. Actually, this is not accurate because of B(H) hysteresis, so that B is not zero when H passes through zero. To illustrate this property, measurements have been run on a transformer initially magnetized ($B_{Init} > 0$) or demagnetized ($B_{Init} < 0$). For this purpose the transformer is driven by a series of square pulses, the secondary being open in order to determine H from the primary current measurement. In a first test, the series have an even number of pulses (as many positive as negative pulses). During one cycle, the B(H) hysteresis is followed, magnetizing and demagnetizing the core successively. In the next series, starting occurs with a negative B_{Init} and the core is not highly saturated (see Fig. 16.20). Fig. 16.21 below is a comparison between simulation and measurement showing the secondary voltage and the primary current (I_{mag}).

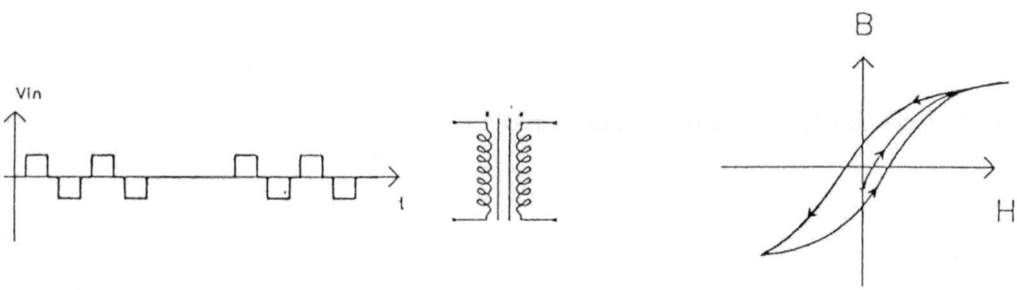

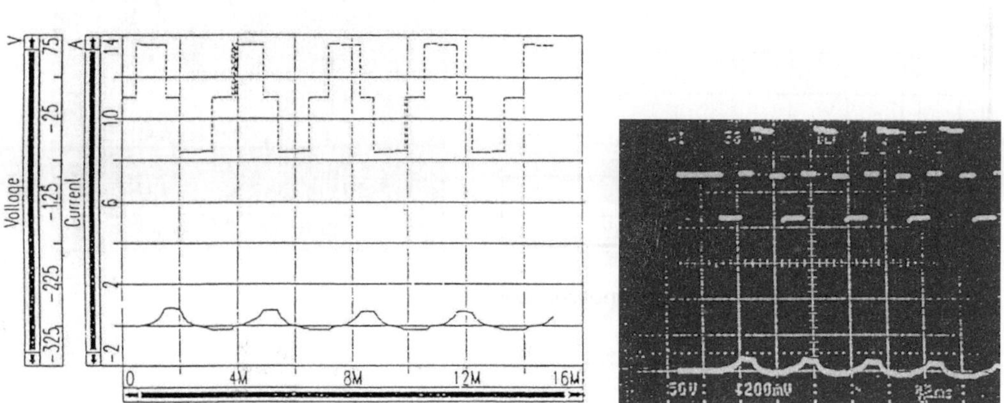

Fig. 16.21 Initial transient period for a demagnetized core

In a second test, the number of pulses during one series is odd. At the end of the cycle, the core is not demagnetized and the hysteresis is not completely followed. In the next series, starting occurs with a core already magnetized and it becomes highly saturated during the first alternation (see Fig. 16.22). Figure 16.23 compares measurement and simulation, and shows that good agreement between the two approaches is achieved.

It should be noticed that during the first alternation, the secondary voltage starts to decrease, showing saturation. Simulation allows to foresee the behavior of a generator at starting and permits prediction of initial operation transients that lead to additional losses, and that are difficult to measure experimentally.

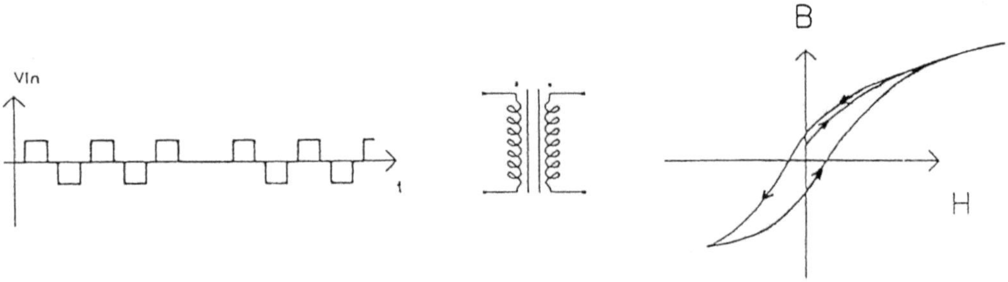

Fig. 16.22 Hysteresis following an initially magnetized core

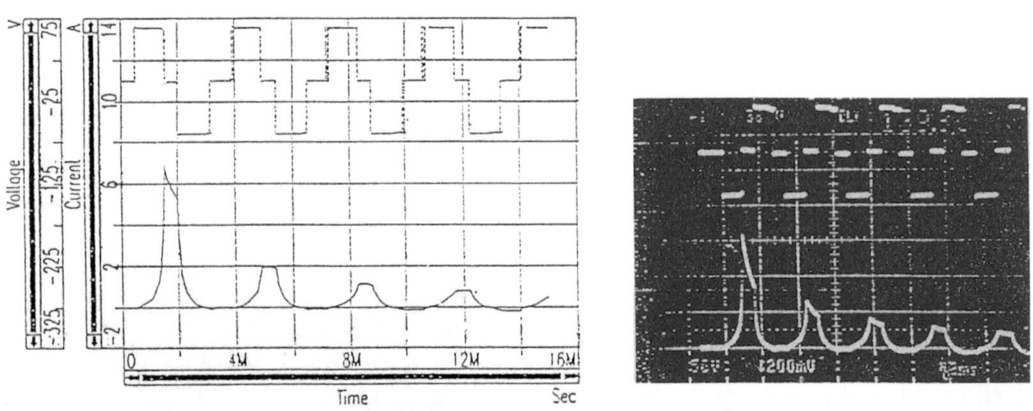

Fig. 16.23 Initial transient period for a magnetized core

16.6 TRANSMISSION LINES

16.6.1 Transmission Lines Used in the Active Sonar Field

The aims of this chapter are to show a way of optimising the transmission line that provides energy to an acoustical array, and to explain how to include this in the whole system simulation. Energy transfer between the emitter and the electroacoustic array is a vital function in the transmission chain. Overall efficiency improvement also requires line efficiency optimisation. That is, for a given line, loading it with a proper impedance can raise its efficiency. However, in our case, things are more complex and optimisation is made under constraints. The designer has to take into account the following parameters:

- size and weight of lines in regard to hydrodynamic constraints,
- reasonable mechanical characteristics.

The goal is, therefore, to make the best use of the available space, the electrical conductors and the generators. The problem could be posed as: "Knowing the possible sizes and weights of conductors and having fixed frequencies, powers and impedances, transmission lines and matchings must then be selected in order to obtain the best efficiency of the transmission chain".

16.6.2 Value of Simulation Tools

The above-mentioned aspects show what can be obtained with computing tools. Indeed, even if experience remains important in this field, calculation is preponderant. From the preceding, several processes can lead us to our goal, namely:

- line electrical characterisation according to size and mechanical aspects,
- line efficiency calculation according to loads and electrical characteristics,
- matched line impedance calculations.

Moreover, thermal aspects should not be forgotten. Indeed, the designer must develop tools intended for this optimisation, but must afterwards take into account these results to simulate the entire system. Following the process depicted in Fig. 16.24, the different steps will be developed below.

16.6.3 Line Electrical Characterisation

This first step is intended to determine the line primary constants (R, L, C, G), starting from size and material considerations.

R: its determination is, of course, based on Ohm's law $R = \rho.l/a$, where ρ is the resistivity, l is the length and a is the cross-sectional area of the conductor. However, also to be taken into account are the variations of ρ with temperature and frequency which are described in literature. The corresponding relations involve Bessel functions related to $(\phi^2.\mu.\rho.f)$, ϕ being the conductor diameter, μ the permeability, and f the frequency. They can be computed, giving good results.

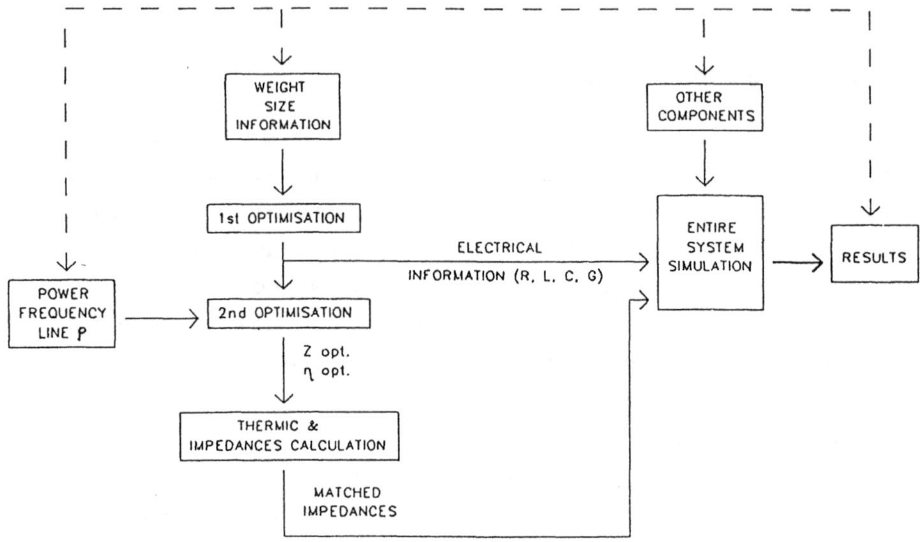

Fig. 16.24 Transmission line optimisation process

L: for different line types, empirical or mathematical relations exist, which are function of μ, ϕ, d, ρ and f.

C: this calculation uses basic relations in case of pairs or coaxial lines and involves ε (insulator permittivity), ϕ (small conductor diameter) and d (distance between conductors).

G: good knowledge of the dielectric material used permits, for given frequencies and operating conditions (voltage, mechanical constraints), to deduce G from C. Dielectric characterisation will also determine the allowable voltage gradient.

It has been seen that each primary parameter of a line can be obtained knowing its constituent parts, its topology and the working frequency. Values of R, L, C and G can then be computed, realising one of the operations depicted on Fig. 16.24. Measurements previously made at GERDSM show good agreement with theory. Indeed, everything necessary to build a line model and use it in a simulator is available. This model can be based on the general equations for lines found in annex 16.1, or in the lumped element transmission line schematic found in annex 16.2.

16.6.4 Line Efficiency Optimisation

In the relations of annex 16.1, one can see that, thanks to primary parameters, it is possible to determine the secondary constants α and β as well as the voltage and current above and below the line, for a given frequency and load. The major aspect here is the efficiency expression: $\eta = Re(V_o . I_o) / Re(V_l . I_l)$

In this way, expressing efficiency as a function of the load impedance (R, X) and canceling its derivative, it is possible to find an impedance for which line losses are reduced. But this optimum process is for low level drives and it is necessary to optimise under power constraints. These limits are the maximum tolerable voltage and heating effects. Only an iterative method applied to the complex load impedance is able to determine the best line efficiency, being given the power, voltage and frequency. The result of these iterations is a complex impedance, which must be seen down the line. Such software has been developed at GERDSM to determine best efficiencies.

16.6.5 Impedance Determination

These calculations, after efficiency optimisation, will allow to know the impedance value 'seen' up the line, and thus to determine matching required for good working conditions in the power amplifier. These calculations are based on line equations found in the annexes and can be carried out using a lumped element transmission line model in case of the whole transmission chain simulation.

16.6.6 Thermal Calculations

In this field, no tool has yet been developed and problems are treated individually. However, a simulation using thermal-electrical equivalents is foreseen.

16.7 CONCLUSION

We have just seen that optimisation of the transmission line is a part of the whole chain optimisation. With use of an analog simulator, cables can be modeled by an equivalent schematic circuit. Future studies using this methodology should enable the simulation code to determine the electrical aspects of the line from its size and material characteristics.

For other components, simulation must allow the design of an entire chain, starting with functional blocks and thus describe the global behavior of the chain in order to define its topology. It must then be possible to refine each block to study precisely its operation and design its practical realisation. Thus a multi-level simulation must ease the processes, starting from a general idea to an ultimate product.

In the more distant future, one should think of a sufficiently "clever" simulator to work with the designer's constraints (available weights or volumes and mechanical constraints), but this relates to the subject of artificial intelligence.

REFERENCES

16.1 *Les transistors de puissance en régime de commutation*, THOMSON-CSF (SESCOSEM, 1975)

16.2 P. ALOISI, *Power switch* - (MOTOROLA, 1987)

16.3 R. CROZE, L. SIMON, J.-P. CAIRE, *Transmission téléphonique, théorie des lignes,* (éditions EYROLLES, 1973).

Annex 16.1

$$V_I = V_O . ch\gamma l + Z_c . I_O . sh\gamma l$$

$$I_I = I_O . ch\gamma l + \frac{V_O}{Z_c} . sh\gamma l$$

$$Z_I = Z_c . \frac{Z_O + Z_c . th\gamma l}{Z_c + Z_O . th\gamma l}$$

$$V_O = V_I . \left(ch\gamma l - \frac{Z_c}{Z_I} . sh\gamma l \right)$$

$$I_O = I_I . ch\gamma l - \frac{V_I}{Z_c} . sh\gamma l$$

$$Z_O = Z_c . \frac{Z_c - Z_I . th\gamma l}{Z_I - Z_c . th\gamma l}$$

$$\gamma = \alpha + j\beta = \sqrt{(R + j.L.\omega) . (G + j.C.\omega)}$$

$$Z_c = R_c + j. X_c = \sqrt{\frac{R + j.L.\omega}{G + j.C.\omega}}$$

$$\eta = \frac{Re\left[V_O . \dot{I_O}\right]}{Re\left[V_I . \dot{I_I}\right]} = \frac{R_O . |I_O|^2}{R_I . |I_I|^2}$$

Annex 16.2

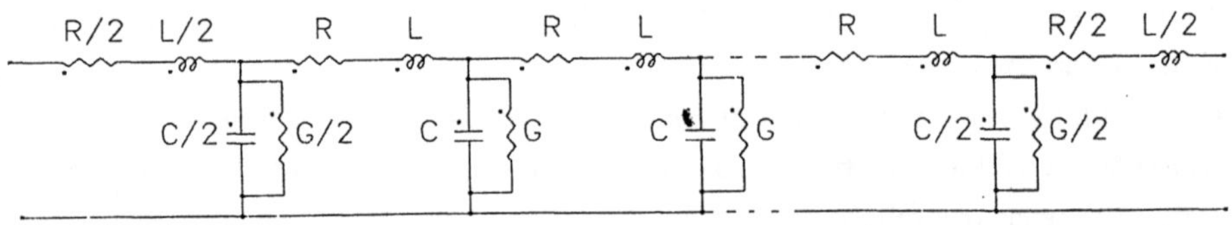

17. Calibration of Underwater Acoustic Transducers at NRL/USRD

A.L. Van Buren and J.E. Blue

Underwater Sound Reference Detachment, Naval Research Laboratory, P.O. Box 568337, Orlando, Florida 32856-8337, USA

At the beginning of World War II, the United States recognized the need for establishing systematic methods for calibrating and evaluating sonar transducers and established the Underwater Sound Reference Laboratories. The Underwater Sound Reference Detachment of the Naval Research Laboratory evolved from this beginning and now serves as the focus for underwater acoustic metrology in the United States. It functions as a "National Bureau of Standards for Underwater Acoustics" by providing two types of reference services for a fee. Calibration, test, and evaluation services and standard transducer loan services cover the frequency range of 0.2 Hz to 2 MHz. This paper describes the calibration, test, and evaluation services. It first covers the measurement facilities for performing the services. Measurement methods for calibrating transducers are then discussed. Next a description of ongoing research in calibration measurement methodology is presented. Finally, some comments are made about calibration system requirements for implementing new measurement techniques.

17.1 INTRODUCTION

Prior to the beginning of World War II, the U.S. Navy's capability for calibrating sonar transducers was virtually non-existent (17.1). The Office of Scientific Research and Development recognized the need for standards for calibrating sonar transducers and in July 1941 entered into a contract with the Bell Telephone Laboratories (BTL) to supply measurement instrumentation and systems and in March 1942 with the Columbia University Division of War Research to operate the Underwater Sound Reference Laboratories (USRL) located in Mountain Lakes, New Jersey and Orlando, Florida. In 1942 the USRL began to study and test reciprocity calibration concepts that had been independently devised by MacLean (17.2) in 1940 and Cook (17.3) in 1941. Standard hydrophones were developed by BTL using piezoelectric crystals of Rochelle salt and ammonium dihydrogen phosphate. BTL developed modified moving-coil sound projectors for the audio-frequency range and piezoelectric crystal sources for the ultrasonic range. This marked the birth of standard underwater acoustic metrology practices in the United States and of what has become the Underwater Sound Reference Detachment (USRD) of the Naval Research Laboratory.

Bobber gives a summary of the state-of-the-art in underwater acoustic metrology at the end of World War II as follows:

At the end of World War II in 1945, it was possible to calibrate a small hydrophone from 2 Hz to 2.2 MHz under ambient environmental conditions. Projectors or sound sources weighing up to a few hundred pounds could be calibrated from about 50 Hz to 140 kHz, with driving powers of 1.5 kW available in the audio-frequency range. When pressure was a variable, the capabilities were limited to 2-100 Hz and 100 psi for small hydrophones only, and to 10-150 kHz and 300 psi for fairly small hydrophones and projectors (less than 100 lbs). Standard hydrophones and projectors were adequate, but far from ideal. Stability with time, static pressure, and temperature in many cases was poor for "standard" purposes. Sound sources generally were cumbersome, and response curves were not smooth and flat as is desirable in calibration work. The measuring systems were not capable of pulsed-sound measurements. Good free-field, or unbounded medium, conditions were generally assumed but seldom obtained. Thus, in spite of great strides forward during World War II, the state of the art in 1945 was still relatively crude.

These capabilities have been improved upon considerably since that time Figure 17.1 shows an aerial view of the USRD on Lake Gem Mary where its main building was completed in 1951. The pier structure on the lake was also completed about that time. This Lake Facility is one of four facilities utilized by the USRD in pursuit of its mission. The lake temperature varies during the year between 15° and 30°C. Water in the test wells is separated from the rest of the lake water and from unwelcome marine life by the use of acoustically transparent swimming pool liners and kept isothermal with a bubbler system. There is nothing unusual about this facility. Its primary use is in calibrating the USRD's loan standards prior to shipment to customers. It is also used for outside customers and for in-house research.

Fig. 17.1 Aerial view of USRD

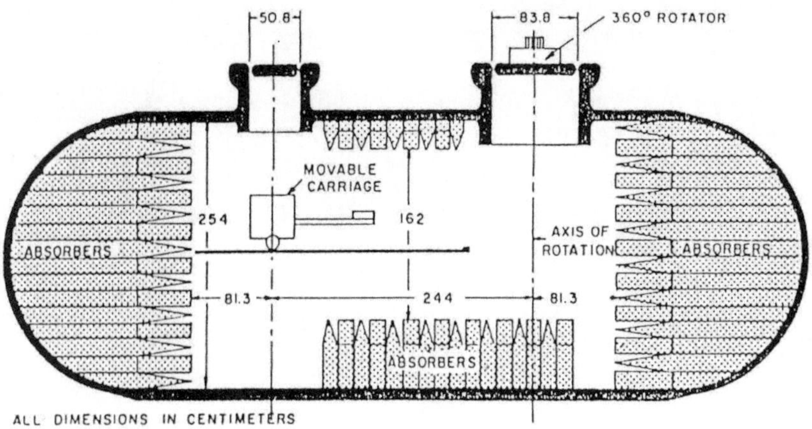

Fig. 17.2 Anechoic Tank Facility

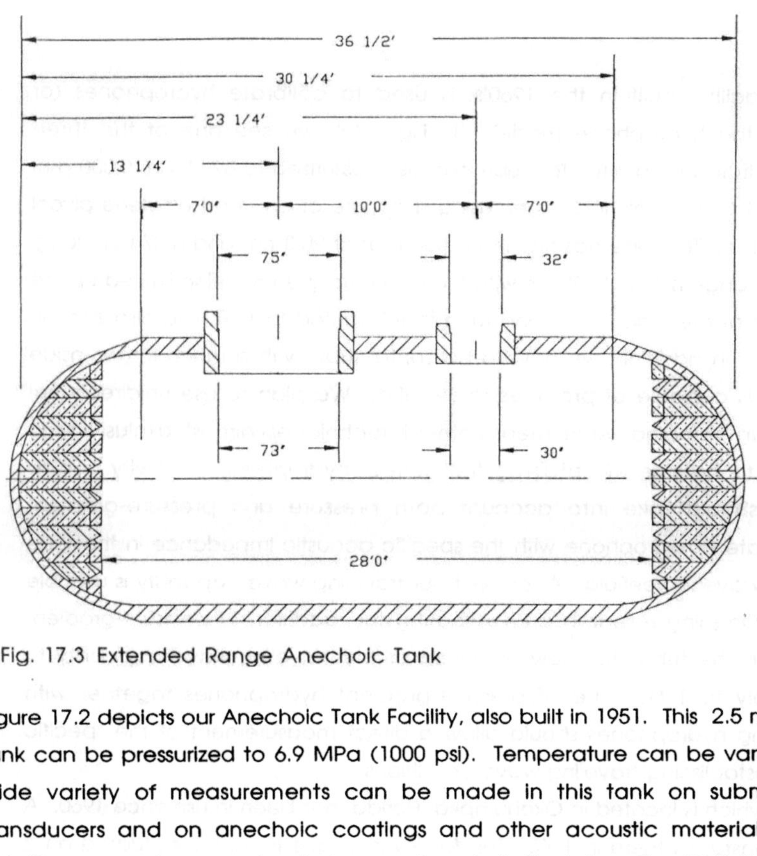

Fig. 17.3 Extended Range Anechoic Tank

Figure 17.2 depicts our Anechoic Tank Facility, also built in 1951. This 2.5 m diameter, 7.6 m long tank can be pressurized to 6.9 MPa (1000 psi). Temperature can be varied from 2° to 40°C. A wide variety of measurements can be made in this tank on submarine and weapons transducers and on anechoic coatings and other acoustic materials. We are presently installing a new larger, extended-range Anechoic Tank (see Fig. 17.3) that can be pressurized to 20.7 MPa (3000 psi).

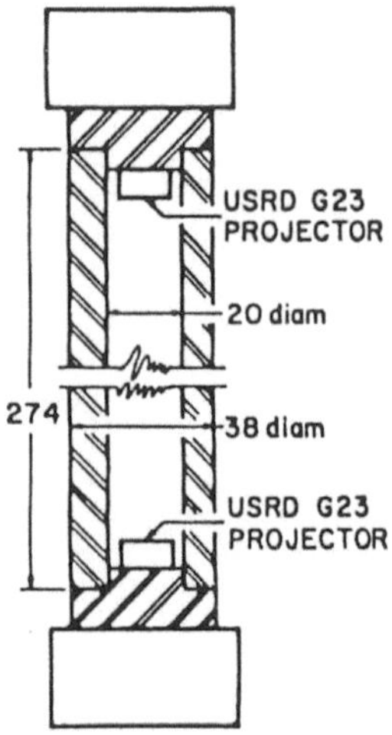

Fig. 17.4 System J

Our Low Frequency Facility, built in the 1960's, is used to calibrate hydrophones (or reciprocal transducers in the hydrophone mode). In Fig. 17.4. we see one of the three calibration tubes that constitute the facility. This tube can be pressurized to 69 MPa (10,000 psi). Temperature can be varied from -2° to 40°C when using a mixture of one part ethylene glycol to two parts water, by volume. The tube has an inside diameter of 20.3 cm and is 274 cm long. The operational frequency range is 1 to 4,000 Hz, with the upper frequency being limited by the first radial resonance mode of the tube. A second tube that is limited to 13.8 MPa also exists in our Low-Frequency Facility. In addition, we have a horizontal tube with a much larger inside diameter of 38.1 cm that is capable of pressures to 69 MPa. We plan to use unidirectional traveling-wave rather than standing-wave measurement techniques almost exclusively in these tubes in the future to be sure of obtaining the correct hydrophone sensitivity. Since hydrophone evaluation should take into account both pressure and pressure-gradient sensitivity, one must calibrate a hydrophone with the specific acoustic impedance in the tube equal to that for a plane wave in freefield. At present our traveling-wave capability is useable down to about 100 Hz. Ongoing research is investigating the addition of pressure-gradient monitoring hydrophones in the tubes to allow extension of the traveling-wave capability to lower frequencies, possibly to 1 Hz. Use of pressure-gradient hydrophones together with existing pressure monitoring hydrophones should allow a direct measurement of the specific acoustic impedance for establishing traveling wave conditions.

Our Leesburg Facility, which is located in Okahumpka, Florida, has been in use since 1960. A permanent structure was installed there in 1965. This facility is unique in that it is situated on a small spring-fed lake 52 meters deep. There is no main well-head. Water seeps in from numerous places around the walls of the cavity in such a way as to produce no measurable flow noise. Flow maintains the spring at about 22°C at all times of the year. The flow is such that below about 2 meters the water is isothermal with depth. Oxygen content of the water is so low at 5 meters of depth and below that fish are not commonly found at the normal calibration depths used in this facility. The main work load for this facility is the calibration of towed-line array sonars.

Measurement methodology used in USRD's four facilities was, for the most part, developed after World War II. These four facilities have been used in developing the standard transducers that are available on a loan basis from the USRD, are used in research for new standard transducer and methodology development, and are used widely by the USRD, and other U. S. Navy laboratories and contractors in their development of sonar transducers and underwater acoustic materials. In addition these facilities and standards are made available to any non-naval U. S. institution or industry with the need for standards or calibration of underwater acoustic devices. Arrangements can be made by NATO countries to utilize these services through agreements with the U.S. Naval Sea Systems Command (NAVSEA) or through U. S. subsidiaries of NATO country companies. This paper discusses how primary underwater acoustic standards are calibrated. The calibration of secondary standards, 'standard' projectors, evaluation of sonar transducers, and calibration systems are also discussed.

17.2 CALIBRATING THE PRIMARY STANDARDS

Even though the USRD has about 1800 transducers (hydrophones, reciprocal transducers, and projectors) of about 40 different types in its standard loan inventory, we only consider hydrophones, i.e., transducers used solely as hydrophones, to be suitable for primary standards. In particular, we only consider those hydrophones that maintain stable sensitivity over large temperature and pressure ranges and over a long time to be suitable for standards. The other transducers that we maintain are either necessary in calibrating the primary standards or in enabling customers to meet some of their specialized requirements for acoustic measurement, test, and evaluation.

There are three methods for calibration in use at the USRD for absolute calibration of primary standard hydrophones. They are:

- free-field reciprocity calibration,
- coupler reciprocity calibration,
- the two-projector null method of calibration.

Free-field reciprocity methods are used in the Lake and Leesburg Facilities down to about 100 Hz. Below 100 Hz lack of a suitable robust reciprocal transducer limits this technique. In the Anechoic Tank Facility the free-field reciprocity method is used down to about 2 kHz. A discussion of free-field reciprocity including phase calibration from Luker and Van Buren (17.4) is reproduced here:

The hydrophone calibration procedure described here is an extension of conventional three-transducer reciprocity calibration (17.1) to include phase. Use of this procedure requires three transducers (a projector P, the hydrophone H to be calibrated, and a reciprocal transducer T that can be used as both a hydrophone and a projector). A series of three projector-hydrophone measurements are made using either P or T as a projector and H or T as a hydrophone. The measurements are made under free-field conditions with the hydrophone located in the farfield of the projector. The three experimental setups are indicated in Fig. 17.5. The input current and output voltage values are complex, i.e., they include both amplitude and phase. The input currents in setups 1 and 2 are chosen to be identical. In this case i_p does not appear in the final expression for the sensitivity of the hydrophone and need not be measured.

For setup 1 the farfield pressure p_{PH} produced at H, which is located d_1 meters from P, is

$$p_{PH} = (i_P S_P d_0 / d_1) \ \exp (jk (d_0 - d_1)), \tag{17.1}$$

where S_p is the transmitting current response of P and d_0 is the reference distance, normally equal to 1 m, at which the transmitting pressure is specified in the definition of S_p. The wavenumber k is equal to ω/c, where ω is the angular frequency in radians per second and c is the sound speed in the surrounding medium in meters per second. The assumed time dependence $\exp(j\omega t)$ has been suppressed for convenience. The open-circuit voltage produced by H is given by

$$e_{PH} = M_H p_{PH} = (M_H i_P S_p d_0 / d_1) \exp(jk (d_0 - d_1)), \tag{17.2}$$

where M_H is the receiving voltage sensitivity of H.

Similarly, we obtain for setup 2:

$$e_{PT} = M_T p_{PT} = (M_T i_P S_p d_0 / d_2) \exp(jk (d_0 - d_2)), \tag{17.3}$$

where M_T is the receiving voltage sensitivity of T. Combining Eqs. (17.2) and (17.3) yields

$$e_{PH}/e_{P_T} = (M_H d_2 / M_T d_1) \exp (jk (d_2 - d_1)). \tag{17.4}$$

Since T is a reciprocal transducer, we have $M_T = JS_T$, where J is the complex spherical wave reciprocity parameter given by Beranek (17.5) as

$$J = (4\pi d_0 / j\omega\rho) \exp (jkd_0), \tag{17.5}$$

where ρ is the density of the surrounding medium. From setup 3 we obtain

$$e_{TH} = M_H \ p_{TH} = (M_H i_T S_T d_0 / d_3) \ \exp (jk (d_0 - d_3)). \tag{17.6}$$

Combining Eqs. (17.4) and (17.6) with the use of Eq. (17.5) produces the following expression for the receiving voltage sensitivity of the hydrophone H:

$$M_H = \{(4\pi e_{PH} e_{TH} \ d_1 d_3) \ / \ (j\omega\rho e_{PT} \ i_T d_2)) \ \exp (j(\omega/c) (d_1 + d_3 - d_2))\}^{1/2}. \tag{17.7}$$

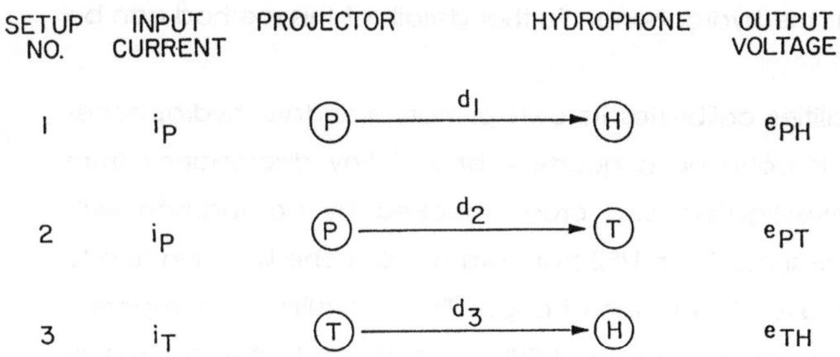

Fig. 17.5 Measurement setups for freefield reciprocity calibration

Standard practice at the USRD and elsewhere in the United States is to ignore the phase calibration and do magnitude calibration only. The distances d_1, d_2, and d_3 in this case are usually all the same. For phase calibration using free-field reciprocity the discussion by Luker and Van Buren continues:

The difficulty in determining the phase of M_H using this method lies in accurately determining both the sound speed and the measurement distances d_1, d_2, and d_3. For example, at 100 kHz in water an error of only 1 mm in any one of the distances gives a phase error of about 12°. However, we can avoid this difficulty by positioning all three transducers P, H, and T in a straight line with H located between P and T. This assures that $d_2 = d_1 + d_3$. Then Eq. (17.7) simplifies to

$$M_H = ((4\pi e_{pH} e_{TH} d_1 d_3) / (j\omega\rho e_{PT} i_T d_2))^{1/2} \tag{17.8}$$

Since the distances and sound speed no longer appear explicitly in a phase term in Eq. (17.8), the accuracy of the phase of M_H calculated using Eq. (17.8) is limited only by the accuracy of the phase measurements of the voltages and current.

Phase calibration experiments (17.4) have proven the feasibility of this method, and new computer-controlled calibration systems being developed at the USRD will have the capability of making on-line complex calibration measurements. These measurements will become routine after the required specialized rigging is further developed and as experience is gained through additional experiments.

In the Low Frequency Facility, reciprocity couplers, such as shown in Fig. 17.6, are used in the absolute calibration of hydrophones over the frequency range from 10 Hz to 8 kHz, the temperature range from -2 to 35° C, and the pressure range from 0 to 105 MPa (15,000 psi). The two-projector null method, as illustrated in Fig. 17.7, has been used in the USRD's low frequency tubes for the absolute calibration of hydrophones from 0.2 to 20 Hz. In this method the two projectors are driven electrically with a provision for controlling the relative phase and amplitude of the two drives. The input to a moving coil projector is adjusted in amplitude and phase until a null pressure is obtained at the hydrophone. Further details of this method can be found in the book by Bobber (17.1).

Each of USRD's four calibration facilities calibrates its own primary standard hydrophones using the appropriate methods. This is done on a quarterly basis. Any discrepancy from previous calibrations is thoroughly investigated and cross checked by comparison with calibrations in other USRD facilities. Semi-annually an H52 standard hydrophone is sent in turn to each of the four facilities for calibration over their range of capabilities. Results are compared, and facility differences are resolved. In this manner the USRD maintains a higher degree of calibration accuracy and confidence than is possible with one facility alone. This confidence is important in setting underwater acoustic metrology standards for the United States.

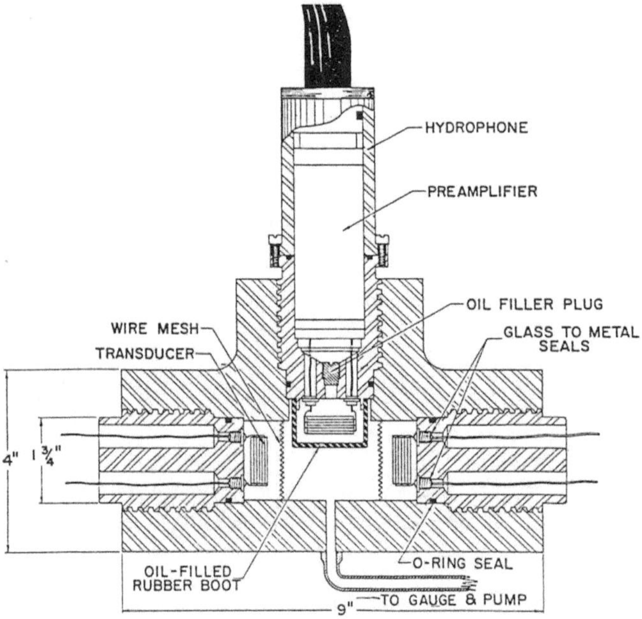

Fig. 17.6 Typical reciprocity coupler

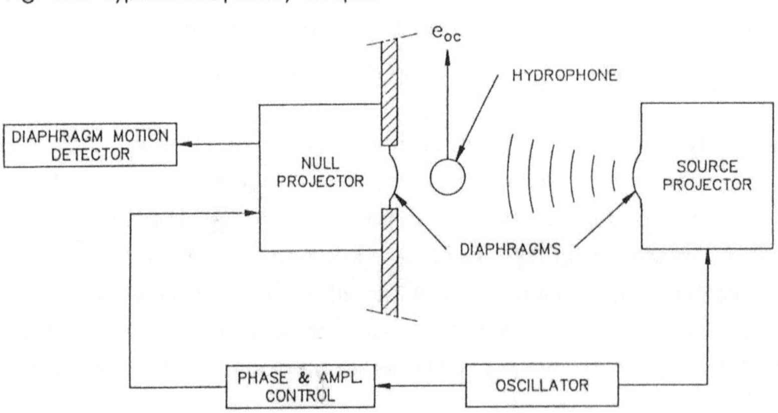

Fig. 17.7 Two-projector null calibration

17.3 SECONDARY CALIBRATIONS OF STANDARD LOAN HYDROPHONES, PROJECTORS, AND RECIPROCAL TRANSDUCERS

For economic reasons, all loan standards available from the USRD are calibrated secondarily using the USRD primary standard hydrophones which are calibrated quarterly. All of the standard types used by the USRD have been carefully developed and individually tested over environmental conditions that more than cover their range of intended use. The USRD Transducer Catalog (17.6) describes most of the standard transducers available to its customers.

The standard loan transducers are routinely calibrated at the USRD Lake or Leesburg Facility. Routine calibration entails using pulsed or continuous wave swept frequency secondary calibration methods at two different separation distances under good freefield conditions. Two separations are used to ensure that the transducers are responding properly to the pressure field. Failure of the transducer response to follow the inverse spreading law for spherical waves may imply improper farfield conditions or an inappropriate pressure-gradient sensitivity of the transducer, among other possibilities. Rigging has been carefully designed so as not to appreciably affect the calibration, and care is taken to see that there are no bubbles adhering to the transducers. Most transducers used in United States calibration facilities are loan standards calibrated by secondary methods. Some facilities perform freefield reciprocity calibration on their own or USRD standards. However, most Naval activities insist on maintaining traceability of their underwater acoustic measurements to the USRD since the USRD maintains traceability of its basic measurements used in setting underwater acoustic standard to the National Bureau of Standards. Just as some measurements done at the National Bureau of Standards result in an 'accepted value' being assigned to a standard, measurements on standard underwater acoustic transducers at the USRD result in 'accepted calibration curves' being assigned to those transducers. In this manner, secondary calibration of standard hydrophones loaned to a wide variety of customers sets underwater acoustic metrology practices for the United States.

17.4 EVALUATION OF SONAR TRANSDUCERS

Facilities at the USRD are also available for calibration, test, and evaluation of sonar transducers and arrays. These facilities have served the Navy well in the development of transducers and arrays for a wide range of applications. In particular, the Low Frequency Facility has aided development of reliable hydrophones for long range surveillance systems. The Anechoic Tank Facility has advanced the development of torpedo and various submarine transducers. The Leesburg Facility has been involved with practically every towed-line array sonar in the Navy. However, the USRD Facilities are not adequate to handle all of the U.S. Navy's sonar calibration, test, and evaluation. Other Navy Laboratories have small calibration facilities such as Dodge Pond and the indoor test pool at the Naval Undersea Systems Center (NUSC) and TRANSDEC at the Naval Ocean Systems Center (NOSC). The Naval Weapons Support Center performs calibration measurements on sample lot units from production runs of fleet sonar transducers. The Navy also maintains three Transducer Repair Facility calibration facilities at Navy shipyards. In addition, to make some of the sonar calibration, test, and evaluation measurements that are not possible in small bodies of water, NUSC maintains a facility with a water depth of about 200 m at Lake Senaca in New York and NOSC maintains a capability at the David Taylor Naval Ships Research and Development Center's Lake Pend O'reille 400 m-deep facility. In general, these facilities provide their own electronic calibration equipment but for the most part rely on techniques developed by and standards supplied by the USRD.

Bobber (17.1) has done an excellent job of covering calibration practices at the USRD up to 1970. In this paper we will discuss a measurement methodology that has been developed since that time or is presently being developed. Much of the newer methodology was developed in order to allow lower-frequency calibration measurements in the multipath-limited environment of small facilities and is facilitated by new digital signal processing techniques involving analog-to-digital converters and computers.

An example of newer digital techniques replacing older analog ones is in immittance measurements. Bobber's (17.1) description of the apparatus and techniques for high level signal measurements is as follows:

> *One technique that has been used successfully is illustrated by the simplified diagram in Fig. 17.8. The transducer voltage or current pulse is sampled by a step-down current transformer consisting of a toroid coil surrounding a single conductor. Simultaneously, the signal in a continuous-wave reference circuit is sampled. The two signals are added and displayed on a cathode-ray oscilloscope. The reference signal contains only the fundamental frequency. The amplitude and phase of the reference signal is adjusted until a null condition in the steady-state part of the pulse is observed on the oscilloscope. Then the reference signal is equal in amplitude and opposite in phase to the fundamental frequency component of the transducer signal pulse. There will be a signal residue in the null because of the harmonic distortion frequencies in the signal. If the harmonic distortion is small, the residue is small and does not introduce measurable error in the nulling technique. If the residue is large, a low-pass filter is used to attenuate the harmonic frequencies.*
>
> *The three transformers should be identical to minimize errors. In practice, the current and reference signal transformers can be one and the same.*
>
> *The reference circuit is a low-signal continuous-wave circuit. The reference voltage and current are readily obtained by conventional calibration methods. Phase difference between the reference current and voltage is controlled and measured by the phase shifter.*

Using this null-balance technique, immittance and sound pressure levels in the water were measured as a function of input power in order to provide an indication of the linearity of a sound projector.

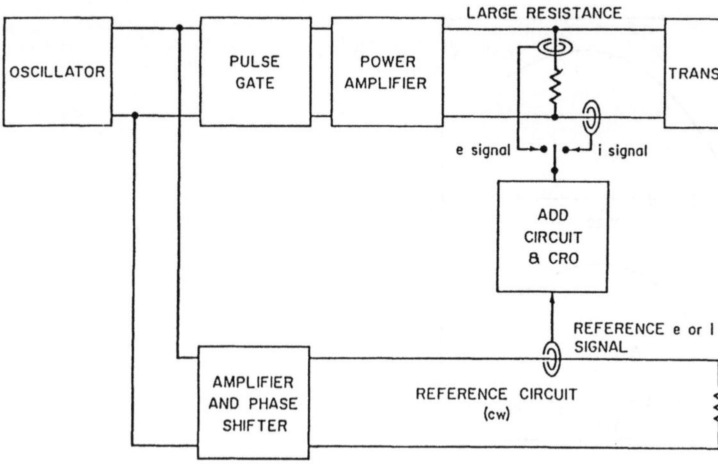

Fig. 17.8 Electronics for analog immittance measurements

In cases where there existed appreciable harmonic distortion, a high degree of accuracy could not be obtained from this null-balance technique. Newer techniques used at the USRD involve digitizing the input voltage, current, and standard hydrophone voltage waveforms and using Fast Fourier Transforms (FFT) in immittance and linearity calculations. These techniques involve sampling of continuous waves or pulsed sinusoids at rates that are multiples of or that are integrally related to the driving frequency thereby assuring a spectral line component in the FFT at the driving frequency. Immittance is then easily calculated by ratios of FFT's.

Use of digital computers in calibration systems has led to consideration of calibration techniques that do not require steady state signal conditions. One such technique investigated at the USRD by Beatty, George, and Robinson (17.7) is the Prony method. For a transducer to be capable of being calibrated by using the Prony method, it must be representable as a linear point or "lumped" device which can be described by a Laplace Transform. Data used in the Prony method must be acquired before the signals are contaminated by boundary reflections. Early work on the Prony method involved reciprocity calibration in the USRD's Lake Facility (17.7). Traditional reciprocity calibration measurements in this facility use pulsed sinusoids to gate out boundary reflection interference and have a low-frequency limit of about 500 Hz. Using two USRD Type J11 reciprocal transducer and a USRD Type H56 hydrophone, the experimental waveforms shown in Fig. 17.9 were acquired and used, via the Prony method, to calculate the amplitude and phase calibration curve shown in Fig. 17.10. These results show excellent agreement with calibration values obtained in USRD's Low Frequency Facility down to about 25 Hz. Work is still ongoing to develop methods for making accurate measurements in the USRD's Anechoic Tank Facility (and indeed in any measurement facility) where there is a requirement to determine the transmitting voltage and current response of low-frequency high-Q sonar projectors at resonance where steady state cannot be reached before wall reflections interfere. (Our so-called Anechoic Tank is not anechoic at low frequencies.)

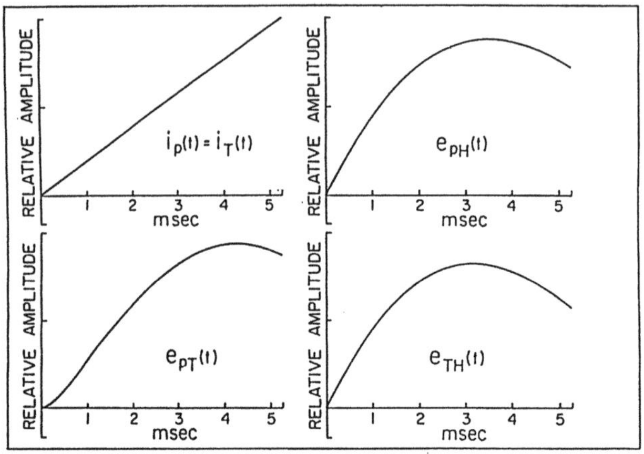

Fig. 17.9 Prony method waveforms

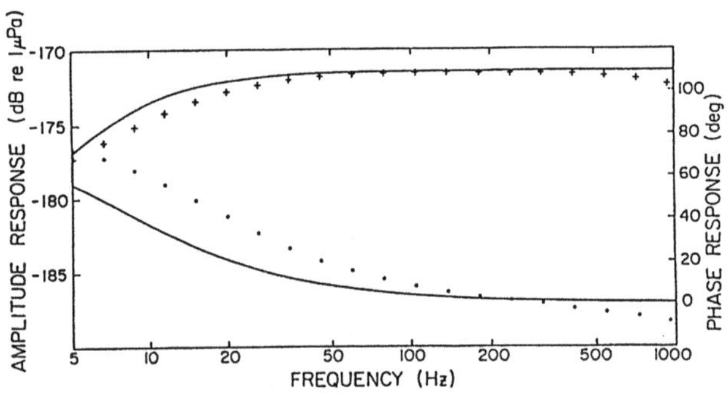

Fig. 17.10 H56 sensitivity (amplitude + phase)

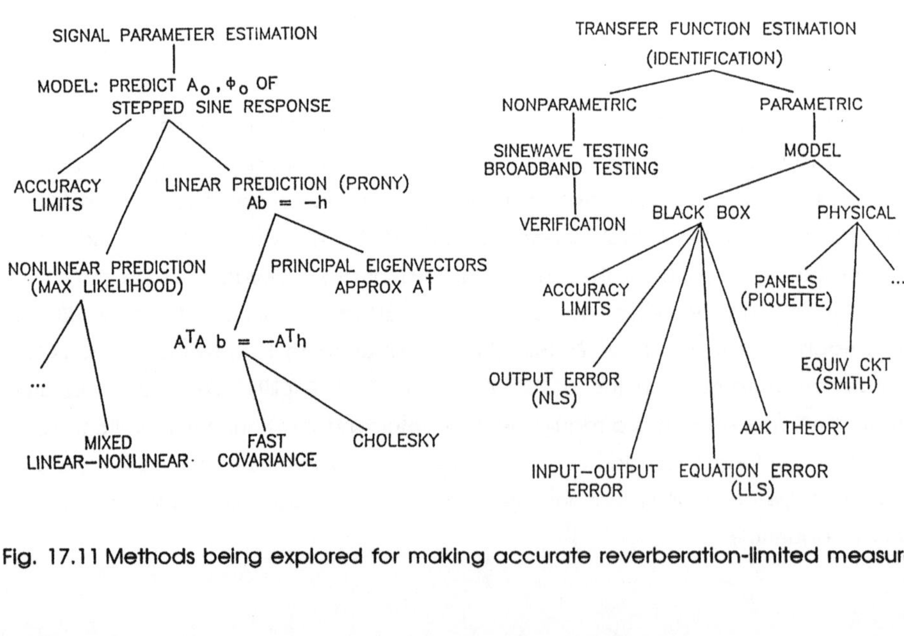

Fig. 17.11 Methods being explored for making accurate reverberation-limited measurements

The wide variety of methods being explored for making these so-called reverberation-limited measurements are depicted in Fig. 17.11 (17.8-17.10). Included are both the stepped-sinusoid input case where the intent is to determine the steady-state amplitude and phase response one frequency at a time and the broad-band transient input case where a significant portion of the transducer transfer function is obtained from a single measurement. It is not expected that one method will be appropriate for calibration of all transducers. Rather a family of methods will be required to handle the broad range of transducer types, Q values, frequency ranges, and boundary reflection conditions.

Sonar array calibration has received quite a lot of attention in the United States. Calibration techniques have been investigated for both line arrays and area arrays. The USRD's Leesburg Facility is involved in developing procedures for calibrating towed-line array sonar in open-water. Measurements on hydrophones in a long line are not always as simple as measurements on standard hydrophones. Towed-line arrays are designed primarily to detect, at sea, a plane wave from a distant target impinging on their acoustically sensitive length. Measurements at the Leesburg Facility usually are made with the line suspended vertically, but establishment of a plane wave over the entire acoustically active length is not possible. An element-by-element calibration is made by suspending the line or a line segment in the water and insonifying, in turn, each element with a plane wave. If the line array is acoustically transparent, and if it is properly rigged, the calibration is performed by the usual straightforward comparison method used for small hydrophones.

Typical calibration results are shown in Fig. 17.12. For most lines, the length of a hydrophone grouping in the line is short enough to allow a plane wave to be established over the hydrophone length. As seen in Fig. 17.13, the rest of the line, however, generally is not in a plane-wave field. It is difficult, if not impossible, to shield acoustically the inactive portion of the line at the low frequencies that are involved. If the line is not acoustically transparent, acoustic excitation of the nonreceiving portion of the line by the non-planar-wave field that exists there may cause some difficulties. For example, longitudinal and transverse modes of vibration of the line may be excited that would not be excited by a plane-wave. These modes can cause non-acoustic pressure fields in the vicinity of the hydrophone elements and can thus significantly affect the calibration. When acoustic excitation of the nonreceiving portion of the line occurs, results such as those shown in Fig. 17.14 are obtained. We note that low-frequency undulations appear in the 1 m data, but that the data at 3 m and 5 m produce relatively smooth, and coincident, results. The shapes of the curves at low frequencies vary from element to element, depending upon element position in the line, but symmetrically placed elements have almost identical low-frequency undulations. When two towed-line array segments are joined, results identical to those in Fig. 17.14 are obtained, which indicates that the low-frequency undulations are due to longitudinal rather than transverse vibrations. This conclusion is further substantiated by the fact that hanging weights varying from 15 to 500 lbs on the lower end of the line does not make much difference in the low-frequency undulations.

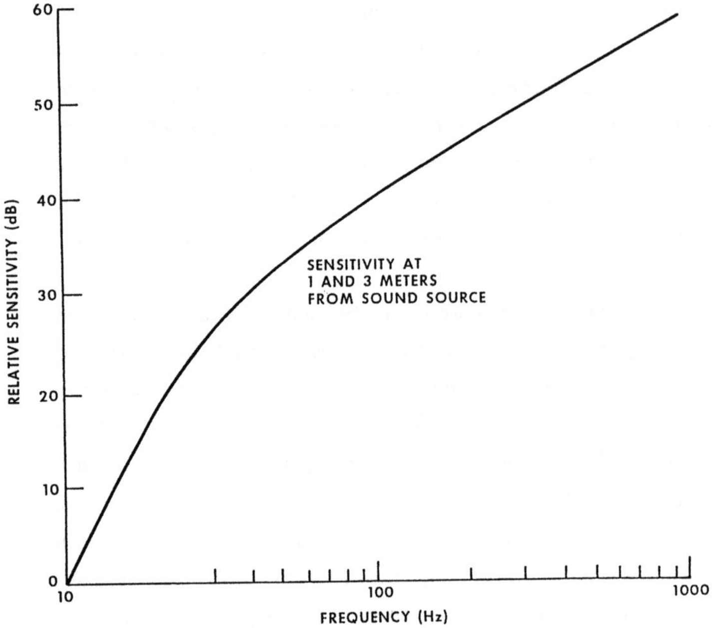

Fig. 17.12 Sensitivity of a towed-line array hydrophone

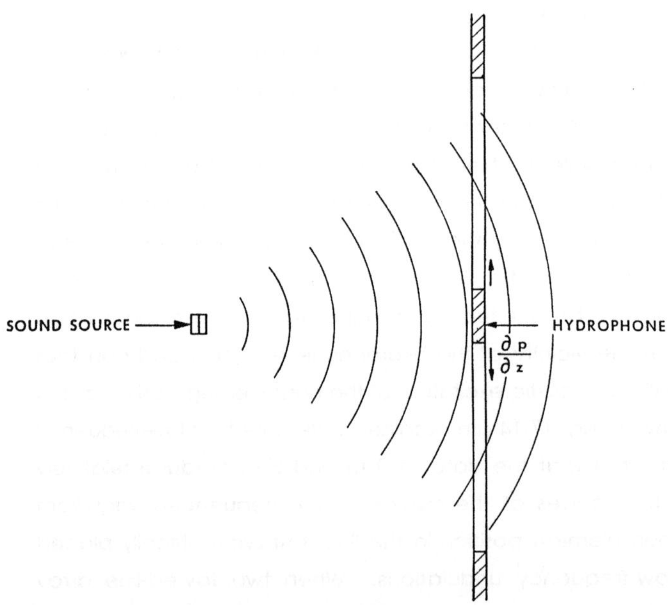

Fig. 17.13 Incident spherical wavefield

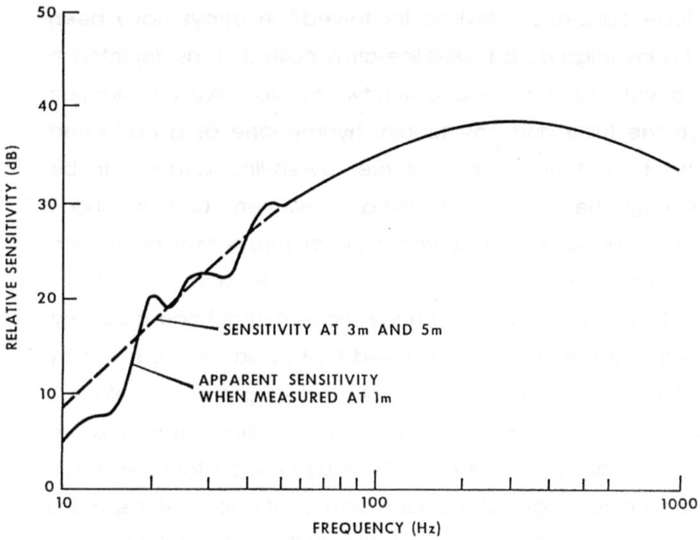

Fig. 17.14 Apparent sensitivity of a towed-line array hydrophone

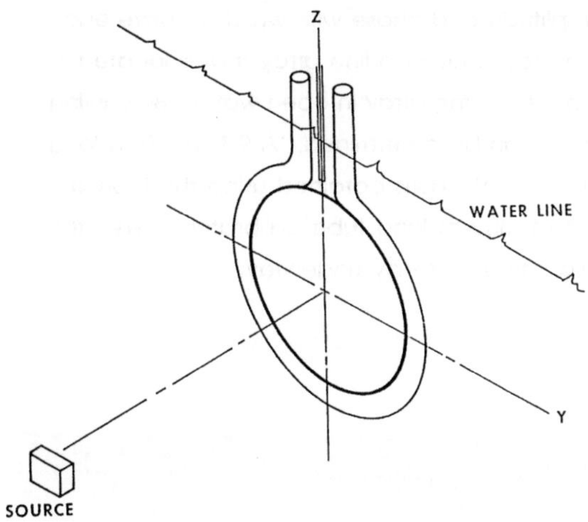

Fig. 17.15 Hoop-rigged towed-line array

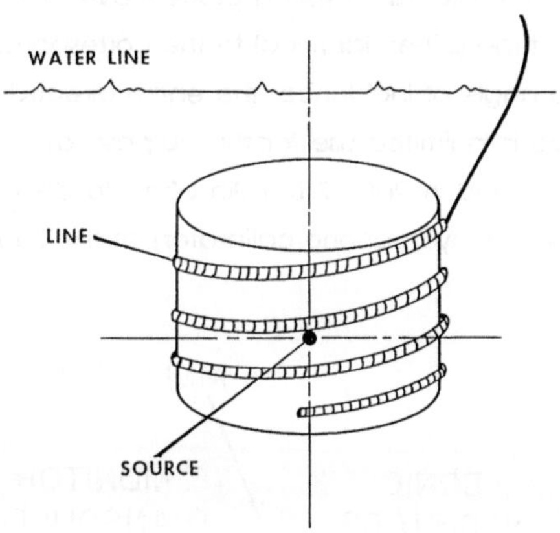

Fig. 17.16 Reel-rigged towed-line array

When the hydrophone grouping length in a towed-line array is too long to allow farfield measurements to be made under free-hanging vertical conditions, the line can be bent around a hoop and the hydrophone group can be calibrated with a sound source aimed at the center of the hoop as shown in Fig. 17.15. If the line is too long for a hoop, it can be coiled on an acoustically transparent reel and calibrated with the sound source suspended in the center of the reel, as shown in Fig. 17.16. Results for lines that can be calibrated free-hanging, on a hoop, or on a reel verify the validity of these techniques.

Efforts to provide specialized tube calibration devices for towed-line arrays have been undertaken. Zalesak and Trott (17.11) investigated a towed-line-array calibrator, as depicted in Fig. 17.17. This calibrator consists of a water-filled open tube with two tubular drivers to produce a nearly uniform sound field within the tube and one tubular hydrophone as a calibrated monitor. Comparison calibration of short elements of some towed-line arrays can be performed by threading the array through the tube and centering the elements (or hydrophone groupings) one by one in the tube. This device works very well for arrays that have short hydrophone groups and that are not excited significantly by the large drop in acoustic pressure at both ends of the tube. However, most towed arrays have hydrophone groupings that are long compared to the length of the uniform sound field that could be produced by this device. Zalesak and Rogers (17.12) developed a method for producing a sound field in a long tube that can be made to appear as a plane-wave incident at any direction to a towed array placed in the tube. Figure 17.18 illustrates this device. The tube wall contains projector-hydrophone pairs located along the entire length of the tube with center-to-center spacing less than about one-sixth of a wavelength. Before setting up the desired sound field at any given frequency, the projectors are driven, one-by-one, and the complex received voltage outputs of all the hydrophones are recorded. This results in a transfer matrix that is inverted. The inverse can then be used to determine the amplitude and phase with which to drive each projector for any desired plane-wave direction. The response of a line array that is located in the tube is then identical to the corresponding response of the array in open water. By varying the angle of incidence, the entire directivity patterns can be determined. A 9.1 m (30 ft) long tube is in limited use for this purpose at the USRD. Recent results obtained using this tube are described in Ref. 17.13. An effort to develop a 91 m (300 ft) long tube calibrator (called the long-line-hydrophone calibrator) at the Leesburg Facility is presently underway.

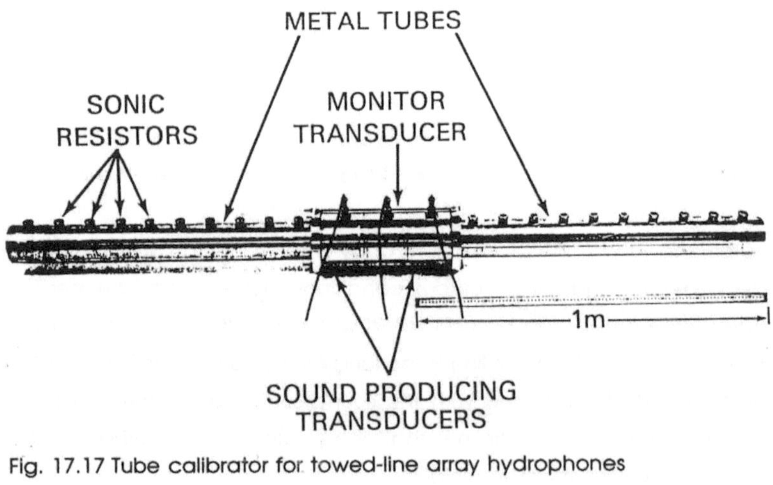

Fig. 17.17 Tube calibrator for towed-line array hydrophones

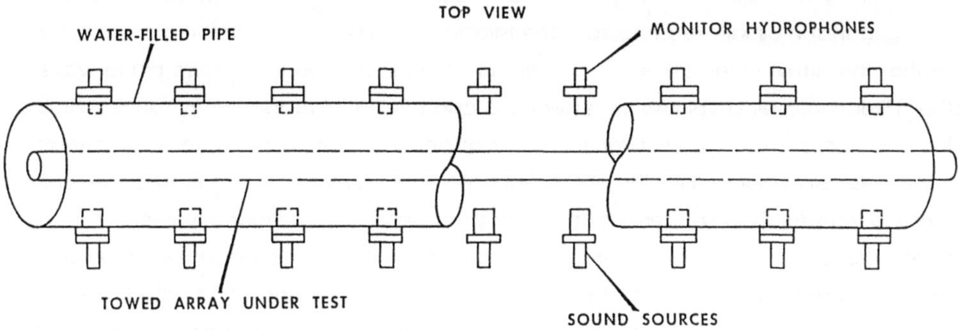

Fig. 17.18 Tube calibrator for towed-line arrays

Problems in calibrating large transducers and transducer area arrays where farfield conditions cannot easily be achieved in compact calibration facilities have received considerable research attention since the early 1960's. One result of this research was the development of two independent methods for obtaining farfield responses from nearfield measurements. Bobber (17.1) reviews both methods, the DRL Method and the Trott Nearfield Calibration Array (NFCA) Method, and discusses their limitations. Some of the limitations of the DRL method have been overcome with the advent of faster and cheaper computers. This Helmholtz type of approach is also finding application in examining the total pressure field from the nearfield to the farfield produced by vibrating objects.

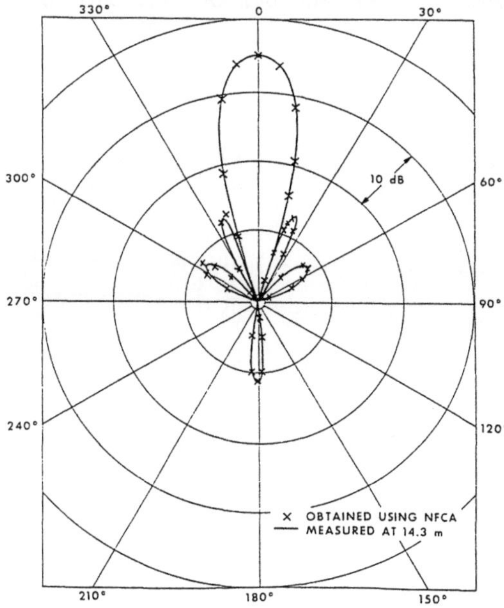

Fig. 17.19 Beam pattern obtained using the NFCA

The Trott NFCA is a large area array of small reciprocal transducer elements that are arranged in a grid that is usually rectangular. The relative acoustic outputs of the elements are selected so that the array, when driven as a projector, produces a nearly uniform plane wave in a specified direction over a volume in its nearfield and over a wide frequency range. As a receiver the NFCA is a plane-wave filter to sound originating from a projector or scatterer located in the nearfield volume, i.e., the NFCA response is proportional to the farfield pressure distribution of the transducer or scatterer for the direction opposite to the plane wave.

Several Trott planar NFCA's were constructed in the late 1960's, but to our knowledge none of them are still operational. Two large planar NFCA's are presently being constructed, one for use as a projector in NRL's target characteristics measurement facility in Washington, D.C., and one for use as both a projector and a receiver in the USRD Leesburg Facility. Van Buren (17.14) investigated cylindrical NFCA's using only a single line array of transducers. The NFCA is synthesized by rotating the test transducer near the line, making measurements of the response of the line at discrete angular positions corresponding to line locations in the cylindrical NFCA, and synthesizing the cylindrical array response using a calculator or computer to perform the calculations. This technique was implemented for the Anechoic Tank Facility using a 1.19 m-long line to achieve an upper frequency capability of about 50 kHz, a factor of 5 higher than previous NFCA's (17.15). Figure 17.19 shows the results of a calibration using this line. More recently, the NFCA concept was extended to a synthetic spherical configuration (Fig. 17.20) that provides full three-dimensional beam patterns of enclosed transducers or scatterers with a single 360° rotation of a semi-circular arc line array of hydrophones (17.16).

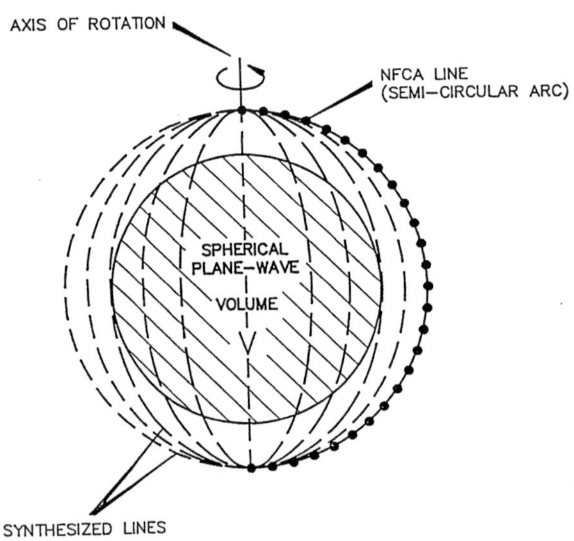

Fig. 17.20 Synthetic spherical NFCA

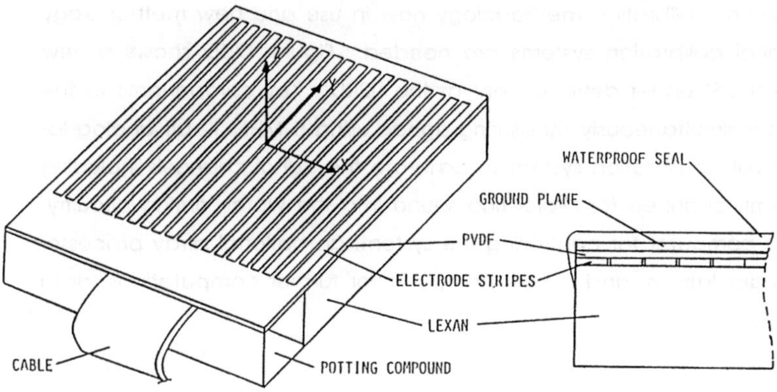

Fig. 17.21 Wavenumber calibrator

With the advent of large area hydrophones designed to reduce flow noise in receiving arrays, the need arises to determine the response of area hydrophones to hydrodynamic pressure fields with non-acoustic wavenumbers. To address this need, we are investigating the use of a piezoelectric polymer array to produce the required wavenumbers evanescently. Figure 17.21 shows such an array using a striped-electrode pattern with independent drive for each stripe. Calculations have shown that no appreciable acoustic energy should radiate to the farfield of the array when the phasing between the drive of neighboring stripes corresponds to wavenumbers above the acoustic region, i.e., non-acoustic wavenumbers. Experimental results obtained with a 12.5 cm x 12.5 cm square prototype wavenumber calibrator show that single wavenumber non-acoustic pressure fields can be obtained (17.17).

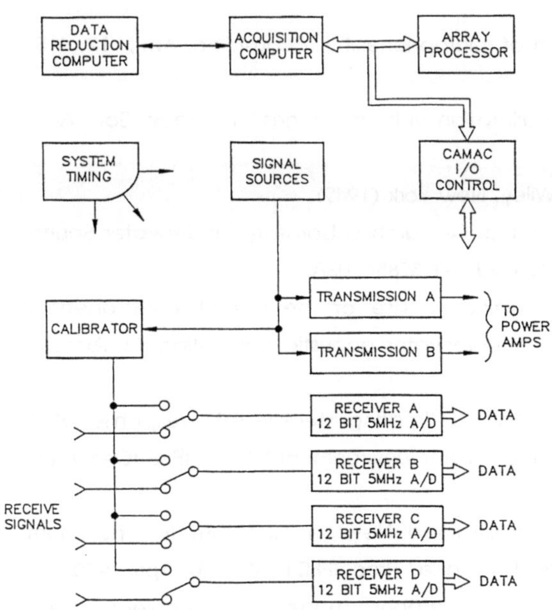

Fig. 17.22 New precision measurement system

In order to better implement calibration methodology now in use and new methodology being developed, new digital calibration systems are needed. Figure 17.22 shows a new precision measuring system (PMS) under development at the USRD. Key components to the system include 4 channels for simultaneously measuring signals using 5 MHz, 12-bit analog-to-digital converters, a built in self- calibration system to calibrate the signal channels including the A/D converters, 2 transmit channels for projecting sound into the water, timing circuitry, signal generation section, a computer for controlling the system, an optional array processor for rapidly handling array calçulations, and a host computer for further computations, data storage, and report generation.

17. 5 CONCLUSION

There has been a recent trend in the United States toward fewer calibration facilities. As calibration methodology becomes more complex, systems more complex and costly, and facility maintenance more costly, we expect this trend to continue.

REFERENCES

17.1 R.J. Bobber, *Underwater Electroacoustic Measurements*, U. S. Government Printing Office, Washington, D. C., July 1970. (Not in print. Available from NTIS).

17.2 W.R. MacLean, "Absolute measurement of sound without a primary standard," J. Acoust. Soc. Am., **12**, 140-146 (1940).

17.3 R.K. Cook, "Absolute pressure calibration of microphones," J. Acoust.Soc. Am., **12**, 415-420 (1941).

17.4 L.D. Luker, and A.L. Van Buren, "Phase calibration of hydrophones," J. Acoust. Soc. Am., **70**, 516-519 (1981).

17.5 L.L. Beranek, *Acoustic Measurements*, Wiley, New York (1949).

17.6 *USRD Transducer Catalog*, April 1982, Naval Research Laboratory, Underwater Sound Reference Detachment, P. O. Box 8337, Orlando, FL 32856, USA.

17.7 L.G. Beatty, J.D. George, and A.Z. Robinson, "Use of the complex exponential expansion as a signal representation for underwater acoustic calibration," J. Acoust. Soc. Am., **63**, 1782-1794 (1978).

17.8 J.D. George, V.K. Jain, and P.L. Ainsleigh, "Estimating steady-state response of a resonant transducer in a reverberant underwater environment," Proc. IEEE ICASSP-88, April 1988.

17.9 J.D. George, and P.L. Ainsleigh, "A constrained Newton algorithm for maximum likelihood modeling of sums of exponentials in noise," Proc. IEEE ICASSP-90, April 1990.

17.10 J. D. George ,and P.L. Ainsleigh, "Stepped sine response modeling using noisy sampled data," NRL Memorandum Report No. 6644, June 1990.

17.11 J.F. Zalesak, and W.J. Trott, "Low-Frequency technique for the underwater calibration of individual elements of a line hydrophone array," J. Acoust. Soc. Am., **62**, 91-101 (1977).

17.12 A description of this method is given in the reference by L.D. Luker, J.F. Zalesak, C.K. Brown, and R.E. Scott, "Automated digital benchtop calibration system for hydrophone arrays," J. Acoust. Soc. Am., **73**, 1212-1216 (1983).

17.13 L.D. Luker, and J.F.Zalesak, "Free-field acoustic calibration of long underwater acoustic arrays in a closed chamber," submitted to J. Acoust. Soc. Am.

17.14 A.L. Van Buren, "Cylindrical nearfield calibration array." J. Acoust. Soc. Am., **56**, 849-855 (1974).

17.15 A.L. Van Buren, M.D. Jevnager, and A.C. Tims, "A 5-to 50-kHz synthetic cylindrical nearfield calibration array," J. Acoust. Soc. Am., **77**, 1927-1932 (1985).

17.16 A.L. Van Buren, "Spherical nearfield calibration array for three-dimensional scanning," J. Acoust. Soc. Am., **85**, 2655-2660 (1989).

17.17 D.H. Trivett, L.D. Luker, S. Petrie, A.L. Van Buren, and J.E. Blue, "A planar array for the generation of evanescent waves," J. Acoust. Soc. Am., in press, (1990).

18. Measurement Methods for Low-Frequency Transducers

C. Giangreco, S. Faure, and J.F. Rossetto

GERDSM -DCAN Toulon, DCN, Le Brusc, F-83140 Six Fours Les Plages, France

This paper describes two methods which are used in GERDSM to calibrate low frequency transducers. The first technique is based on cross correlation functions. The feasibility of the technique has been demonstrated for transducers with low quality factor Q and its extension to projectors with high quality factor Q is described. A second method which reduces the transient part of the acoustic pressure by controlling the headmass velocity of the transmitting transducer has been tested successfully on actual projectors.

18.1 INTRODUCTION

Many transducers operate at low frequencies, below 1500 Hz. So that their performance can not be measured using pulse sound technique in a water tank of reasonable size. A technique based on the cross correlation function has been tested. Limitations and future projects are discussed. A second method which reduces the transient part of the acoustic pressure generated by the projector is also described and is very useful in calibrating resonant transducers with high quality factor.

18.2 CORRELATION TECHNIQUE

18.2.1 Principle of the Method

The transfer function measured during the calibration of transducer is described on Fig. 18.1, where X(f) is the input signal and Y(f) is the output signal.

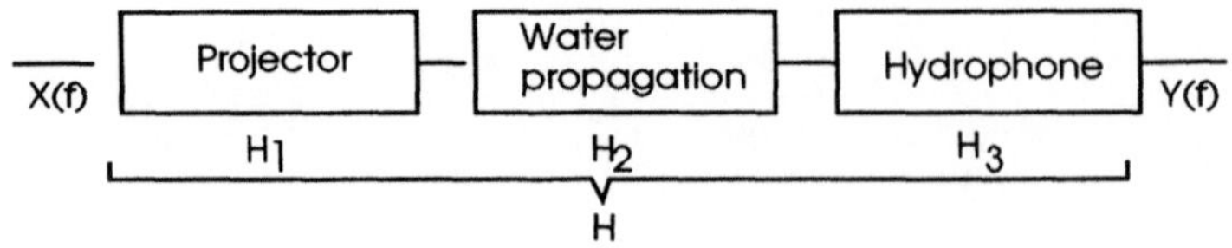

Fig 18.1 Estimation of H

Thus, $H(f) = H_1(f)\, H_2(f)\, H_3(f)$ where $H_2(f)$ and $H_3(f)$ are known.

An estimation of H(f), $\hat{H}$ (f) can be obtained by:

$$\hat{H}(f)=\frac{G_{xy}}{G_{xx}} \tag{18.1}$$

where: G_{xx} is the power spectral density of the input signal and G_{xy} is the cross spectral density between the input and output signal.

The cross correlation function $\Gamma_{xy}(\tau)$ is obtained by applying an Inverse Fourier Transform to G_{xy}. This function is defined by:

$$\Gamma_{xy}(\tau) = \int_{-\infty}^{+\infty} x(t)\, y(t+\tau)\, dt \tag{18.2}$$

As the water tank is a confined environment, the output signal can be expressed by:

$$y(t) = \sum_{l=0}^{\infty} y_l\left(t - \frac{d_l}{c}\right) \tag{18.3}$$

where c is the sound speed of the water, $y_0(t)$ is the direct echo, d_o is the distance between the transducers and $y_j(t)$ are the echoes from the boundaries of the tank.

Substituting y(t) from (18.3) into (18.2) we obtain:

$$\Gamma_{xy}(t) = \Gamma_{xy0}\left(\tau - \frac{d_0}{c}\right) + \sum_{l=1}^{\infty} \Gamma_{xyl}\left(\tau - \frac{d_l}{c}\right) \tag{18.4}$$

It can be seen that the cross correlation function is the sum of the cross correlation function between the input signal and the direct signal which arrives at $t = d_o/c$ and the cross correlation between the signal input and the boundary echoes which arrive at d_l/c.

By applying a window on this function it is possible to remove the effect of the boundary echoes if the peaks are narrow enough. Then the Direct Fourier Transform is applied to obtain the free G_{xy}^F (without the echoes of the tank). So the projector transfer function can be found by:

$$H_1(f) = \frac{G_{xy}^F(f)}{G_{xx}(f) \times H_2(f) \times H_3(f)} \tag{18.5}$$

The duration of the intercorrelation function peak must be shorter than $(d_l\text{-}d_o/c$. So we must use an input signal with an autocorrelation function with narrow peaks. Therefore the signal must have large bandwidth in the frequency domain to apply this technique. In practice we use a Gaussian Random Signal (GRS) or aLinear Frequency Modulated Signal (LFMS).

These signals are adequate if the transducer does not reduce the bandwidth in the frequency domain of the input signal. Otherwise, improvements are necessary and will be discussed later.

18.2.2 Measurements Results

A known hydrophone B&K 8103 was first measured by this technique to test its feasibility. The transducers are located in the tank as shown in Fig 18.2. With a bandwidth in the frequency domain (50 Hz - 10 kHz) two different input signals were tested a GRS with 300 averages and a LFMS with 25 averages. Before doing a calibration using this technique, it is very important to measure the coherence function $C_{xy}^2(f)$ which is defined by:

$$C_{xy}^2(f) = \frac{|G_{xy}(f)|}{G_{xx}(f) \times G_{yy}(f)} \qquad (18.6)$$

where | | is the modulus.

This parameter is an indicator of the linearity of the measuring system. The coherence function of the L F M S (Fig. 18.3) is better than the coherence function of the GRS (Fig. 18.4). These curves show that the measurement will be valid only above 600 Hz. This is due to the projector which generates low acoustic pressure levels below 600 Hz. The cross correlation functions are represented in Figs. 18.5 and 18.6. The effect of the boundary echoes can be seen. It is interesting to note that the cross correlation peaks are shorter when G R S is used. The results are plotted on Fig. 18.7. Curves 18.7 a & b are obtained after windowing the cross correlation function as indicated in Figs. 18.5 and 18.6.

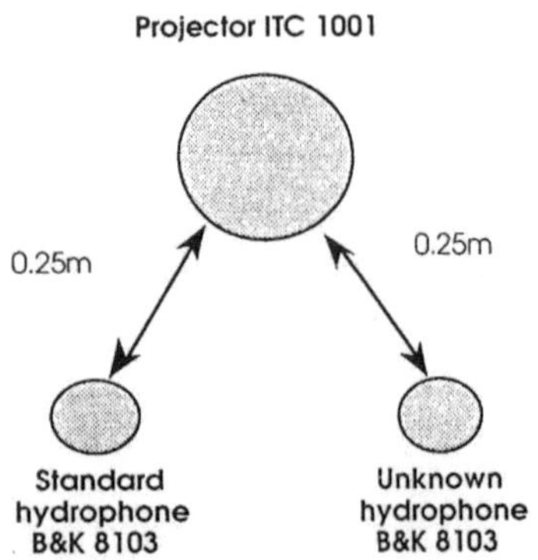

Fig. 18.2 Set up of the experiment

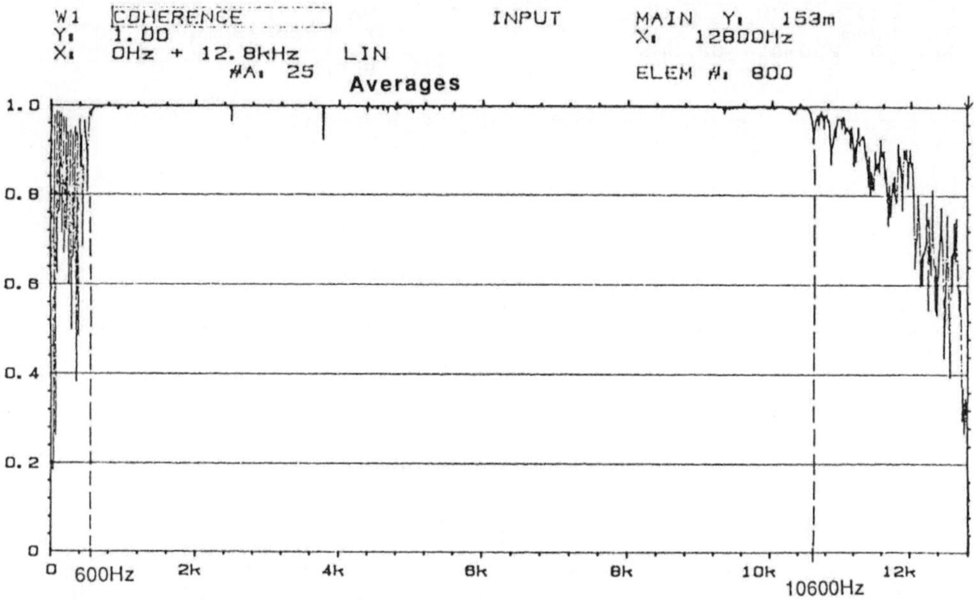

Fig. 18.3 Coherence function of the LFMS

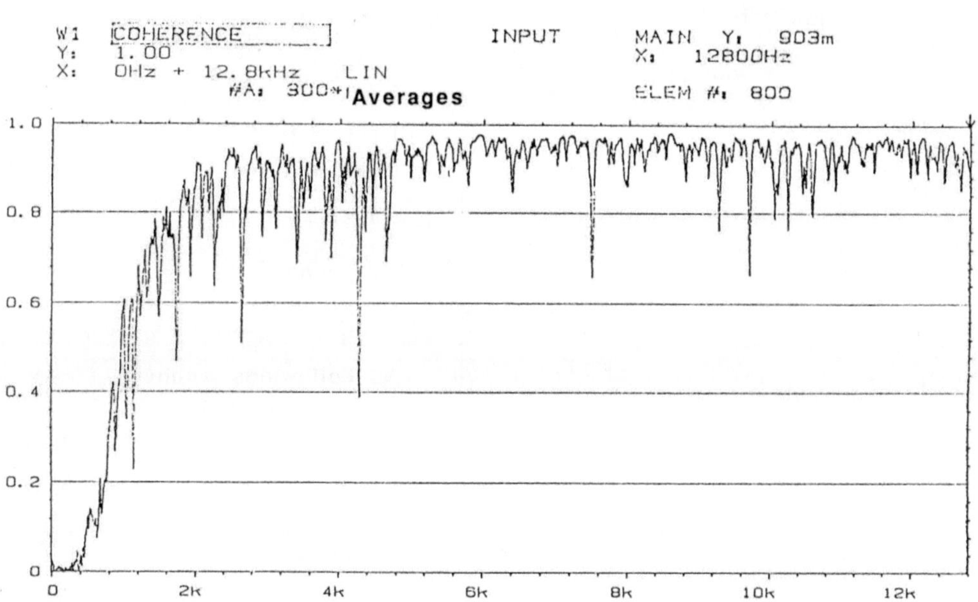

Fig. 18.4 Coherence function of the GRS

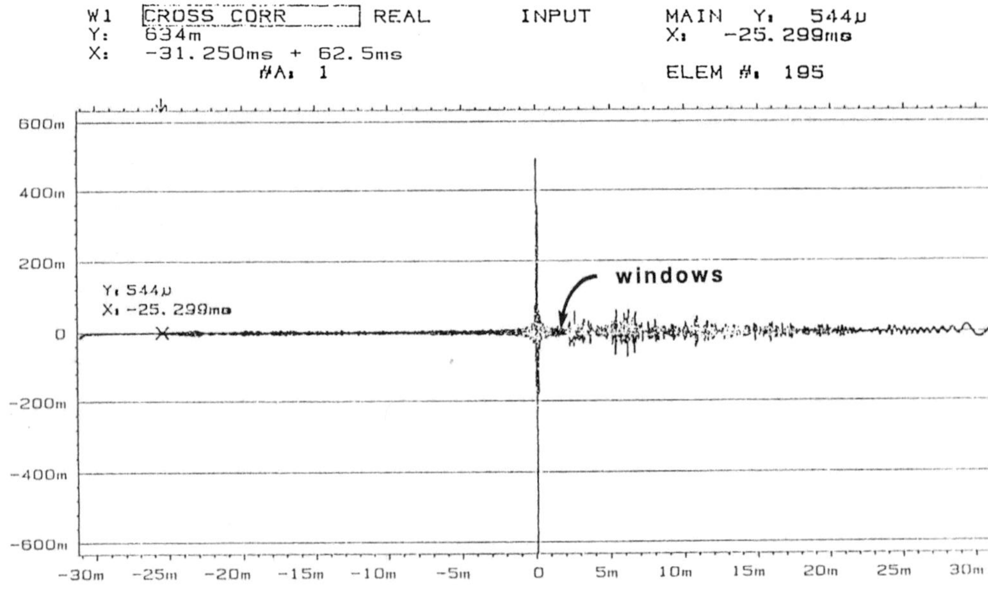

Fig. 18.5 Cross correlation function of the LFMS

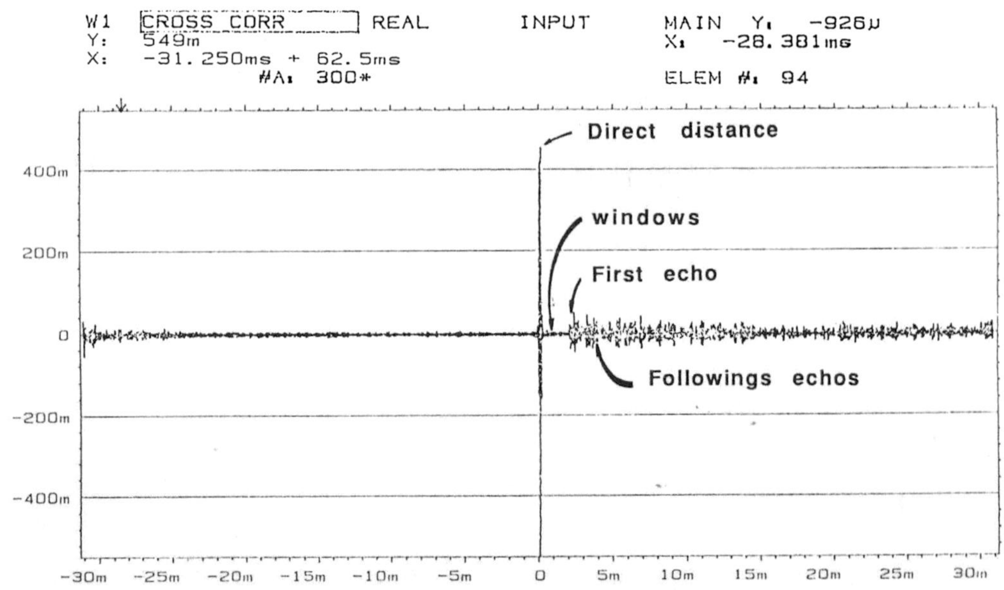

Fig. 18.6 Cross correlation function of the GRS

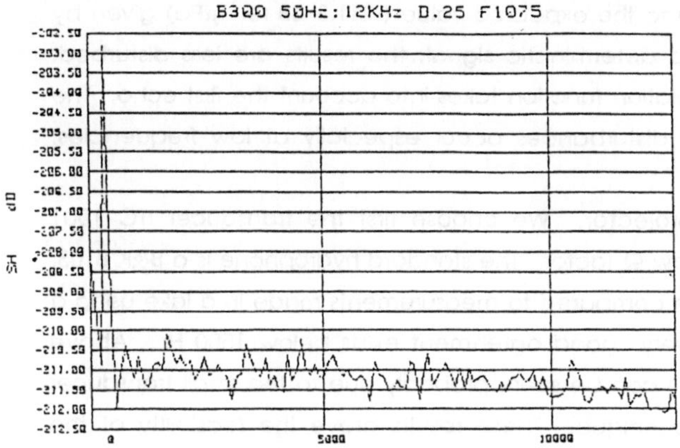

Fig. 18.7a Window of the cross correlation function of Fig. 18.5

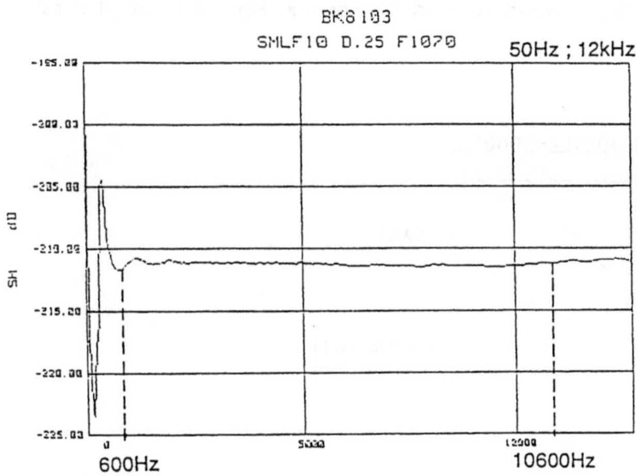

Fig. 18.7b Window of the cross correlation function of Fig. 18.6

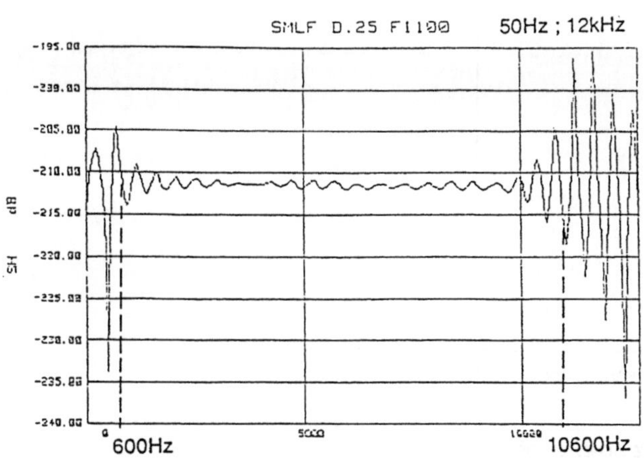

Fig. 18.7c Window of the cross correlation function with the first echo

These results are close (less than 1 dB) to the expected value (-211.3 dB re. 1μPa) given by the constructor. Because the LFMS is a deterministic signal, the results are less disturbed. Curve c is obtained when the cross correlation function takes into account the first echo. The input signal is LFMS. As can be seen, disturbances occur especially at low frequencies, showing the advantage of the technique.

The next step was to calibrate a projector. We choose first the transducer ITC 1001 because it is omnidirectional and has a low Q factor. The standard hydrophone is a B&K 8103. The results obtained by this technique are compared to measurements made in a lake using a pulsed sound technique (Fig. 18.8). A very good agreement exists below 1500 Hz. Above 1500 Hz a small deviation (less than 1 dB) can be seen, probably due to the small transducer separation used with the correlation technique. These results prove the feasibility of the technique when the transducers used have low Q factors. When a projector has high Q factor the peaks of the cross correlation function widen as can be seen in Fig. 18.9. So it is not possible to window.

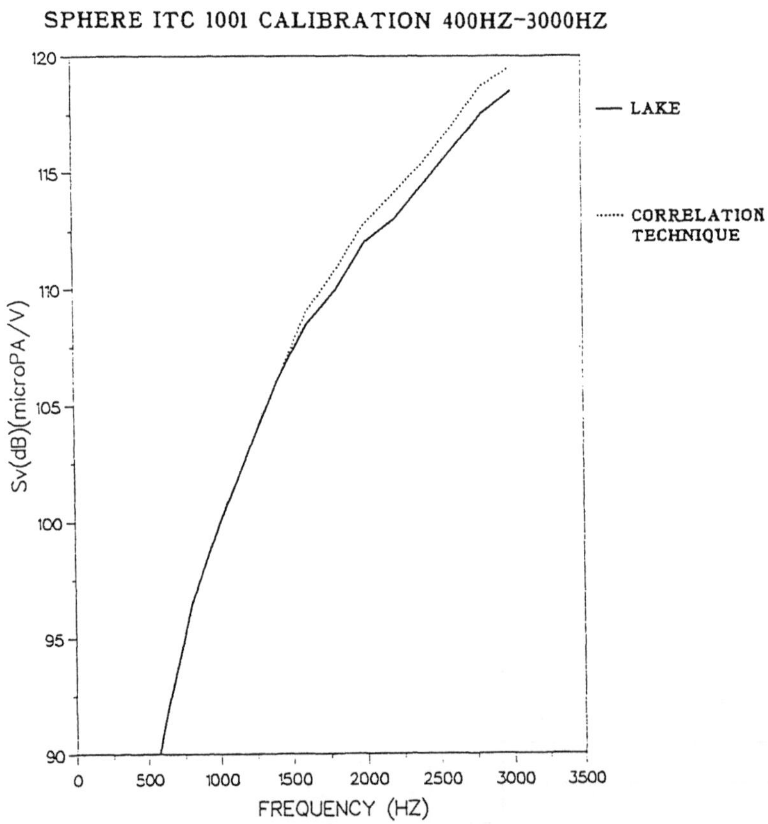

Fig. 18.8 Sphere ITC 1001: 400 Hz - 3000 Hz calibration

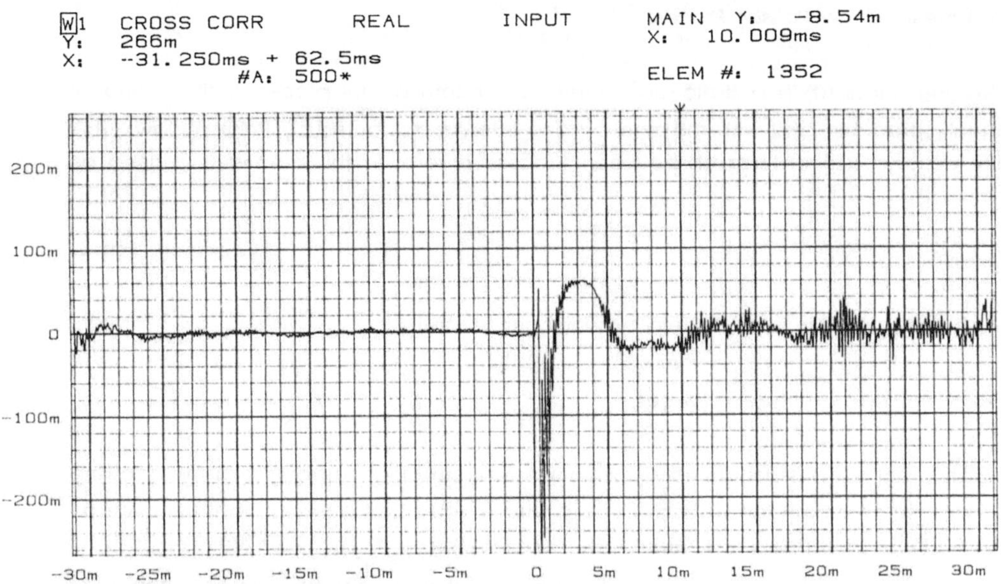

Fig. 18.9 Cross correlation function for a projector with a high Q factor

18.2.2.1 Resonant Transducers

As a matter of fact in this particular case, the input signal must take into account the characteristics of the projector. The method which is currently developed uses the idea of E. Kasiwagi (18.1). The input signal can be a FM signal modulated by the random signal in the form of a binary Poisson process.

A FM signal is described by:

$$x(t) = a_0 \cos (\omega_c t + \mu(t)) \qquad (18.7)$$

where a_0 is the amplitude and $\omega_c/2\pi$ the carrier frequency. The phase $\mu(\tau)$ is related to the modulating signal r(t) by:

$$\frac{d\mu(t)}{dt} = \Delta\omega \; r(t) \qquad (18.8)$$

where $\Delta\omega$ is the frequency deviation per volt.

A FM signal of this type may be easily made to form both the symmetric or asymmetric concave power spectra with respect to its center frequency, by adjusting its parameters as it is shown in Ref. 18.1. This procedure is applicable to compensate for the narrow-band gain characteristic of the high-Q underwater acoustic transducers and to extend the effective bandwidth for the measurement. The two types of binary Poisson processes are described hereafter.

18.2.2.2 Symmetric Binary Poisson Process

The modulating signal r(t) is a stationary symmetric random binary process with +1 and -1 values equally probable and with the zero-crossing interval distributed by Poisson's law. The probability density $P_o(t)$ of intervals between successive zeros in this process is defined as follows:

$$P_o(t) = \lambda\, e^{-\lambda t} \tag{18.9}$$

where λ is the expected number of zero-crossings per second.

18.2.2.3 Asymmetric Binary Poisson Process

The modulating signal r(t) is a stationary asymmetric random binary process with the following characteristics: It takes both the +1 and -1 values with unequal probability, so that the average lifetime of r(t) in these states are different, $1/\lambda_1$ and $1/\lambda_2$, respectively. Both lifetimes of r(t) in states +1 and -1 follow Poisson's distributions. Lengths of adjacent intervals between zero-crossings are statistically independent. Then, the probability density of the lengths of intervals during which r(t) is +1 or -1 may be defined respectively as:

$$P_o^+(t) = \lambda_1\, e^{-\lambda_1 t} \text{ when r(t) is +1} \tag{18.10a}$$

$$P_o^-(t) = \lambda_2\, e^{-\lambda_2 t} \text{ when r(t) is -1} \tag{18.10b}$$

The asymmetric process is more useful in practice because the transducer transfer function is usually asymmetric with respect to the resonance frequency. E. Kasiwagi (18.1) used an electronic approach to generate an approximate binary Poisson process which is not easy to use. So we chose to generate the binary Poisson process using an arbitrary waveform synthesizer shown on Fig. 18.10.

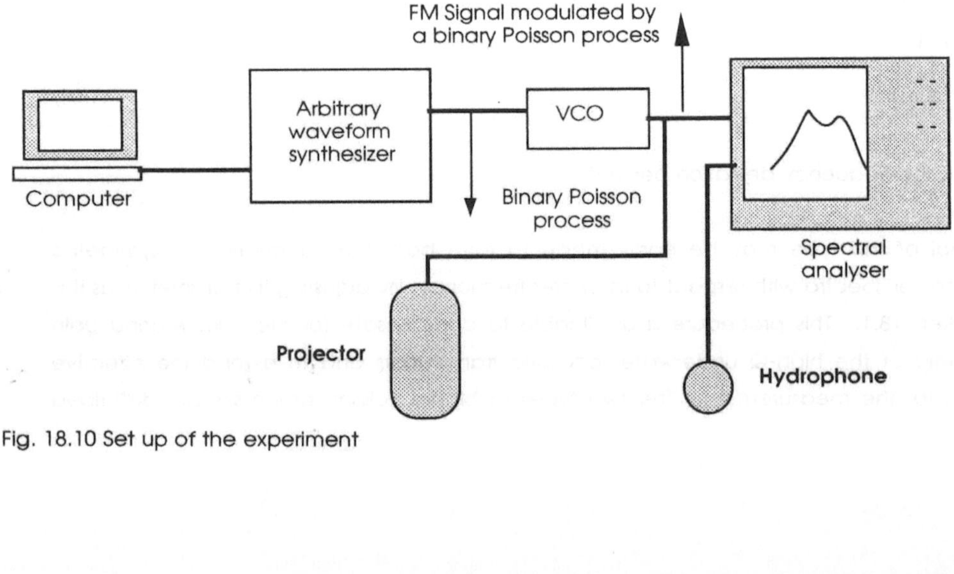

Fig. 18.10 Set up of the experiment

18.3 MEASUREMENTS USING CONTROLLED MOTIONAL CURRENT

This method reduces the transient part of the acoustic pressure generated by the projector under test.

18.3.1 Principle of the Technique

A typical piezoelectric transducer can be described around its resonant frequency by the analog circuit represented Fig. 18.11.

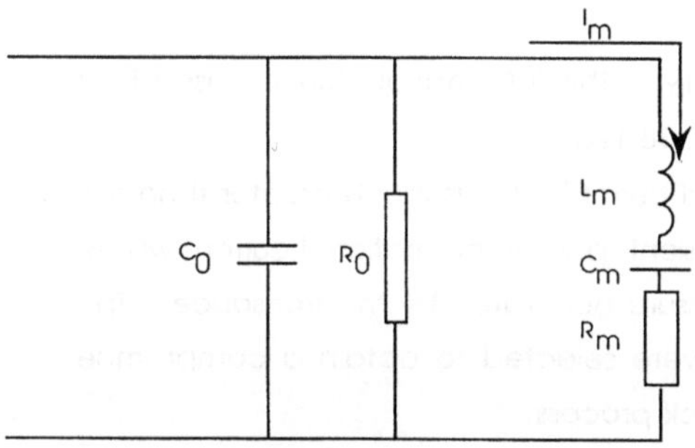

Fig. 18.11 Analog circuit for a piezoelectric transducer around its resonant frequency

Where C_0 is the electrical blocked capacitance which is easily measured in practice, R_0 the dielectric resistance, often neglected because of its high value and L_m, C_m, R_m the components of the motional branch of the analog circuit.

The idea used here is to control the pressure waveform in water by controlling the head-mass velocity of a transmitting transducer. The information on this velocity is given by the measurement of a motional current. The motional current of a transducer is measurable by using an additional parallel capacitor and a current transformer as illustrated in Fig. 18.12a.

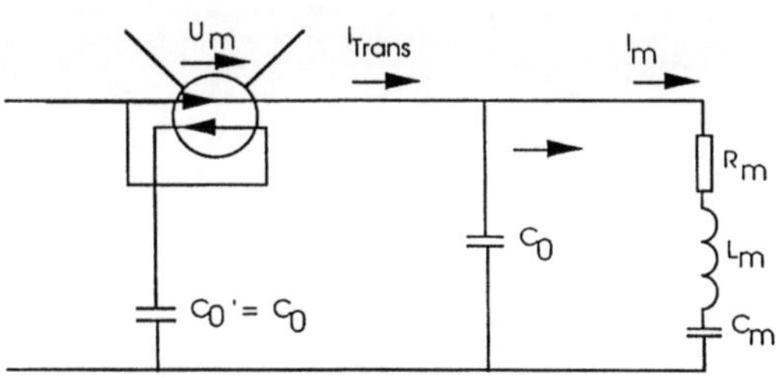

Fig. 18.12a Measurement of the motional current

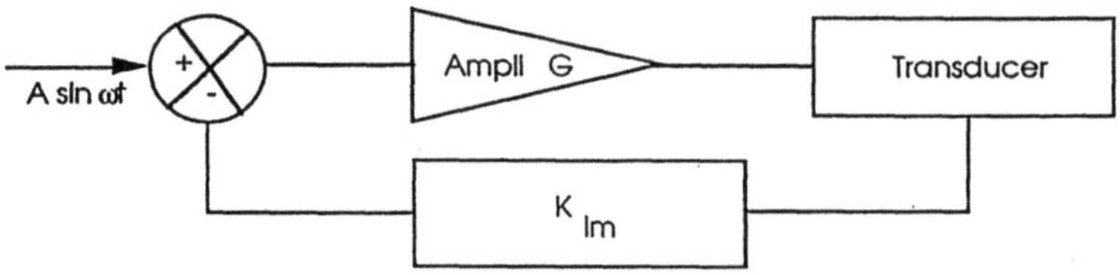

Fig. 18.12b Feedback process

When the capacitance C_0' is equal to C_0. The current in the toroid coil is

$$I_m = I_{trans} - I_{C_0'} \tag{18.11}$$

So the output voltage of this transformer is $U_m = K\ I_m$. This information (U_m) is used in a feedback process, diagrammatically represented in Fig. 18.12b.

The transducer will operate with a constant motional current if GK is much greater than the motional impedance of the transducer. Hence the transient part of the motional current will be reduced and it will be the same for the acoustic pressure generated by the transducer. The values of the electronic components of this device were selected to obtain a compromise between the stability and the steepness of the feedback process.

18.3.2 Measurements

At first, this technique was tested on electronic components to simulate the analog circuit of a piezoelectric projector at a resonance frequency of 960 Hz and a Q factor close to 10. The following components were selected:

$C_0 = C_0' = 1.8\ \mu F$

$C_m = 0.347\ \mu F$

$L_m = 3.8\ H$

$R_m = 47\ \Omega$

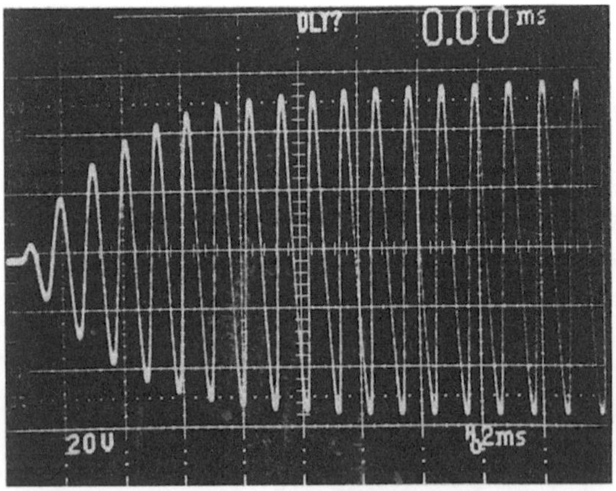

Fig. 18.13 Voltage across the resistance R_m

Figure 18.13 shows the voltage across the resistance R_m when a classical voltage feedback amplifier is used to excite the projector. So, it is an image of the motional current which is also an image of the acoustic pressure. It can be seen that the steady state of the signal is reached after 10.8 ms only, the frequency of the excitation is 960 Hz. This time (duration) is too long to do a proper measurement in a typical acoustic tank where the boundary echoes come 5 ms after the beginning of the excitation pulse. Then, the 'simulated' transducer is excited at the same frequency by a current controlled amplifier which works as shown on . 18.14. In this case the steady state is obtained after 2 ms, so an acoustical measurement can be made.

These measurements were conducted over a frequency range around the resonance frequency and in each case the motional current I_m was constant, thus demonstrating the feasibility of the method.

After these simulations a real projector was tested. The so-called TZ 800 transducer with a resonance frequency around 800 Hz and a Q factor equal to 10 was chosen. The set up of the experiment is represented diagrammatically by the arrangement in Fig. 18.14.

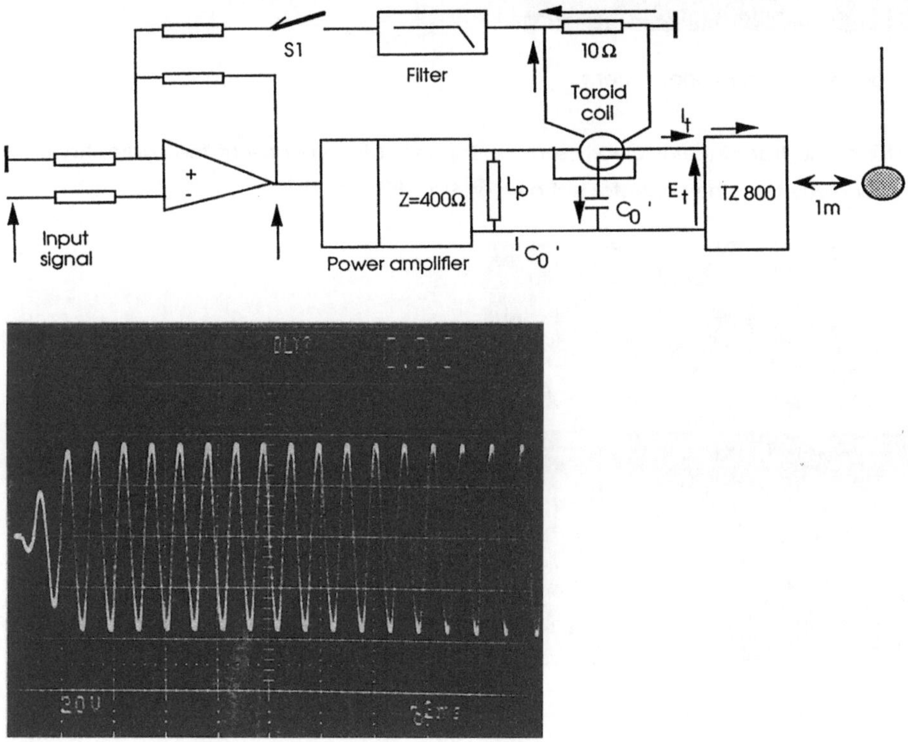

Fig. 18.14 Set up of the experiment

The filter is used to increase the stability of the feedback process.The switch S1 separates two functions. When it is closed, the device works as a current controlled amplifier. The device works as a voltage feedback amplifier when the switch S1 is open. First we analyze the behavior of the transducer when it is excited by the voltage controlled amplifier. Figure 18.15 shows the oscillogram of the motional current.

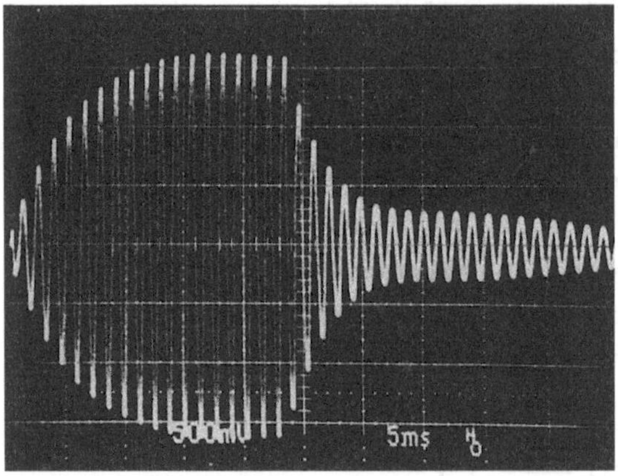

Fig. 18.15 Oscillogram of the motional current

The high-Q factor of the transducer is clearly shown. The receiving voltage of the hydrophone located at 1 meter away from the projector is shown on Fig. 18.16.

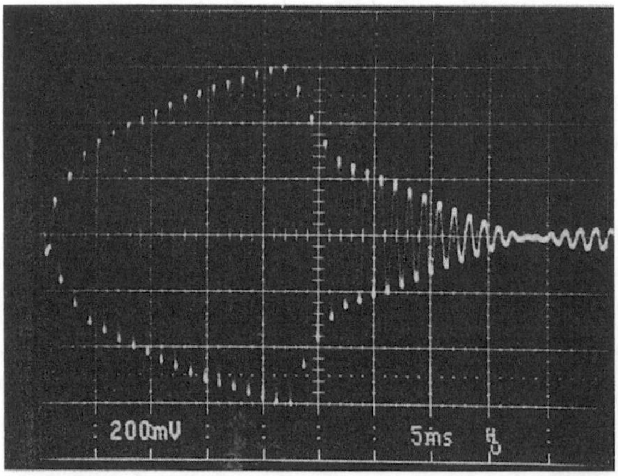

Fig. 18.16 Receiving voltage of the hydrophone located at 1 meter away from the projector

The steady state is not reached before the arrival of the boundary echoes in the tank. So that good measurements are not possible. This experiment was conducted again with the switch S1 closed. It is interesting to note the high level which occurs at the beginning of the transducer current pulse (Fig. 18.17) to reduce the transient part of the acoustic pressure.

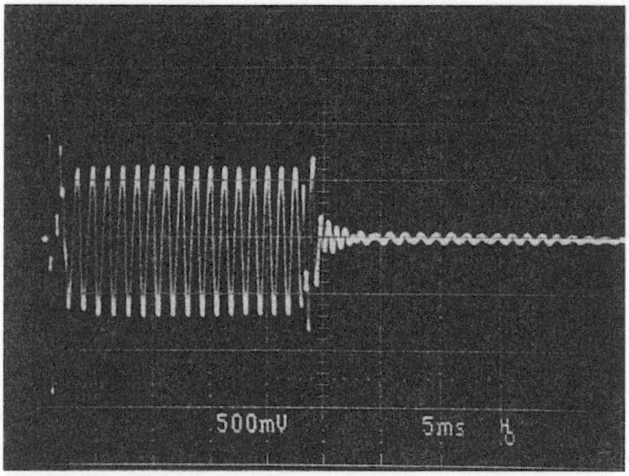

Fig. 18.17 Transducer current pulse

As can be seen in Fig. 18.18 the motional current steady state is reached at the third period of the signal. The receiving voltage of the hydrophone confirms this result (Fig. 18.19). The acoustic measurement can be made during 2 periods before the arrival of the disturbances due to the boundaries of the tank.

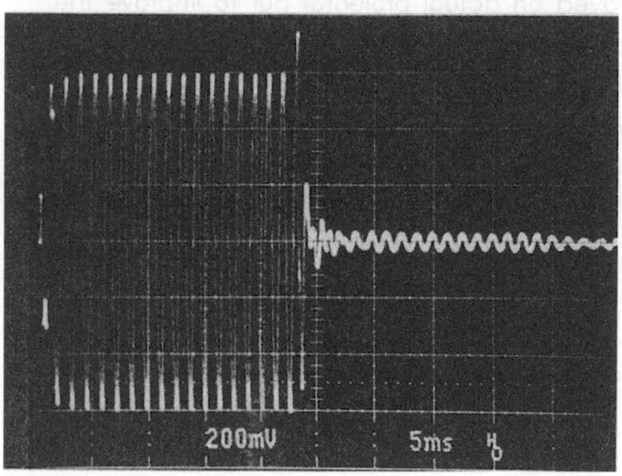

Fig. 18.18 Motional current

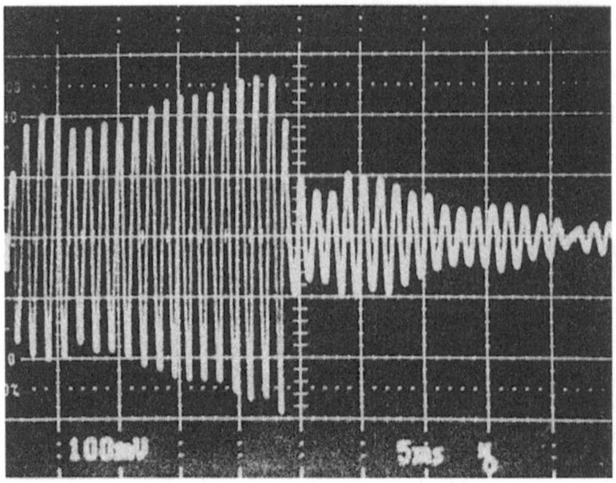

Fig. 18.19 Receiving voltage of the hydrophone

18.4 CONCLUSION

Two methods which allow the calibration of low frequency transducers have been described. The first technique is based on the cross correlation function and the feasibility of the technique has been shown on transducers with low Q factor. If a projector with a high Q factor is tested using this technique, the input signal must take into account the characteristics of the projector. An input signal which is a FM signal modulated by a random signal in the form of a binary Poisson process is currently being tested and results from tests on real projectors using this input signal will be published later. The second method reduces the transient part of the acoustic pressure. The idea used here is to control the headmass velocity of the projector. The feasibility of this technique has been proved on actual projector but to improve the efficiency of the technique the following should be done:

the feedback process must be improved,

- an analog circuit of the piezoelectric transducer which takes into account the transient part of the signal must be developed.

REFERENCE

18.1 E. Kasiwagi, "FM signal-correlation for underwater acoustic measurement," Acustica, **37**, (1977).

19. Poster Session

P.1 Static Properties of Terfenol-D Rods

L. Sandlund, and M. Fahlander

Feredyn AB, Fyrisvallsgatan 11, S-752 28 Uppsala, Sweden.

Very high magnetostrictive strains are observed in Terfenol-D. The material can produce high forces, and give fast precision motions without backlash. The high magnetomechanical coupling factor means that it can efficiently convert electric energy into mechanical energy and vice versa. The unique properties of Terfenol-D makes it favorable where high energy densities and low frequencies are needed e.g. in powerful long-range sound projectors.

In this poster paper a versatile test rig to investigate static properties of Terfenol-D rods is described. The rig is based on a watercooled solenoid with a magnetic flux feedback. Also, experimental data are presented showing the overall magnetostriction as well as the magnetic flux density as a function of the magnetizing field at different mechanical loadings.

P.2 Research on the Origin and Qualification of Losses in a Tonpilz Transducer

M.M. Amadei[*], P. Gonnard[*], L. Eyraud[*], and D. Boucher[**]

[*]Laboratoire de Génie Electrique et Ferroélectricité, INSA,
F-69621 Villeurbanne CEDEX, France

[**]GERDSM -DCAN Toulon, DCN, Le Brusc, F-83140 Six Fours Les Plages, France

In a previous study, we had worked on the characterization of a transducer lent by the GERDSM: a low frequency Tonpilz for power emitting. It is made of 7 independent pairs of ceramics, a nodal electrode, a tail mass in brass and of a head mass in aluminium with a rubber housing (to ensure the waterproofness). In each pair the piezoceramics are organised with opposed polarities and electrically driven with parallel connection. The stack is stuck at every interface and on the external surface of the ceramic stack and is held in position thanks to a prestress of about 200 bars on the ceramics. Its electrical, mechanical and thermal characterization has been made in air at its fundamental resonance and under nominal mechanical conditions.

Various problems have thus been stressed:

- saturation of the radiating face acceleration according to the voltage,
- important losses, about 54 W,
- a maximum heating of 38°C near the nodal electrode in free natural convection.

Under the above working conditions, the losses of ceramics (active elements of this transducer) are very low and have been widely increased by the 4 W value. This is why we had to look for other kinds of losses by carrying out various assemblies.

1) Without rubber: The transducer vibrates much better: the nominal mechanical conditions are reached at 40 V instead of 100. However, a transducer deterioration is observed and after 50 minutes, the current and radiating face acceleration fall to the same value as that of the original transducer. This means that another element occurs in time much more than rubber.

2) Without glue: We have mounted a transducer the same way as previously with rubber on the headmass but without glue. The effect is instantaneous. At 100 V, the study in time shows that the current keeps on increasing until it stabilizes at a value nearly twice as important as that of the transducer with glue: 725 mA, instead of 397. The nominal speed of the headmass is reached for 80 V. Furthermore, no saturation of the acceleration according to tension occurs. The glue was then responsible for all the degradations of the initial transducer.

3) Without rubber or glue: This manipulation was necessary to estimate the effects of the rubber without those of glue. 39 V are enough to reach the nominal mechanical conditions. Only 20 W instead of 54 are dissipated. The thermal study reinforces these results: the maximal heating is 24°C on the nodal electrode.

34 W have already been saved. The only aim of the manipulation without the rubber was to evaluate its losses in air. In water, the rubber is necessary for the waterproofness of the assembly. Glue, on the contrary, is not necessary: it would be better to take it off, for its effects are deplorable. New studies are trying to find out the origin of the other 20 watts: from now on, they show that the electrodes utilized as well as the value of the prestress on ceramics play a significant role.

P.3 Software for Analyzing Magnetostrictive Material

G. Reyne

LEG ENSIEG, Rue de la Houille Blanche,
BP 46, F-38402 Saint Martin d'Hères, France

This is a demonstration of the programs Flux2 and Flux3 with an emphasis on their application to the three-dimensional study of the polarization of Terfenol rods. Other examples stress the significance of the 2D-3D modeling for non linear magnetostrictive systems. According to requests, various works on magnetomechanical coupling as well as the vibratory aspect could be provided. The demonstrations will be illustrated with graphic displays, explanations and posters.

P.4 Rare-earth Low Frequency Tonpilz Transducer

M. Letiche

Thomson Sintra Activités Sous Marines, Route des dolines,
Parc d'activité de Valbonne, B.P. 38, F-06561 Valbonne CEDEX, France

The opportunity of developing low frequency transmitting transducers with reasonable size and weight is offered by the incentive properties of rare earth-iron alloys (Terfenol-D).

First, the characteristics (coupling factor, Young's modulus, permeabillity, magnetostriction, etc) of most of these materials available on the world market have been measured. The results underscore the advantages of manufacturing low frequency transducers using this material in comparison with those using PZT ceramics.

A tonpilz transducer, using four non grain oriented Terfenol-D rods, has been manufactured and calibrated in a water tank. Its acoustic properties are displayed. The most important properties are the following: low frequency, low mechanical factor, high transmitting response.

Taking into account its small size and low weight (about 8 kg with its housing), this transducer could be favorably compared to a piezoceramic transducer. The high power measurements have shown a good linearity in its response and a high power response is expected after optimization.

P.5 Compensated Flextensional Transducer Cooling System

G. Grosso

ERAMER, ZI du camp Laurent, F-83500 La Seyne sur Mer, France

Our class IV flextensional transducers are made up with:

- a compact engine group, made of six piezoelectric ceramic stacks lined up on the same plane, with an steel guide at each end and in the middle, a central steel electrode to make its insertion into the shell possible,
- a GRP shell (elliptical shape) with 2 integral metallic inserts,
- 2 stainless-steel plates closing the transducer,
- an elastomere membrane for waterproofness,
- the prestress in the ceramic stack is obtained using the shell. A defined force is applied to the minor axis to increase in size the major axis which is smaller than the ceramic stack. Then, the drive is inserted, the force released so the prestress is applied,

The main novelties are:

- Integration in the transducer of a simple design cooling system,

- setting up outside and under the transducer, of the great depth compensation device ERAMER, which is not bulky and independent.

Therefore, the transducer has been designed with internal filling rooms made of aluminium fitting exactly the internal cross section of the shell, so that the functional space is available, and the rest of the space is occupied. The role of these filling rooms is:

- to reduce the internal space of air to a maximum,
- to be used as an internal gap and frame for the whole transducer,
- to make it possible to set up several natural convection cooling pipes made of stainless steel, that cross the transducer from one side to the other,
- to act as a cooling system,
- to avoid excessive bending of the shell (protection against shocks).

The advantages of the device are:

- possible use of the transducer at high power,
- efficient cooling of the ceramic stacks,
- simple design (natural convection) and not very bulky.

The advantages of adapting the great depth compensation system of ERAMER to flextensional transducers are:

- working at great depths (1000 m or more),
- the bending of the shell with depth is far less significant, constant and independent from immersion,
- significant reduction of the shell internal stresses (no need to compensate for the prestress loss due to immersion), reduction of the static and dynamic fatigue of the material and avoiding of glass fibres delamination,
- no prestress loss on the ceramic stack according to the immersion of the transducer,
- since there is no bending of the shell due to immersion, a space is available to set up the cooling system,
- this system is compatible with the design of an array made of flextensionals.

P.6 High Depth Gas Compensation Device

G. Grosso

ERAMER, ZI du camp Laurent, F-83500 La Seyne sur Mer, France

A high depth gas compensation device has been designed and developed by ERAMER, and can be adapted to all kinds of transducers. It is under an international patent.

This device is made up with:

- two stainless steel bottles with an elastomere bladder,
- a pressure release system mounted on one of these bottles, composed of a sliding waterproof piston in the body of the pressure release system, with a significant end

diameter, sea side, so as to obtain a satisfying sensitivity and of a return spring, bottle side,

- waterproof and flexible pipes resistant to hydrostatic pressure, with a waterproof joint at the ends,
- the transducer to be balanced under pressure.

At surface level the bladder of the first bottle without pressure release system is deployed at the atmospheric pressure. The second bottle is pumped up at a several bar pressure (10b) called transfer pressure, the effects of which are:

- to repel the piston,
- to compress the spring,
- to close the pressurized gas passage of that second bottle towards the transducer and to open the passage of the first bottle.

Underwater and before transfer pressure, the transducer is balanced by the first bottle, so that there is no overpressure in the transducer at the surface.

When going down, (transfer pressure) some gas stays in the first bottle, the external pressure inside the pressure release system is equal to the internal pressure of the second bottle (10b). The spring repels the piston towards the bottle, which involves:

- the closing of the first bottle,
- the opening of the second bottle,

From this immersion, the transducer is balanced by the second bottle with an overpressure of 10b, that will enable it to go down to the maximum immersion.

At great depths, the bladder of the first bottle stops far before maximum immersion. The pressure release system stays in balance until maximum immersion. When it comes back to the surface, the bladder of the second bottle expands until it stops in its metallic housing. Then, we are back to transfer pressure.

When going up, (transfer pressure), the bladder of the second bottle is blocked, thus the internal pressure is steady whereas the external pressure keeps on decreasing, repelling the piston towards the sea, closing the second bottle and opening the first one.

To the surface, the transducer is balanced again, without overpressure, by the first partially deployed bottle.

Notes

It is possible to use more than two bottles if greater immersions have to be reached. The working principle remains the same; a bottle without any pressure release system at atmospheric pressure, other bottles equiped with a pressure release system pumped up at different pressures. It is possible to actuate the piston by means of an electromagnet controlled by a pressure release system set on the pumped bottles.

Advantages

- unlimited immersion made possible for all types of transducers,
- the gas volume being between 5 and 15 times less important than for a standard compensation. The more significant the desired immersion, the more important the increase in this ratio,
- the overpressure of surface prepumping is possible (simplified design of the transducers),
- the balancing of the transducers (low and very low frequencies) with a high volume of residual gas without generating too high bladder volumes is possible,
- automatic working of the device under hydrostatic pressure,
- the use of simple design transducers is possible (expense and weight savings),
- the balancing of a transducer stack with a single device is possible (expense and weight savings),
- the geometry of the device enables easy integration in a cylindric or plane array,
- processing and control of the pressure as well as of the gas hydrometry are easier for an array stacks.

P.7 Technique for Characterising Low Frequency Transducers Using Test Tank Measurements

J. Delany

Dowty Maritime, Bridport Road, Greenford, Middlesex UB6 8UA, England

Reverberation effects tend to limit the usefulness of test tanks for evaluating low frequency sonar transducers. Impedance measurements generally exhibit a distorted character compared to that observed under free field conditions. In particular, a well behaved device may indicate a shift in its resonance and appear multi-resonant.

This behavior may be understood in a general manner using an equivalent circuit approach. In particular, any reverberation signal present during the measurement may be represented by an equivalent "reverberation" impedance element in series with the radiator impedance. The frequency response of the "reverberation" impedance is dependent on the test configuration and nature of the boundary surfaces, and is generally complicated. However, using a cylindrical test tank it has been observed that the reverberation effect can produce a high frequency modulation to the transducer mechanical impedance.

The above observation forms the basis for a filtering technique performed with a desktop computer on data obtained with an HP 4194 impedance analyser. Results are presented for an 800 Hz flextensional transducer measured under test tank conditions and in open water. The "filtered" response is shown to provide at least a first order approximation of the free field response.

The technique has proved useful with other low frequency devices, including benders. It has allowed initial transducer development and evaluation to be carried out under laboratory conditions rather than necessitating open water measurements. In particular the 'filtered' responses have been found to provide a good basis for equivalent circuit representations of various transducers. These have proved useful for evaluating acoustic performances, understanding array interaction effects and developing drive system designs.

P.8 Class III Flextensional Transducers

D.F. Jones

Defence Research Establishment Atlantic, P.O. Box 1012, Dartmouth, Nova Scotia, B2Y 3Z7, Canada

The number of papers appearing in the open literature on flextensional transducer technology is increasing. The driving force behind this trend is the growing interest, primarily from naval research laboratories, in low-frequency, long-range, active sonar systems. One of the various flextensional classes, the Class IV transducers have been the most widely studied. At DREA however, there has been a considerable amount of research done on Class III transducers such as the barrel-stave projector (D.F. Jones and G.W. McMahon, J. Acoust. Soc. Am. **82**, Suppl. 1 S75 (1987)). In this paper, a detailed discussion of the measured acoustic performance results for Class III dual-shell projectors is given. The frequency range of interest is between 1 and 2 kHz.

P.9 Reciprocity Calibration of a Flextensional Transducer from Electrical Impedance Measurements in Water and in Air: the Delta-Z Reciprocity Calibration Method

T.K. Janacek and S.R. Baker*

Department of Physics, Code PH/BA, Naval Postgraduate School, Monterey, California 93943, USA

The Delta-Z Reciprocity Calibration Method is a procedure by which the open- and short-circuit receiving sensitivities, and the transmitting current and voltage responses of a reversible underwater acoustic transducer may be obtained from measurements of its input electrical impedance in water and in air. The method has previously been applied to obtain the open-circuit receiving sensitivity of several piezoelectric sonar transducers. The application of the Delta-Z Method to obtain the transmitting response of a small flextensional transducer is described in this paper. The results show that using the Delta-Z Method, it is possible to obtain an engineering estimate of the transmitting response of a low-frequency piezoelectric sonar transducer from impedance measurements made in a small tank, i.e. without the need of a deep-water test facility. *(Work supported by NAVSEA, NRL-USRD, and the Naval Postgraduate School.)* *LT, USN.*

P.10 Numerical Modeling of Arbitrarily Dense Active Sonar Arrays

K.A. McLean [a], *S.R. Baker* [b], *C.L. Scandrett* [c], *D.R. Canright* [c], *and E. Kuntsal* [d]

(b) Department of physics, Code PH, (c) Department of Mathematics, Code MA, Naval Postgraduate School, Monterey, CA 93943, USA

A procedure has been developed by which the performance of an arbitrarily dense volumetric array of sonar transducers may be computed. The method is an extension of the so-called T-matrix method, which has been extensively applied to all types of scattering problems, to the problem of coupled electroacoustic transducers. The acoustic field is represented in terms of its radiation eigenfunctions, the coefficients of which are found self-consistently by imposing an impedance-matching condition at the structure-fluid boundaries. No restriction is placed on the arrangement of the transducers, so that multiple-scattering is rigorously included. An analytical dynamical description of each transducer is derived using a finite-element computer code. A description of the method and some preliminary results will be presented. *(Work supported by NUSC, NRL-USRD, and the Naval Postgraduate School.). (a)LT, USN. (d) Present address: International Transducer Corporation, Goleta, CA, USA.*

P.11 Development of a Resonant Pipe Projector

J. Ellis, Sparton of Canada Ltd., London Ontario, Canada.
B. Fanning, FanTech Consulting Service, Dartmouth, Nova Scotia, Canada
G.W. McMahon, Defence Research Establishment Atlantic, P.O. Box 1012, Dartmouth, Nova Scotia, B2Y 3Z7, Canada

A high power, low frequency projector has been developed and built for oceanographic research applications. Driving excitation for the Resonant Pipe Projector is provided by a piezoceramic/steel composite ring which is capable of a maximum source level of 206 dB/μPa and has an electroacoustic efficiency of 85 percent, at the cavity resonance of 260 Hz. A tuning method is outlined which allows the resonance to be adjusted by the addition of compliant material to the cavity. Performance is well predicted by MAVART, a finite element modeling program which includes the acoustic radiation solution. Scaling to other frequencies is readily achievable. If broadband moderate power is required, the projector can be utilized between cavity resonant frequencies. The maximum depth of operation is essentially unlimited due to the free-flooding construction.

P.12 Empirical and Analytical Evaluation of Small Arrays of High Power Flextensional Transducer Elements

F. A. Tito
Naval Underwater System Center, Code 2131, New London, CT, 06320, USA

Several small active arrays composed of flextensional transducer elements were evaluated at the Naval Underwater Systems Center (NUSC) Seneca Lake Facility. These arrays included: (1) a 4x3 array on an air backed flat plate fixture; (2) a 4x3 free field array on frame fixture; and (3) a 1x6 array on a simulated rigid baffle fixture (6 pairs , with each pair mounted back to back). This paper deals with the impedance and power distributions across the arrays, and source level measurements for the various configurations. The results showed that some elements exhibited negative power for various steering conditions in the 4x3 arrays especially as the array was steered off broad-side. Maximum element negative power was found at the maximum steered angle. Modeling analysis was also performed for the various array configurations and measured source level performance compared well with the analysis.

P.13 Design Model of Large, Uniform Conformal Arrays of Bender bar and Flextensional Transducers

G. Brigham, Raytheon Co., P.O. Box 360, West Main Road,
Portsmouth RI, 02871 USA
B. McTaggart, Naval Underwater System Center, Code 213,
New London, CT, 06320, USA

The single element design technology of flextensional and bender bar transducers is several decades old and well-known. However, the use of either type of transducer in a conformal array presents new and formidable design problems. When the array is very large, the effects of edge diffraction on the outer elements can be ignored to allow generation of a zeroeth ordered wave-guide design model to estimate radiation loading at any steering angle. Highly eccentric shelled flextensional and bender bar transducers in this paper are included in one common format. Since both types are largely flexural, they are resonant at frequencies where the interelement spacing is much smaller than a wavelength in water and only the low frequency inertial and plane wave volume flow components of loading need to be determined. Several array geometries have been studied and their reactive loadings calculated. This paper shows both the theoretical and numerical results of various mass loadings as a function of array and element geometry.

P.14 Very Low Frequency Emitting Sonar Array Design

E. Sernit, and T. Mazoyer

Thomson SIntra Activités Sous Marines, Route des dolines,
Parc d'activité de Valbonne, B.P. 38, F-06561 Valbonne Cedex, France

Nowadays, submarine have become so quiet that their detection by passive acoustics is less and less efficient. So, it appears there is a real need to design powerful very low frequency acoustical emitters in order to activate these passive systems, thus providing long range detection active SONAR. Designing such antennas, difficult technological as well as complex modeling problems arise:

- regarding the very low frequency domain, the array dimensions are limited by the environment, typically to the order of the wavelength,
- high sound levels that are required imply strong electro-mechanical demands in materials. In case of transducer arrays, a very precise knowledge of acoustical behavior is needed for each source to estimate the whole system performances,

A design process is presented, including software tools, that has been developed in the laboratories at Thomson Sintra ASM. There are two steps:

Step 1: <u>computation of antenna acoustical interactions</u>

To obtain precisely the coupling effects between elementary transducers <u>and</u> what we call environment (structures, baffles etc.) we adopt a numerical three-dimensional modeling technique, Helmholtz Integral Equations and Boundary Finite Elements. This is the only means to take into account real geometries and baffle conditions, because of the complex structure of acoustical near fields close to small radiating bodies. After that, a fast post-processor gives us: antenna directivity patterns as a result of elementary far field combination, power distribution map of the array, by using *in antenna* mutual acoustical impedances.

Then refering to sonar specifications (directivity, etc.) and technological constraints (power distribution, etc.), one may have a first-step optimization of the system shape.

Step 2: <u>Transducer design.</u>

Acoustical results are coupled to an equivalent circuit for transducer behavior modeling. The operator then chooses the technology for elementary sources, the electrical modularity of the array and the electrical supply type (U, I or Power). At that step we optimize the equivalent circuit parameters by computing sound level and electrical impedances, and comparing them to the required ones. Furthermore, computed electro-mechanical powers provide the technological limits for the system modeled. Let us emphasize the very low computation costs involved in this step. This allows to foresee the computing of many different variations.

The last part of the poster deals with future developments: extensions to other applications (different kinds of transducers) as well as implementation of automatic optimization procedures. Our goal is to go on to a computer aided design tool.

P.15 Evaluation of Sonar Array Directivities Using Elementary Column Diagrams

J. Garcin

GERDSM -DCAN Toulon, DCN, Le Brusc, F-83140 Six Fours Les Plages, France

Electrical weighting is often needed to get the required directivity diagrams of multicolumn active array. Simple geometrical considerations or more sophisticated models can be used to estimate these values. But, due to the complexity of real environment and strong acoustical interactions in compact low frequency arrays, the experimental diagrams are often significantly different from those expected.

Because such measurements need an important equipment set-up for phase and module weighting and a lot of time, the number of trials is limited and the optimal shading is generally not reached.

To improve the result, we tried a linear superposition process performed in two steps. First in a lake, we have measured the complex radiation patterns (magnitude and phase) of elementary groups with a minimal equipment set-up (no delay lines or other weighting apparatus. All the transducers being in short circuit except those of interest.) Second, to obtain the directivity corresponding to a set of weighting, we use a micro-computer to sum these elementary diagrams with the desired coefficients. The results take into account environmental effects and acoustical interactions. This operation is done quickly and can be automated to sort the best weighting.

Three main limitations of the method can be seen:

- phase measurement requires a very stable distance between array and hydrophone,
- the error on the final directivity is the sum of the elementary errors,
- transducers must be linear.

Despite the restrictions mentioned above, our predicted and measured directivities and transmitting levels have turned out to be in good agreement.

Subject Index

MIX
Papier aus verantwortungsvollen Quellen
Paper from responsible sources
FSC® C105338

If you have any concerns about our products,
you can contact us on
ProductSafety@springernature.com

In case Publisher is established outside the EU,
the EU authorized representative is:
Springer Nature Customer Service Center GmbH
Europaplatz 3, 69115 Heidelberg, Germany

Printed by Libri Plureos GmbH
in Hamburg, Germany